Conservation Monitoring in Freshwater Habitats

Clive Hurford · Michael Schneider
Ian Cowx
Editors

Conservation Monitoring in Freshwater Habitats

A Practical Guide and Case Studies

Broads Authority
The Broads – a member of the National Park family

Länsstyrelsen
Västerbotten

CSIC
Estación Biológica de Doñana
Consejo Superior de Investigaciones Científicas

Cyngor Cefn Gwlad Cymru
Countryside Council for Wales

Springer

Editors
Clive Hurford
Conservation Monitoring Team Leader
Countryside Council for Wales
Woodside, Orielton, Pembroke
UK

Michael Schneider
Large Carnivore Specialist
and Conservation Officer
Västerbotten County Administration
Umeå
Sweden

Ian Cowx
Professor in Applied Fisheries Biology
University of Hull
International Fisheries Institute
Hull
UK

ISBN 978-1-4020-9277-0 e-ISBN 978-1-4020-9278-7
DOI 10.1007/978-1-4020-9278-7
Springer Dordrecht Heidelberg London New York

Library of Congress Control Number: 2009941850

Printed on acid-free paper

Springer is part of Springer Science+Business Media (www.springer.com)

Acknowledgements

This book has been possible only because of the cooperation and support of very many friends and colleagues, and initially our thanks must go to the authors who have contributed chapters to the book. Thanks also to the bodies that supported the studies underpinning many of the examples presented.

Thanks are due to Dan Guest for preparing maps, and to Terry Rowell, Dan Guest and Chris Millican for helping with the thankless task of copy-editing and commenting on the draft content. Similarly, Nigel Holmes and Richard Lansdown were always a source of inspiration, advice and support. Our gratitude is also extended to the publishing editors and technicians at Springer NL, particularly Catherine Cotton and Ria Kanters, for their encouragement, support and patience throughout the writing and collation phases of the publication process.

In the Countryside Council for Wales we would like to thank: David Parker, Hazel Drewett and David Allen for supporting the book and for helping to secure funding for the colour sections; the employees of CCW who participated in training events and data collection, namely Sam Bosanquet, Charlotte Gjerlov, Dan Guest, Jon Hudson, Chris Lawrence, Tracey Lovering, Jon Turner, Karen Wilkinson and Julian Woodman; and Helen Jones and Delyth Hughes of Training Branch for supporting both the collaboration with our colleagues in Sweden and our freshwater monitoring training programme.

In the Environment Agency we would like to thank Richard West, Julie Gething and Gemma Dickson for providing guidance on benthic invertebrate identification, specifically for the data collection phase of the Western Cleddau monitoring case study.

In Sweden, we thank: Björn Jonsson of the Västerbotten County Administration for ensuring that the collaboration with CCW was possible, and for his continuing support; Kajsa Berggren of the County Administration for her interest in chapter illustrations; Johnny Berglund of the County Administration for the benefit of his time and expertise and for taking us safely through troubled waters in ringed seal habitat; Per Risberg at the Swedish Environmental Protection Agency, Tero Härkönen at the Swedish Museum of Natural History, Malin Hemmingsson at the Swedish National Board of Fisheries and Mattias Karlsson, formerly at the County Administration, for a fruitful co-operation during the production of the draft action plan for the ringed seal in Sweden; Torbjörn Lövbom of the Swedish Hunters'

Association for sharing his experience of seal hunting in the drift ice of the Bothnian Bay; and Jonas Grahn and Emma Widmark of the County Administration for discussing Natura 2000 issues and for having great patience during the creation of the action plan for the ringed seal, respectively.

At Doñana Natural Space, we thank: Dr. J.J. Chans for his interest and support, and Ricardo Díaz-Delgado for coordinating the contributions from Spain.

At Swansea University, we thank: Dan Forman, Charles Hipkin and Penny Neyland for their help and support during the time that the CCW monitoring team was working with them on the water vole habitat monitoring project.

We wish to acknowledge the financial support of the Countryside Council for Wales, Västerbotten County Administration and The Broads Authority: this ensured that the case studies could be produced in colour. We would also like to acknowledge Doñana Natural Space for making staff time available for contributing to the book and for covering the costs of interpreting the chapters from Spain.

Finally, we would both like to thank our partners Chris and Doris, and Michael's kids Ronja and Matthis, for their endless patience, particularly during the period of collating this book.

Contents

Contributors

Johan Ahlström
Liming Administration and Environmental Assessment Officer,
Västerbotten County Administration, SE-901 86 Umeå, Sweden
johan.ahlstrom@lansstyrelsen.se

Juan A. Amat
Department of Wetland Ecology, Doñana Biological Station, CSIC, Seville, Spain

Iban Ameztoy
Technician of Remote Sensing and GIS Laboratory (LAST),
Doñana Biological Station, CSIC, Seville, Spain

David Aragonés
Technician of Remote Sensing and GIS Laboratory (LAST),
Doñana Biological Station, CSIC, Seville, Spain

José Luis Arroyo
Specialist Technician of Birds Section, Monitoring Team,
Doñana Biological Station, CSIC, Seville, Spain

Sam Bosanquet
Recorder for Hepatics, British Bryological Society & Phase II Lowland Habitat Surveyor, Countryside Council for Wales, Cnwc y Llwyn, Brechfa, Carmarthen, Carms, UK

Miguel Ángel Bravo-Utrera
Head of Limnology Section, Monitoring Team, Doñana Biological Station, CSIC, Seville, Spain

David Broughton
Macrophyte Specialist and Consultant Ecologist, Scott Wilson Ltd.,
West One, Leeds, UK

Javier Bustamante
Department of Wetland Ecology and Head of Remote Sensing & GIS Laboratory (LAST). Doñana Biological Station, CSIC, Seville, Spain

Ulf Carlsson
Västerbotten County Administration, Umeå, Sweden

Richard Chadd
Environmental Monitoring (Analysis and Reporting) Officer, Environment Agency of England and Wales, Anglian Region (Northern Area), Spalding, Lincolnshire. UK

Alfredo Chico
Specialist Technician of Birds Section, Monitoring Team, Doñana Biological Station, CSIC, Seville, Spain

Eduardo Costas
Department of Animal Production, Veterinary Faculty, Complutense University, Madrid, Spain

José Luis del Valle
Specialist Technician of Birds Section, Monitoring Team, Doñana Biological Station, CSIC, Seville, Spain

Ricardo Díaz-Delgado
Head of Landscape and Remote Sensing, Monitoring Team & Technical Head of Remote Sensing and GIS Laboratory (LAST), Doñana Biological Station, CSIC, Seville, Spain

Chris Dyson
Countryside Council for Wales, Llandrindod Wells, Powys, UK

Juan Manuel Espinar
Ranger of Doñana Natural Space, Centro Administrativo El Acebuche, Matalascañas, Huelva, Spain

John Ferguson
Fish Ecology Division, National Marine Fisheries Service, Seattle, USA & Department of Wildlife, Fish and Environmental Sciences, Swedish University of Agricultural Sciences, Umeå, Sweden

Dan Forman
Tutor in Conservation Ecology, Conservation Ecology Research Team (CERTS), Department of Pure and Applied Ecology, Institute of Environmental Sustainability, Swansea University, Swansea, UK

Luis García
Specialist Technician of Birds Section, Monitoring Team, Doñana Biological Station, CSIC, Seville, Spain

Emili García-Berthou
Institut d'Ecologia Aquàtica, Universitat de Girona, Girona, Spain

Héctor Garrido
Specialist Technician of Birds Section, Monitoring Team, Doñana Biological Station, CSIC, Seville, Spain

Andy J. Green
Department of Wetland Ecology, Doñana Biological Station, CSIC, Seville, Spain

Dan Guest
Countryside Council for Wales, Swansea, UK

Jane Harris
Kepwick Ecological Services, East Carleton, Norwich, UK

Jonathan P. Harvey
Hull International Fisheries Institute, University of Hull, UK
j.p.harvey@hull.ac.uk

Bob Haycock
Countryside Council for Wales, Pembroke, UK

George Hinton
Analysis and Monitoring, Natural England, Peterborough, UK

Charles Hipkin
Plant Biology and Ecology, Department of Pure and Applied Ecology, Institute of Environmental Sustainability, Swansea University, Swansea, UK

Nigel Holmes
The Almonds, Warboys, Huntingdon, Cambs, UK

Fernando Ibáñez
Specialist Technician of Birds Section, Monitoring Team, Doñana Biological Station, CSIC, Seville, Spain

Mats E. Johansson
Västerbotten County Administration, SE-901 86 Umeå, Sweden

Lars Karlsson
National Board of Fisheries, Älvkarleby, Sweden

Andrea Kelly
Head of Conservation, Broads Authority, Norwich, UK

Richard V. Lansdown
45 The Bridle, Stroud, Glos. UK

Stefan Larsson
Department of Ecology and Environmental Sciences, Umeå University, Umeå, Sweden

Hugues Lefranc
Head of Database and Data Quality, Monitoring Team, Doñana Biological Station, CSIC, Seville, Spain

Kjell Leonardsson
Department of Wildlife, Fish and Environmental Sciences, Swedish University of Agricultural Sciences, Umeå, Sweden & Department of Ecology and Environmental Sciences, Umeå University, Umeå, Sweden

Victoria López-Rodas
Department of Animal Production, Veterinary Faculty, Complutense University, Madrid, Spain

Hans Lundqvist
Department of Wildlife, Fish and Environmental Sciences, Swedish University of Agricultural Sciences, Umeå, Sweden

Manuel Máñez
Head of Birds Section, Monitoring Team, Doñana Biological Station, CSIC, Seville, Spain

Antonio Martínez
Specialist Technician of Birds Section, Monitoring Team, Doñana Biological Station, CSIC, Seville, Spain

Penny Neyland
Conservation Ecology Research Team (CERTS), Department of Pure and Applied Ecology, Institute of Environmental Sustainability, Swansea University, Swansea, UK

Jan Nilsson
Department of Wildlife, Fish and Environmental Sciences, Swedish University of Agricultural Sciences, Umeå, Sweden

Richard A. Noble
Hull International Fisheries Institute, University of Hull, Hull, UK

Regino Núñez
Advanced Technology Systems (SATEC), Seville, Spain

Andrew D. Nunn
Hull International Fisheries Institute, University of Hull, Hull, UK

Johan Östergren
Department of Wildlife, Fish and Environmental Sciences, Swedish University of Agricultural Sciences, Umeå, Sweden

Daniel Palm
Department of Wildlife, Fish and Environmental Sciences, Swedish University of Agricultural Sciences, Umeå, Sweden

Tim J. Pankhurst
PLANTLIFE, Cambridge University Botanic Gardens, Cambridge UK

Peter Rivinoja
Department of Wildlife, Fish and Environmental Sciences,
Swedish University of Agricultural Sciences, Umeå, Sweden

Rubén Rodríguez
Specialist Technician of Birds Section, Monitoring Team,
Doñana Biological Station, CSIC, Seville, Spain

Héctor Rodríguez-Pérez
Department of Animal Production, Veterinary Faculty, Complutense University, Madrid, Spain

Terry A. Rowell
21 Cadnant Park, Conwy, UK

Ignacio Serrano
Department of Ecology and Environmental Sciences,
Umeå University, Umeå, Sweden

Jane Southey
Macrophyte Specialist and Principal Consultant Ecologist,
Scott Wilson Ltd., Unit 33, Mansfield i-Centre, Oakham Business Park, Hamilton Way, Mansfield, UK

Justin Steventon
Senior Development Lead, Steventon Consulting, Redmond, WA, USA

Ryan J. Taylor
Hull Inernational Fisheries Institute, University of Hull, Hull, UK

Chris Uttley
Water Policy Officer, Countryside Council for Wales, Maes y Ffynnon, Ffordd Penrhos, Bangor, Gwynedd

Montserrat Vilà
Doñana Biological Station, CSIC, Seville, Spain

Richard West
Environment Agency of England and Wales, Wales Region (South East Area), Cardiff, UK

Section 1
Setting the Scene

Chapter 1
Conservation Monitoring in Freshwater Habitats: An Introduction

Clive Hurford

Introduction

This book focuses on the role of monitoring in the conservation of freshwater habitats and species. The chapters are grouped in the following sections:

1. Introductory
2. Biological indicators
3. Rivers – threats and monitoring issues
4. Rivers case studies
5. Lakes and wetlands case studies
6. Doñana Natural Space: an integrated programme of surveillance, monitoring and management

Part I sets the scene for the rest of the book. This chapter provides a brief overview of the protocols for developing efficient and reliable conservation monitoring projects. The chapter on planning management highlights the need for well-informed and transparent decision-making, and is followed by an overview of the links between the Habitats Directive and the Water Framework Directive.

Part II comprises a series of short chapters that discuss the role of different groups as biological indicators in freshwater habitats, i.e. mammals, birds, invertebrates, plants, phytoplankton and invasive species. Reliable monitoring projects need a clear focus, and these chapters draw attention to species, or co-occurring species (Grime 2001), that may be suitable indicators of condition in freshwater habitats. For example, Grime *et al.* (1990), Ellenberg (1979, 1988) and Ellenberg *et al.* (1991) can be useful places to start for identifying vascular plant habitat indicators.

C. Hurford
Countryside Council for Wales, Woodside, Orielton, Pembroke, SA71 5EZ, UK
e-mail: clive.hurford@btinternet.com

C. Hurford et al. (eds.), *Conservation Monitoring in Freshwater Habitats: A Practical Guide and Case Studies*, DOI 10.1007/978-1-4020-9278-7_1,

Part III is an eclectic mix of chapters that looks initially at the threats faced by rivers, and then at some of the methods currently used to record river vegetation. This section also contains chapters highlighting the problems associated with observer variation; probably the most important threat to the reliability of a monitoring project.

Parts IV and V provide a series of monitoring case studies in freshwater habitats, some focussing on habitats and others on species. These chapters are intended to reflect the current situation with regard to the development of monitoring projects in rivers, lakes and wetlands. Perhaps surprisingly, given the history of targeted water quality monitoring and the development of targeted monitoring projects for terrestrial habitats (e.g. Hurford and Schneider 2006), there are very few examples of targeted monitoring projects for freshwater habitats and species. As such, most of the case studies demonstrate how existing survey and surveillance can be used to develop site-specific targets for management and monitoring: this is an essential stepping stone to effective conservation management in freshwater environments.

Part VI comprises a series of chapters describing the integrated system of surveillance and monitoring that feeds back into the conservation management of Doñana Natural Space: probably the most important wetland area in Europe.

Survey, Surveillance and Monitoring

In this book, we use the term monitoring to mean assessing the habitats or species against clearly defined and measurable conservation aims. This is broadly in keeping with the definition of Hellawell (1991), which described monitoring as:

'Intermittent (regular or irregular) surveillance carried out in order to ascertain the extent of compliance with a predetermined standard or the degree of variation from an expected norm'.

This definition distinguishes monitoring from the more traditional forms of ecological investigation such as:

- *Survey*, which is typically a one-off descriptive exercise, perhaps to describe the habitats on a site or to map the distribution of a species
- *Surveillance*, which is a replicable survey, often used to detect trends in habitats, populations and environmental change
- *Experimental management*, which tests the effects of different management practices
- *Environmental impact assessment*, which assesses the likely effects of a development or incident
- *Research*, which is carried out to increase our knowledge about a species or habitat, perhaps through ecological modelling, population viability analysis and demographic studies

As in the terrestrial environment, most data collection from freshwater habitats to date falls into the survey, surveillance or research categories. The critical difference between these exercises and a monitoring project is that a monitoring project will clearly identify when we need to make a management response.

A Model for Conservation Management and Monitoring

Monitoring (as defined by Hellawell) is essentially a tool of practical conservation management, and Fig. 1.1 shows a simple, but effective, model for nature conservation management and monitoring.

The need for clear decision-making is implicit in this model. First we must decide what would represent a favourable state for the key habitat or species, and then we must decide when to intervene if the state is (or becomes) unfavourable. A third, often overlooked, but equally important, decision concerns when we would consider the habitat or species to have recovered; this is unlikely to be the same point that we became concerned about it. This decision not only has resource implications, it can also have major implications for other habitats and species (prey species are an obvious example). All of these decisions are essential to the development of an efficient and effective monitoring project.

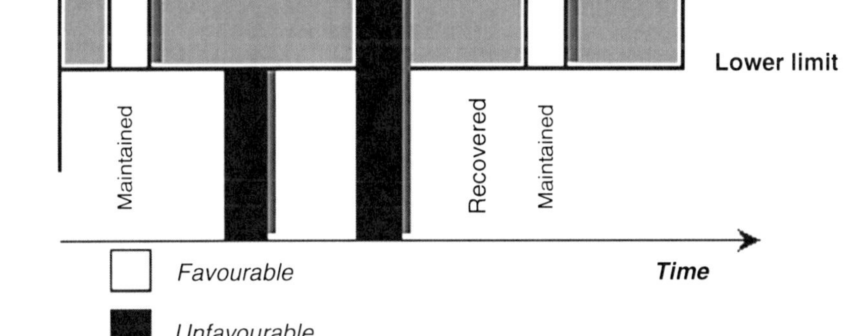

Fig. 1.1 This shows a model for conservation management and monitoring (from Brown 2000). Key decisions are needed to identify (**a**) when the habitat or species is in a favourable state (in the band between the lower and upper limits) and (**b**) when it has recovered from being in an unfavourable state

Developing a Monitoring Project

For a monitoring project to be effective, it must consistently provide the right result. To achieve this, the project should:

- Monitor against site-specific targets that take account of what we believe it is possible to achieve, given the various pressures on the site
- Focus on a small number of carefully selected attributes, drawing on site-specific research, survey and surveillance data
- Express attributes in concise and measurable terms, and test them for measurability
- Use sampling methods tested to assess how much time is needed to collect the data and to check that trained surveyors will deliver consistent and reliable monitoring results
- Use field surveyors familiar with both the attributes and the sampling method

Key Steps in Developing a Monitoring Project

There are several steps in developing a monitoring project; these are outlined below:

1. Identify the priority for conservation on your site (or for each of the different management units of the site).
2. Collate the relevant knowledge from existing research, survey and surveillance on the key habitats or species on your site.
3. Develop a conservation strategy.
4. Develop site-specific condition indicators.
5. Decide where to monitor.
6. Collect the monitoring data.
7. Feed back into site management.
8. Safeguard the monitoring data.

The following sections provide a brief overview of the steps outlined above.

Identifying the Priority for Conservation

Site managers often find it difficult to formally prioritise habitats or species for conservation on their sites, but it is an essential first step in the development of a strategy for conservation management and monitoring.

It could be that managing for one species will deliver the range of habitats available for many other species on a site, perhaps salmon *Salmo salar* could fit into this category on some rivers. If this was the case, then both the management and monitoring could focus on the *S. salar* population. For example, it would be reasonable to assume that if a river is supporting a strong *S. salar* population, it will also be capable of supporting a strong bullhead *Cottus gobio* population.

Life is rarely that simple, however, and it is more likely that different parts of a site will be important for different habitats and species. In this case, we need to think about how the different parts of a site contribute to sustaining the important habitats and species that use it. At this point, there is nothing to be gained from going along the philosophical route of agonising over why one species or habitat is more important than another: the annexes of the Habitats Directive and regional red data lists give clear guidance on this.

Collating Existing Research, Survey and Surveillance Information

This step is self-explanatory. As soon as we have decided which habitats and species we should be managing for, we have a focus for our literature search. Background information on the requirements of the key habitats and species on our sites, coupled with site-specific survey or surveillance data, increases our confidence to make difficult decisions.

Going through this process also draws attention to shortfalls in knowledge and allows us to identify topics for future research.

Developing a Conservation Strategy

Having decided what the conservation priorities are, and taken account of the existing knowledge, we then need to decide what we want to achieve and where we want to achieve it.

The best way to work through these issues is to adopt a map-based approach: the chapter on site unitisation (Chapter 16) describes one way of doing this for a river system. Another approach would be to produce a condition map for the site (Hurford and Schneider 2006). In the latter case, having decided which habitat or species is the priority in any particular section of the site, map the current extent of the habitat (or potential habitat – if appropriate) and how much of it currently meets the criteria for being in good condition. This, of course, requires us to have developed criteria for recognising when the habitat is in good condition. However, having produced this map, we can then assess whether the current extent of habitat is sufficient, and whether enough of it is in good condition. If not, we should also be in a position to decide whether there is the potential to increase the extent of the habitat and where it would be possible to improve its condition. This will identify key areas for both management and monitoring.

Developing Site-Specific Condition Indicators

We use the term 'condition indicators' to describe the suite of attributes and targets that we will use as evidence for the condition of the habitat (or species). In effect, the condition indicators are a form of ecological shorthand to help us recognise when the key habitat is in a state of high conservation value. Typically, the condition indicators, which should be applied at the management unit level, will comprise:

- A target for the overall extent of the broad habitat
- A target for the extent of good quality habitat
- Unambiguous definitions for both the broad habitat and good quality habitat

The condition indicators should incorporate knowledge both from research and from site-based survey and surveillance projects. Critically, we should use existing information to help us decide where the key habitats and species should be on our sites, and what would represent optimum condition for them. This information should then be presented in a concise and transparent form.

Table 1.1 is an example of a condition indicator table for humid dune slack habitat at Kenfig SAC in South Wales: this type of table is a template for facilitating the

Table 1.1 A condition indicator table for the humid dune slack habitat at Kenfig NNR in South Wales. Note the site-specific habitat definitions, these minimise the opportunity for observer variation during the monitoring exercise. The requirement for obviously >25% open ground in the embryo slack definition is the only subjective element in the table

Condition indicator table	The *humid dune slack* habitat at Kenfig will be in optimum condition when	
Habitat extent	Lower limit	Extent of humid dune slack identified on remote sensing image in 1999
Habitat quality	Lower limit	In Section 1 (see map) >30% of dune slack sampling points in Area Y And >45% of dune slack sampling points in Area Z are either embryo or successionally-young slack vegetation And In Section 1 outside of Areas Y and Z >7 dune slacks of >0.5 ha, and >3 dune slacks of >0.25 ha have vegetation where >70% of sampling points are either successionally young or orchid-rich slack vegetation
	Site specific habitat definitions	
Dune slack vegetation	Moist vegetation on level ground between sloping dunes, typically with *Salix repens* present	
Successionally-young dune slack vegetation	Bare soil and thalloid liverworts present, and at least four of the following present: *Carex viridula* ssp. *viridula*, *Juncus articulatus*, *Anagallis tenella*, *Samolus valerandi*, *Eleocharis quinqueflora*, *Ranunculus flammula*, *Liparis loeselii*, within any 50 cm radius And None of the following present: *Phragmites australis*, *Molinia caerulea*, *Calamagrostis epigejos* within any 1 m radius	
Embryo slack vegetation	Obviously >25% open ground with *Salix repens* forming clonal patches And At least two of *Carex arenaria*, *Sagina nodosa* or *Juncus articulatus* present within any 1 m radius	
Orchid-rich dune slack vegetation	At least 2 of the following present: *Epipactis palustris*, *Dactylorhiza incarnata*, *Gymnadenia conopsea*, *Pyrola rotundifolia*, in any 50 cm radius And None of the following present: *Phragmites australis*, *Molinia caerulea*, *Calamagrostis epigejos* within any 1 m radius	

critical decisions in a conservation management project. Targets for habitat extent are best dealt with by reference to site habitat maps or remote images. It is important to differentiate between the area of the broad habitat and the area of habitat that must meet the criteria for optimum condition: the best examples of habitats rarely have a uniform structure or composition. The lower part of the table is for site-specific habitat definitions, i.e. any terms that we use in our targets for extent and quality must be defined unambiguously here.

We should take every care to ensure that the information in the condition indicator table is unambiguous and expressed in terms that allow objective assessment. Failure to do this will result in an unreliable monitoring project, where the monitoring result is an artefact of observer bias.

The Importance of Unambiguous Definitions

The definition in Box 1.1 below was developed initially for SERCON (System for Evaluating Rivers for Conservation) and was intended to differentiate between the bank and the channel: this is an attempt to make the attributes objective. This definition has been repeated in other documents subsequently.

These definitions are entirely subjective and will result in unknown levels of observer variation. There is simply no way that a surveyor could know the section of river well enough to make these judgements. In this instance, a simple solution would be to provide a list of (a) aquatic, (b) emergent and (c) terrestrial (or bank) species and to deal with the species accordingly.

Box 1.1 The definition of a 'bank' developed for use in SERCON river evaluations

'At the sides of the river all parts of the substratum are included which are likely to be submerged for more than 85% of the year. The 'bank' can be usefully defined as *that part of the side of the river (or islands) which are submerged for more than 50% but less than 85% of the time.* In general terms, therefore, 'river' records are reserved for those macrophytes occurring in the region of the river which is rarely uncovered, and those shallow sections *which have an upper limit that may be exposed for a maximum of 50 days in any year.* 'Bank' records are for those plants that occur above the limit of the 'river' plants, and *are thus out of the water for more than 50 days in any one year*, yet will be submerged, or partially so, during mean flow periods. The upper limit of the 'bank' *excludes all the areas which are submerged during the 150 days of each year* when river flows are at their highest. Such estimates have to involve guesswork, but estimates of submergence levels do allow better interpretation of the data and clearer insights into the ecology of individual species and communities at different sites.'

Box 1.2 All of the italicised terms in this target (and definition) for siltation in the JNCC rivers guidance are open to observer interpretation and will lead to inconsistent monitoring results

'Target: No *excessive* siltation. Channels should contain *characteristic* ranges of substrate types for *unmodified* rivers.

Definition: For river types characterised by *extensive Ranunculus* beds, there should be a *predominance* of *clean* gravels, pebbles and cobbles, with *relatively low cover* by *silt-dominated* substrates. Maximum fines content *should not be too great* to prevent establishment of new plants. Fines are defined as particles <0.83 mm.'

The example in Box 1.2 is taken from the JNCC Common Standards Monitoring guidance, and comprises a target and definition.

Targets and definitions of this type are commonplace in guidance documents and present a real barrier to the development of reliable monitoring projects and consistent recording. To be of any practical value, the targets must be defined in unambiguous terms.

Deciding Where to Monitor

When we have decided what we want our management to achieve and where, deciding where to monitor is relatively straightforward. Random sampling may be appropriate for research or statistical purposes, but is not appropriate for conservation monitoring. If we know which sections of a river, lake or wetland are most likely to support a habitat or species, these are the places that we should monitor.

For example, the Kenfig condition indicator table (Table 1.1) refers to 'Areas Y and Z', these are the youngest areas of dune slack development on the site: if these do not meet the criteria for successionally young dune slack vegetation then nowhere else will either. We could have set an alternative target for 'n %' of the dune system to be successionally young dune slack vegetation and then set up a stratified random sampling programme to answer the question, but this would have been ignoring everything that we have learnt about the site, and would have resulted in a sampling programme where most of our time was spent collecting data from non-target habitat.

The equivalent in rivers might involve targeting areas that we know are suitable for the habitat or species, rather than sampling indiscriminately at 5 km intervals along the length of a river.

Critically, we should be using existing information from research, survey and surveillance to select the most appropriate places to monitor, i.e. the locations most likely to provide answers to the questions in the condition indicators.

Collecting the Monitoring Data

To some degree at least, the methods of data collection will be determined by what we are monitoring. As a general rule, however, the data we collect should have a clear focus and should be collected using objective methods. The critical questions for a conservation manager centre on whether the key habitats and species on their site are in an acceptable state, and if not, what to do about it. Monitoring should answer the first of these questions as efficiently as possible.

For example, if there is a target for a migratory fish to be able to spawn 50 km upstream in a river, the obvious place to start sampling is in suitable spawning habitat at least 50 km upstream. If the fish are spawning there, we can assume that they can also spawn in suitable habitat downstream of this location. Similarly, for monitoring short-lived species at least, if there are always enough adults, we can make some informed assumptions about survivorship.

It is always worth remembering that, for practical conservation purposes, we simply need to know whether what we are doing is delivering what we want. Provided that this question has been clearly defined, the answer should be either ‘yes’ or ‘no’.

Feeding Back into Management

This is self-explanatory. The purpose of conservation monitoring is to feed back into the management of the habitat or species. As such, the site managers should be informed of the monitoring result as soon as it is known.

Safeguarding the Monitoring Data

The condition indicators, the monitoring data and any GIS and GPS files associated with a monitoring project should be bundled together and saved in several locations. Most organisations responsible for managing sites of conservation will have site files, i.e. folders where all of the information on a site is held. This is an obvious place to store the monitoring data. The site manager should also have a copy if the data, as should the monitoring surveyor. Critically, all of the files should be stored together in several different locations.

The Naturalness Issue

Perhaps surprisingly, for what is probably the most cultural of all habitats, the term ‘naturalness’ occurs commonly in the literature on freshwater conservation, and is often seen as the overriding conservation goal. This pursuit of naturalness undoubtedly has noble roots, but it is unlikely to deliver effective conservation, because we don't seem to know how to define it or, by association, how to achieve it.

In my experience, the word 'naturalness' appears at a very early stage in any discussion with freshwater specialists on what would represent a favourable state for a watercourse. In fact, in the CSM guidance for rivers (JNCC 2005) the word 'natural' occurs 99 times in one form or another:

- 'Natural' occurs 36 times
- 'Naturalness' occurs 34 times
- 'Naturally' occurs 10 times
- 'Semi-natural' occurs 9 times
- 'Naturalised' occurs 6 times (only once in relation to alien species)
- 'Unnaturally' occurs 2 times
- 'Near-natural' occurs 2 times

Rarely, if ever, in the guidance is there any attempt to define what these terms mean. Nor is there any attempt to explain the rationale behind trying to manage for and monitor something that we can't define.

There is, of course, a case for protecting the less modified rivers, but this is not guaranteed to deliver effective nature conservation: it will only serve to let processes take their course (natural or not). Some of the rivers of northern Scandinavia, for example, are amongst the least modified in Europe (physically at least): they are also amongst the worst affected by acidification (see Chapter 22). If naturalness is our aim, this presents us with the dilemma of whether to interfere in an attempt to restore the biological diversity of these rivers, or simply let nature take its course? The Scandinavian countries have opted to restore the biodiversity of their lakes and rivers.

There are also those who feel that the human population is a component of the natural world. If that is the case, then the current situation is natural for now, as it reflects the needs of the human population and the relative importance that the human population attaches to nature conservation. It would be wise to hold on to this thought when we are making our management decisions.

We can be reasonably certain that the human population will continue to want water for drinking and everyday luxuries like taking baths and showers, washing clothes, cooking, flushing toilets and recreation. We can also assume that water will be needed to manufacture cars, steel, synthetic fibre, paper, newspapers, and to service agriculture, etc. (Giller and Malmqvist 2005), and that our watercourses will come under increasing pressure for new hydropower schemes.

Against this background, the argument in favour of pursuing naturalness seems a little optimistic, if not misguided. Perhaps a more realistic approach would be to develop strategic plans, outlining which sections of which rivers are important for which habitats and which species, and include the optimal periods for water abstractions.

There is no doubt that the concept of 'naturalness' provides an interesting topic for philosophical debate, but whether it has any place in the practical delivery of nature conservation is far more debatable. Ironically, in deciding that our overriding aim is to return a river to its natural state, we are simply making another human intervention, one that will benefit some species and disadvantage others.

Finally…

In collating a selection of case studies for the book, we invited contributions representing a wide range of topics and approaches. Consequently, a few case studies describe a traditional approach to collecting trends data; some describe a generic 'one size fits all' form of condition assessment, while others describe the development of site-specific monitoring projects that incorporate early warning systems for conservation management.

Similarly, we chose to invite contributions from practitioners with alternative viewpoints, who consequently recommend different approaches. By doing this, we hope that the book will give readers an overview of the options available and provide them with a well-informed basis for making decisions about their own monitoring projects.

We also hope that this book will serve to promote constructive debate on a subject area that is in the formative stages of development.

References

Brown A (2000) Habitat monitoring for conservation management and reporting. 3: Technical guide. Life-nature project no LIFE95 NAT/UK/000821. Integrating monitoring with management planning: A demonstration of good practice in Wales. Countryside Council for Wales, Bangor

Giller PS, Malmqvist B (2005) The biology of streams and rivers. Oxford University Press, Oxford

Grime JP, Hodgson JG, Hunt R (1990) The abridged comparative plant ecology. Chapman & Hall, London

Grime JP (2001) Plant strategies, vegetation processes and ecosystem properties, 2nd edn. Wiley, Chichester

Ellenberg H (1979) Zeigerwerte der Gefäßpflanzen Mitteleuropas. Scripta Geobotanica 9:1–122 (in German)

Ellenberg H (1988) Vegetation ecology of Central Europe, 4th edn. Cambridge University Press, Cambridge

Ellenberg H, Weber HE, Düll R, Wirth V, Werner W, Paulisson D (1991) Zeigerwerte von Pflanzen in Mitteleuropa. Scripta Geobotanica 18:1–248 (in German)

Hellawell JM (1991) Development of a rationale for monitoring. In: Goldsmith B (ed) Monitoring for conservation and ecology. Chapman and Hall, London, pp 1–14

Hill MO, Mountford JO, Roy DB, Bunce RGH (1999) Ellenberg's indicator values for British plants. ECOFACT vol 2, Technical Annex. ITE/DETR. HMSO, Norwich

Hurford C, Schneider M (eds) (2006) Monitoring nature conservation in cultural habitats: A practical guide and case studies. Springer, Dordrecht, the Netherlands

JNCC (2005) Common standards monitoring guidance for rivers. Joint Nature Conservation Committee, Peterborough

Chapter 2
Options for Planning Management

Terry A. Rowell

Introduction

The approach to monitoring set out in this book depends on clear decisions being made about the objectives for habitats and species within a conservation site. These decisions should be made by site managers as a fundamental part of the general decision-making about site management; the development of the site management plan. The link between monitoring and management planning is well-established (Alexander and Rowell 1999; Rowell 2006). It should be taken as read that a management plan is only of conservation benefit if it is based on clear decision-making, and is actually put into practice.

But the position of site management plans in conservation management is paradoxical. The literature has always told us that they are important or even essential (e.g. Duffey *et al.* 1974; Green 1985; Furniss and Lane 1992) but, informally or anecdotally, they are often referred to as "shelfware" – impractical documents that are put away and forgotten about. Conservation organisations have produced many guides to management planning (e.g. Nature Conservancy Council 1983, 1988; Ramsar Convention Bureau 1993, 2002; Countryside Council for Wales 1996, 2003; Eurosite 1999, 2004; English Nature 2005), and no textbook on conservation management is complete without a chapter on how to write a plan (e.g. Sutherland and Hill 1995; Meffe *et al.* 1997). If you work for a conservation organisation, then they will probably insist that you use one of these guides.

The vast range of guidance on management planning presents a problem for the site manager. In general, these guides do not agree with one another. A review of nine planning guides aimed at UK protected sites showed that definitions of key terms were functionally inconsistent, and that rules and other guidance that expanded

T.A. Rowell
21 Cadnant Park, Conwy, LL32 8PR, UK
e-mail: tarowell@tiscali.co.uk

C. Hurford et al. (eds.), *Conservation Monitoring in Freshwater Habitats: A Practical Guide and Case Studies*, DOI 10.1007/978-1-4020-9278-7_2,

those definitions only enlarged the differences between them (Rowell 2009a). Further analysis of the same guides has shown that key decision-making processes, for objectives and management actions, are also inconsistent between guides. These processes are usually incomplete, often flawed, and sometimes absent altogether (Rowell 2009b). It is apparent that when planning guides are revised there is a tendency to change plan format, the definitions of key terms, and the decision-making processes – all without any explanation of why this has been done.

While there may be several possible explanations of why this happens, it is clear that experts do not agree on how to do management planning. It is also tempting to conclude that, possibly, no-one really knows the best way to make these important decisions, especially as there seem to be no documented cases of guidance or plans being tested. In a similar vein, some writers on conservation management conclude that we simply do not know whether our management systems work (Alexander 2008), though this does not sit easily with the frequent successes publicised by conservation organisations – perhaps these are in spite of our management systems.

So, the purpose of this chapter is not to recycle management planning guidance into yet another set of instructions. Rather, I will try to provide some suggestions that may benefit anyone facing these difficult, yet critical, decisions.

Learn to Plan

If you attend a conservation planning course in the UK, you will probably find that it reflects some particular brand of management planning guidance, and we should expect it to have similar problems. One remedy for this is to learn to plan from a better established discipline – one where plans have to be actioned and where poor quality or over-costly outcomes will not be tolerated. While there is no single way to plan in the business or IT sectors, there are accepted standards such as Prince 2. A simple place to start that covers all general aspects of planning is the free *Project Management Primer* (Jenkins 2006), and a wide variety of other free resources is available on the internet, such as those at www.projectsmart.co.uk. Courses in project management are now appearing on the agendas of some UK conservation training centres.

In general, there are no significant differences between a site management plan and any other type of project plan. They consist of a sequence of big decisions – what exactly are we trying to do, what are our quality criteria, what do we need to do in general terms to achieve these – followed by a series of many rather smaller decisions. These smaller decisions are just as important as the bigger ones, as they detail who will do the work, when they will do it, what equipment will be needed, and so on. General planning advice (e.g. Jenkins 2006) will help you work through these smaller decisions, and keep track of the work as it proceeds. But the big decisions need a different strategy.

A Strategy for the Bigger Planning Decisions

Understand Your Site

It may seem trite to say this, but unless the site is understood, you cannot hope to plan its management. You must understand why the habitats are where they are, what processes operate, and how they inter-relate and interact with the activities of people. Similarly, the autecology of the important species will provide clues to how best to manage them. Lastly, it is essential to have a good understanding of how people interact with the site, and who they are. If possible, find out how history has affected the site and its management in the past.

Some of this you can find from reading but, if your site is well known and important then there may simply be too much material; the well-known reserve at Wicken Fen in Cambridgeshire has a bibliography running to well over 800 separate publications (Rowell *et al.* 2007). This is the time to start involving others who, through long experience of this or similar sites, can guide you. Spread your net widely, and do not expect experts to agree all the time. There is, currently, an attempt to build an evidence base for conservation management, so you should consult www.conservationevidence.com and the website of the Centre for Evidence-based Conservation (www.cebc.bangor.ac.uk). The latter has carried out systematic reviews of some freshwater management issues (e.g. Stewart *et al.* 2006, 2007).

Draw a series of maps of your site with as much information as possible about its current situation, such as the locations of habitats, populations, watercourses, adjacent land-uses, and so on (see Nature Conservancy (1988) for simple examples, and Brown (2000) for more complex landscape block diagrams).

Understanding your site is not the same thing as filling in information under each heading in the "Description" which is usually the first section of any management plan. The Description does not necessarily result in understanding, though it would still be useful to summarise what you have discovered.

Nest Your Conservation Plans

The site management plan will not be the only plan for your site. There are bound to be a series of higher level directives, conventions, laws, policies, strategies and plans that affect the site, such as the EU Habitats Directive, the UK's Biodiversity Action Plans, catchment management plans, and so on. Log all of these that you know about, then ask if there are any others. Try to sort out what they mean for your site in terms of issues and priorities. For instance, many freshwater sites will have to take into account the Ramsar Convention. This requires the maintenance of ecological character, and you will have to plan not only how to do this, but also how to monitor unacceptable change (Davis and Brock 2008).

If the list of issues and priorities is long, then drawing a diagram will help sort out the relationships. This will be useful later when you try to prioritise habitats and species for management.

However important you believe your management plan to be, there are likely to be others written specifically for your site. Fire plans and Health & Safety plans are the most likely candidates, but it is not unusual for separate management plans to be written for the people-related aspects of a site, such as access and recreation. Developing a list of these now, and working out their relationships with the conservation plan, will help avoid duplication and ensure that the various plans interact properly.

Consider Your Options

Most management planning guides mention management options in some form, either based on features (e.g. Nature Conservancy 1988; Countryside Council for Wales 1996; Eurosite 1999) or so-called factors (Ramsar 2002; Countryside Council for Wales 2003; Alexander 2005); factors are discussed below. In some guides, the full list of options for managing any habitat is "active management, limited intervention, or non-intervention". This is a basic start, but assumes that other decisions have been made first, such as re-establishment of a lost habitat, or expansion of an existing one.

Sorting out your options at each decision point is crucial. The key decision points in most management plans are:

1. *What are the possible futures for the site?* Scenario planning, or setting out plausible futures (Peterson *et al.* 2003), is not normally included in management planning guidance. Possible futures could involve changing the proportions of habitats, re-establishing a lost habitat or process, or reintroducing a species.
2. Given 1, *what are the candidates for management?* These are usually habitats and species, but could include processes, especially in dynamic freshwater habitats. Whatever your final list of features to be managed, you will almost certainly need to prioritise.
3. *What are the objectives?* Objectives are normally written for the products of decision point 2. This book gives several examples of measurable objectives, and basic options should have been set at decision point 1. Here, you need to consider options for what aspects of individual features might be measurable, and the exact level that you will set as representing optimal condition, the basis for your monitoring.
4. *What actions will secure the objectives?* Many managers do not use scientific information to inform decisions (Pullin *et al.* 2003), but you should consult those evidence bases referred to earlier, conservation management handbooks (e.g. Ward *et al.* 1994; Crofts and Jefferson 1999; Benstead *et al.* 1999), and experts with specific knowledge of your habitats and species. There will often be a range of options – Brooks and Stoneman (1997) and Tait *et al.* (1988) give examples

for various habitats. When there is no clear winner amongst these options, you might like to use more than one in an experimental approach, if you have the space and resources. Some management planning guides seem to insist that you can only manage factors rather than the features themselves (e.g. Alexander 2005). Others point out that you also need to consider how you can make the features more resilient to impacts (Kapos *et al.* 2008). While I feel you should think about both, it's clear that increasing resilience can be at least partly dealt with at the objective-setting stage, by increasing population size, for instance.

5. *How do we future-proof our management?* Some management planning guides suggest that, if all factors are within predefined limits, then a feature will be safe into the foreseeable future (Countryside Council for Wales 2003; Alexander 2005). This can only be the case if security of management (e.g. funding, staffing, control over the management of surrounding land) are specifically dealt with at every site. So you should always develop options for future-proofing, and ensure that you are clear about the advantages and deficiencies of each.

Face Up To Risks and Assumptions

Each option that you consider, particularly those at decision points 4 and 5, will carry some level of risk. Logging risks may help you decide between options, and should clarify the risks arising from the decisions you finally take. There are simple ways of doing this. Rate the probability of each risk occurring as high, medium and low, then rate its impact similarly. Those risks with higher probability and impact should be avoided, or a contingency plan developed if avoidance is not possible.

Similarly, each decision will carry a greater or lesser degree of assumption. For example, a decision to graze in a particular month might be based on information from sites in a different biogeographical zone, so that the effects at your site are assumed. So long as assumptions are made explicit, they can be managed as risks. If your assumption is proved correct, then it can be removed from your log. Hopefully, this discussion makes it clear that you should log all decisions that you make in the running of your site, record the reasons for your decision, and any assumptions and risks.

A Word About Factors

Most management planning guidance advises that you should record constraints and factors, though they tend to use these terms in different ways. Some guides use the term *factor* to cover anything that might affect a feature you have decided to manage, not just negatively, but positively too (e.g. Countryside Council for Wales 2003; Alexander 2005). According to these guides, factors interact in complex ways, such as:

1. Habitats and species may themselves be factors for other habitats and species. Management will change the condition of features, and the future effect of this on other features needs to be considered.

2. The effects of factors may differ in combination with other factors, so they must be considered singly and in combination, with the possibilities of synergies borne in mind.
3. The effects of factors (and, presumably, combinations of factors) may change over time.
4. Factors can be positive or negative in effect (and, presumably, can be both, depending on intensity; e.g. water table may be negative when too high or too low, but beneficial when correct).

It would be easy to get bogged down in these complexities, and the guides that raise them give no suggestions about how to find resolutions. We have to look further afield, and bear in mind that we must focus on the factors and interactions most likely to have impacts on our priority habitats or species. Margoluis *et al.* (2009) recommend the use of simple diagrams to conceptualise these sorts of relationships. Recent management planning software includes a module to help you develop these conceptual models (see Miradi, based on Open Conservation Standards developed by the Conservation Measures Partnership (www.miradi.org)).

Conclusions

Management planning is important, but only if it results in plans that *can* be implemented, and *are* implemented. Too many management planning guides are muddled and impractical. The advice given here may help towards better plans where all options are considered, the reasons for decisions are clear, and assumptions and risks are properly accounted for.

References

Alexander M (2005) The CMS management planning guide. Conservation Management System Consortium, Talgarth

Alexander M (2008) Management planning for conservation; a theoretical basis and practical guide. Springer, Dordrecht, the Netherlands

Alexander M, Rowell TA (1999) Recent developments in management planning and monitoring on protected sites in the UK. Parks 9:50–55

Benstead P, Jose P, Wade M, Joyce C (1999) European wet grassland: guidelines for management and rehabilitation. Royal Society for the Protection of Birds, Sandy

Brooks S, Stoneman R (1997) Conserving bogs: The management handbook. The Stationery Office, Edinburgh

Brown A (2000) Habitat monitoring for conservation management and reporting. 3: Technical guide. Life-Nature Project No. LIFE95 NAT/UK/000821. Integrating monitoring with management planning: a demonstration of good practice in Wales. Countryside Council for Wales, Bangor

Countryside Council for Wales (1996) A guide to the production of management plans for nature reserves and protected areas.Countryside Council for Wales, Bangor

Countryside Council for Wales (2003) Management planning for SSSIs, Natura 2000 sites and Ramsar sites. Countryside Council for Wales, Bangor

Crofts A, Jefferson RG (1999) The lowland grassland management handbook. English Nature, Peterborough

Davis J, Brock M (2008) Detecting unacceptable change in the ecological character of Ramsar wetlands. Ecol Manage Restor 9:26–32
Duffey E, Morris MG, Sheail J, Ward LK, Wells DA, Wells TCE (1974) Grassland management and wildlife management. Chapman and Hall, London
English Nature (2005) NNR management plans: A guide. English Nature, Peterborough
Eurosite (1999) Eurosite toolkit: Management planning. Eurosite, Tilburg, the Netherlands
Eurosite (2004) Eurosite management planning toolkit, complementary guidance: A handbook for practitioners. Eurosite, Tilburg, the Netherlands
Furniss P, Lane A (1992) Practical conservation: Water and wetlands. Hodder & Stoughton, London
Green B (1985) Countryside conservation: the protection and management of amenity ecosystems. George Allen & Unwin, London
Jenkins N (2006) A project management primer. Available at: www.nickjenkins.net/prose/projectPrimer.pdf (accessed on 18 April 2009)
Kapos V, Blamford A, Aveling R, Bubb P, Carey P, Entwistle A, Hopkins J, Mulliken T, Safford R, Stattersfield A, Walpole M, Manica A (2008) Calibrating conservation: New tools for measuring success. Conserv Lett 1:155–164
Margoluis R, Stem C, Salafsky N, Brown M (2009) Using conceptual models as a planning and evaluation tool in conservation. Eval Program Plan 32:138–147
Meffe GK, Carroll CR, Contributors (1997) Principles of conservation biology. Sunderland, MA, USA: Sinauer Associates
Nature Conservancy Council (1983) A handbook for the preparation of management plans. Nature Conservancy Council, Peterborough
Nature Conservancy Council (1988) Site management plans for nature conservation. Nature Conservancy Council, Peterborough
Peterson GD, Cumming GS, Carpenter SR (2003) Scenario planning: A tool for conservation in an uncertain world. Conserv Biol 17:358–366
Pullin AS, Knight TM, Stone DA, Charman K (2003) Do conservation managers use scientific evidence to support their decision-making? Biol Conserv 119:245–252
Ramsar Convention Bureau (1993) Guidelines on management planning for Ramsar sites and other wetlands (first adopted as an annex to resolution 5.7, 5th meeting of the conference of the contracting parties, Kushiro, Japan, June 1993). Ramsar Convention Bureau Gland, Switzerland
Ramsar Convention Bureau (2002) New guidelines for management planning for Ramsar sites and other wetlands (adopted by Resolution VIII. 14). Ramsar Convention Bureau Gland, Switzerland
Rowell TA (2006) The relationship between monitoring and management. In: Hurford C, Schneider M (eds) Monitoring nature conservation in cultural habitats: A practical guide and case studies. Springer, Dordrecht, the Netherlands, pp 379–384
Rowell TA (2009a) Management planning guidance for protected sites in the UK; a comparison of formats and definitions in nine guides. J Nat Conserv 17:99–111
Rowell TA (2009b) Management planning guidance for protected sites in the UK; a comparison of formats and definitions in nine guides. J Nat Conserv 17(3):99–113
Rowell TA, Friday L, Warrington S (2007) Wicken Fen bibliography. Available at: www.wicken.org.uk/research_bibilography.htm (accessed on 18 April 2009)
Stewart GB, Bayliss HR, Showler DA, Pullin AS, Sutherland WJ (2006) Does the use of in-stream structures and woody debris increase the abundance of salmonids? Systematic Review No. 12. Collaboration for Environmental Evidence
Stewart GB, Bayliss HR, Showler DA, Sutherland WJ, Pullin AS (2007) What are the effects of salmonid stocking in lakes on native fish populations and other fauna and flora? Systematic Review No. 13a. Collaboration for Environmental Evidence
Sutherland WJ, Hill DA (1995) Managing habitats for conservation. Cambridge University Press, Cambridge
Tait J, Lane A, Carr S (1988) Practical conservation: site assessment and management planning. Hodder and Stoughton, London
Ward D, Holmes N, Jose P (1994) The new rivers and wildlife handbook. Royal Society for the Protection of Birds, Sandy

Chapter 3
The Water Framework Directive and the Habitats and Birds Directives

An Overview of the Legal and Technical Relationship

Chris Uttley

Background

The Water Framework Directive (WFD) (2000/60/EC) was established by the European Union (EU) as "a framework for community action in the field of water policy". It contains a variety of provisions describing how member states of the EU should manage their surface and ground waters, and can essentially be split into two different types of regime within a single framework. These regimes can be described as (i) a new regime for determining the ecological and chemical quality of surface and ground waters, and achieving the relevant default objective, and (ii) a framework for implementing existing EU legislation (in the field of water policy) and integrating this with the new 'ecological' regime.

The new framework has been the subject of much EU guidance and discussion, particularly through the Common Implementation Strategy established by the European Commission and, through Government advisory groups, has been the focus of most legal and technical work in the UK to date. However, the provisions of the WFD, as they apply to areas protected under the Habitats (Directive 92/43/EEC) and Birds Directive (79/409/EEC) have been less well discussed (Withrington 2005). These provisions are of interest for a number of reasons. Firstly, in the UK, the Habitats and Birds Directives are the only Directives listed within the WFD that are not implemented primarily by the same competent authorities that are responsible for the WFD or the other Directives captured within the field of water policy. They, therefore, represent an institutional challenge to implementation. Secondly, they are the only other Directives within the framework that have their

C. Uttley
Water Policy Officer, Countryside Council for Wales, Maes y Ffynnon, Ffordd Penrhos, Bangor, Gwynedd LL57 2DW, UK
e-mail: c.uttley@ccw.gov.uk

C. Hurford et al. (eds.), *Conservation Monitoring in Freshwater Habitats: A Practical Guide and Case Studies*, DOI 10.1007/978-1-4020-9278-7_3,

own ecological objectives and outcomes. They establish a different way of determining and judging ecological quality, based upon criteria described in the Habitats Directive and the way these have been applied by the relevant agencies in the UK. Thirdly, the Habitats and Birds Directives establish a strong precautionary legal framework for the EU, which in many cases goes beyond that created by the WFD. The WFD provides opportunities for member states to establish longer deadlines, less stringent objectives and allows for deterioration of ecological status under a wider range of conditions than those contained in the Habitats Directive. This difference in level of precaution presents legal and economic challenges to member states.[1]

The following sections describe the provisions of the WFD as they apply to the Natura 2000 network of sites, and the final section summarises these and describes challenges that remain, particularly for interpretation in relation to monitoring requirements and the application of the provisions for species in the Natura Directives.

Which Natura 2000 Protected Areas Are Relevant for the WFD?

Article 6 of the WFD requires Member States to establish a register of all Protected Areas lying within each river basin district that have been designated as requiring special protection under specific Community legislation for the protection of their surface water or groundwater or for the conservation of habitats and species directly depending on water.

Of course all species and habitats depend on water for their survival, but it is possible to identify varying degrees of dependency, which determine the management and monitoring priorities for different species and habitats. Many species and habitats are wholly aquatic, whereas others rely upon saturation or inundation by water (fresh and saline) for their existence. The UK Technical Advisory Group[2] developed a list of water-dependent species and habitats from Annex I and II of the Habitats Directive and the Birds Directive. Species and habitats were segregated according to three types of dependency. This dependency determines (in part) the relevance of each species or habitat to the WFD, although no distinction is made in the Directive itself. Table 3.1 illustrates the definitions used in the guidance document (UK TAG 2003).

In the UK, any Natura 2000 protected area designated for one or more water dependent species or habitat has been entered onto the Register of protected areas.

[1]Water Framework Directive, Article 4 (3, 4, 5,6,7).

[2]UK Technical Advisory Group, (UK TAG) formed from the UK environment and conservation agencies to advise Government on technical aspects of WFD implementation. www.wfduk.org.

Table 3.1 Ecological criteria for identifying those Natura Habitats and Species that are directly dependent on the status of water

Natura 2000 SPECIES	Natura 2000 HABITATS
Aquatic species living in surface waters as defined in Article 2 of the Water Framework Directive (e.g. bottle-nose dolphin *Tursiops truncatus*, freshwater pearl mussel *Margaritifera margaritifera*)	Habitats which consist of surface water or occur entirely within surface water, as defined in Article 2 of the Water Framework Directive (e.g. oligotrophic waters; estuaries; eelgrass beds)
Species with at least one aquatic life stage dependent on surface water (i.e. species that use surface water for breeding; incubation, juvenile development; sexual maturation, feeding or roosting – including many Natura bird and invertebrate species)	Habitats which depend on frequent inundation by surface water, or on the level of groundwater (e.g. alluvial alder wood, blanket bog, fens)
Species that rely on the non-aquatic but water-dependent habitats relevant under 2.b and 2.c in the habitats column of this table (e.g. Killarney fern)	Non-aquatic habitats which depend on the influence of surface water – e.g. spray, humidity (bryophyte-rich gorges)

What Are the Relevant Standards and Objectives for Natura 2000 Protected Areas in the WFD?

As described in the opening paragraphs, the WFD is both a "framework" for ensuring the implementation of existing EU legislation and a new way of evaluating the ecological status of waters. Article 4 of the WFD describes the environmental objectives for the different categories of water and for protected areas.

The environmental objectives for surface and groundwater bodies are contained within Article 4 (1a & 1b). These objectives apply only to water as defined within Article 2 of the Directive. In contrast, although many Natura 2000 protected areas may contain, or may be, water bodies (rivers, lakes and estuaries), they may also contain water not identified as a 'water body', or in many cases, areas of land. Water dependent habitats include wetlands, such as fen, blanket bog and saltmarsh. Otters *Lutra lutra*, marsh fritillary *Eurodryas aurinia* butterflies and many bird species occupy and use areas of land, wetland and water bodies. The environmental objective for these areas of water and land are established in Article 4 (1c).

Article 4 (1c) states:

> In making operational the programmes of measures specified in the river basin management plans, for protected areas, Member States shall achieve compliance with any standards and objectives at the latest 15 years after date of entry into force of this Directive, unless otherwise specified in the Community legislation under which the individual protected areas have been established.

Article 4 (2) states:

> Where more than one of the objectives in paragraph 1 relates to a given body of water, the most stringent shall apply.

Article 4 (1c) therefore applies to any Natura 2000 protected area listed on the register of protected areas. Of particular interest is the use of the phrase "...shall achieve compliance with any standards and objectives...". This formulation differs from other parts of the WFD, which requires member states to protect, enhance and restore with the *aim* of achieving good surface or groundwater status. It implies that member states needs to have achieved all the relevant standards and objectives for Natura 2000 protected areas before 2015. This arguably goes further than the Habitats Directive itself, which also includes the formulation of "aiming" to achieve Favourable Conservation Status.

Whilst the Habitats and Birds Directives set no statutory 'standards' that have to be achieved, they do require the setting of 'conservation objectives' for each habitat and species for which the Natura 2000 site is designated. Conservation Objectives are set in order to determine if the aims and objectives of the Directives are being met, to ensure that measures taken actually maintain or restore features to favourable conservation status, and finally, to help determine the impact of any plans and projects on the features of the site. Conservation objectives are also used to assess whether a designated habitat or species is in a "favourable condition" on each Natura 2000 area, and contributing to Favourable Conservation Status at the member-state scale.

The UK has interpreted the obligations of Article 4 (1c) as requiring all water dependent species and habitats for any Natura 2000 protected areas listed on the register, to be in Favourable Condition, or if not, to have identified the reason "why not" as being clearly unrelated to water management. This requires that water-dependent features are meeting their conservation objectives (UK TAG 2008).

In the UK, conservation objectives have been developed as expressions of intent for the features, with a list of targets for a variety of attributes, such as water quality, population characteristics or physical habitat. Whilst these targets are not themselves standards, they are used as part of a 'weight of evidence' approach to determine whether or not a particular species or habitat is achieving favourable condition.

For Natura 2000 protected areas, it is these conservation objectives and the targets contained within them that form the "standards and objectives" referred to in Article 4 (1c).

Application of the Derogations

Article 4 (3, 4, 5, 6, 7) of the WFD contains the relevant provisions allowing member states to achieve less stringent objectives, extend deadlines or allow the status of water bodies to deteriorate (the derogations).

Article 4 (paragraphs 8 & 9) establish the conditions attached to the use of the derogations. These two paragraphs effectively prevent the use of any of the derogations (apart from time extensions), on Natura 2000 sites, since in practice, the only reason a member state can allow a Natura 2000 protected area to deteriorate is to allow a development that is of "imperative overriding public interest". (Habitats Directive Article 6 (4)).[3]

There has been some debate regarding the extension of the deadline for achieving Natura 2000 standards and objectives. Article 4 (4) states that the deadlines in paragraph 1 (2015) may be extended by member states for the purpose of phased achievement of the objectives for *bodies of water, provided that no deterioration occurs in the status of the affected body of water....*

Article 4 (4) specifically refers to *bodies of water*, not protected areas. Since the WFD contains specific obligations for protected areas, it can be argued that since this provision does not specifically refer to them, it was not intended to apply to them. In addition, for member states to make use of extended deadlines, they must ensure that no deterioration in the status of the water body occurs. For some impacts, notably diffuse pollution of lakes, proliferation of alien and invasive species and sediment pollution, preventing deterioration over the extended timescale will be extremely difficult, without some form of measures to address the cause of the impact.

The current UK position (in April 2009) is that where Natura 2000 protected areas are coincident with water bodies, the deadline for achievement of standards and objectives for those sites may be extended beyond 2015 provided the conditions in Article 4(4) are met. For sites not coincident with water bodies, the deadline cannot be extended. Thus for fens, bogs, saltmarsh and other wetlands, a strict deadline of 2015 for achieving the conservation objectives to contribute to Favourable Conservation Status applies.

The Programme of Measures

Article 11 requires the establishment of a "programme of measures" to achieve the environmental objectives of the WFD. The programme of measures is in effect a list of management actions that a member state will take at a national, regional and local level to achieve the various objectives, including those for Natura 2000 water dependent protected areas.

---

[3]Article 6 (4) of the Habitats Directive states "If, in spite of a negative assessment of the implications for the site and in the absence of alternative solutions, a plan or project must nevertheless be carried out for imperative reasons of overriding public interest, including those of a social or economic nature, the Member State shall take all compensatory measures necessary to ensure that the overall coherence of Natura 2000 is protected. It shall inform the Commission of the compensatory measures adopted. Where the site concerned hosts a priority natural habitat type and/or a priority species, the only considerations which may be raised are those relating to human health or public safety, to beneficial consequences of primary importance for the environment or, further to an opinion from the Commission, to other imperative reasons of overriding public interest."

These actions may include legislative changes, new regulations, or individual and localised management actions required to control the impacts identified on a specific water body or protected area.

The programme of measures for Natura 2000 sites must consist of measures required by the Habitats Directive itself. For example, the review of all significant consents to ensure they are not having an adverse effect on the protected area. It should also include controls, or remedies, for unregulated impacts, such as diffuse pollution, alien species and drainage. Impacts to the physical morphology of waters are a good example. With the exception of modifications relating to flood defence, there are few mechanisms by which water bodies and protected areas can have their morphological integrity restored. If this is resolved, it is likely to be one of the most significant benefits arising from the WFD.

Monitoring for Natura 2000 Water Dependent Protected Areas in the WFD

Whilst there are detailed provisions within the Habitats Directive (Article 11) requiring the monitoring and surveillance of features, the WFD itself contains requirements for additional monitoring of water dependent Natura 2000 protected areas.

Annex V (1.3.5) of the WFD describes these additional requirements. It states that:

> Bodies of water forming these areas shall be included within the operational monitoring programme referred to above where, on the basis of the impact assessment and the surveillance monitoring, they are identified as being at risk of failing to meet their environmental objectives under Article 4. Monitoring shall be carried out to assess the magnitude and impact of all relevant significant pressures on these bodies and, where necessary, to assess changes in the status of such bodies resulting from the programmes of measures. Monitoring shall continue until the areas satisfy the water-related requirements of the legislation under which they are designated and meet their objectives under Article 4.

This strongly implies that additional monitoring of protected areas is required as part of the WFD. However, it is currently unclear whether additional monitoring is required, and, if this is the case, whether the monitoring programmes implemented by the competent authorities in the UK (for both the WFD and Habitats Directives) have the capacity to take account of this. Taking into account the potential economic implications of decisions made with respect to judgements on Favourable Condition and Ecological Status, and the deadline established by the WFD for water dependent Natura 2000 protected areas, the level of resource dedicated to monitoring will probably need to increase during the next phase of river basin planning.

In addition, many wetland habitats are still relatively poorly understood. Their functioning, in terms of irrigation and nutrient requirements, are still the subject of detailed study. A significant level of resource is required to develop a detailed understanding of the measures required on a site-by-site basis. These site investigations are essential if the right type and intensity of measures are to be implemented.

Summary and Remaining Challenges

It is clear that the authors of the Water Framework Directive created specific obligations for member states to ensure that the water requirements of Natura 2000 protected areas are provided for. The regime they created, if implemented correctly, should fully integrate the requirements for water dependent species and habitats into a River Basin Management Plan.

The WFD creates the requirement to achieve any standards and objectives for protected areas before December 22nd 2015, but this can only happen with concerted efforts from member states Governments' and the competent authorities responsible for implementing both the Water Framework Directive and Natura Directives. Many protected areas will of course take considerably longer to recover ecologically from years of degradation, but the WFD should ensure that measures to restore habitats and species are implemented within a planned framework before 2012.

Many new mechanisms will be required to tackle impacts from drainage, diffuse pollution and invasive species. Deriving the true financial cost of this work will illustrate the gap between what is currently funded and the level of resource actually needed to bring these sites into favourable condition.

Although the UK has produced a body of guidance on this issue, there has to date been little or no consideration of how the WFD can provide a framework for achieving those provisions of the Habitats and Birds Directives relating to species and habitat protection outside of protected areas. Reporting on the status of freshwater species and habitats outside the protected sites series will need further consideration in the second cycle of River Basin Planning.

Finally, it is not yet clear that the monitoring programmes implemented by the competent authorities in the UK give effect to all the provisions for monitoring of Natura 2000 protected areas contained within the WFD. Certainly we know that in many cases, detailed site-specific investigations may be required to ensure that the right measures are adopted. In many cases, the resource for this work remains to be found.

References

UK TAG (2003) UK technical advisory group on the water framework directive – Guidance on the identification of natura protected areas. Available at http://www.wfduk.org/tag_guidance/Article_06-07/TAG2003%20%20WP4a%20%28PR03%29 (accessed on 25 April 2009)

UK TAG (2008) UK technical advisory group on the water framework directive – Guidance on determining whether Natura 2000 Protected Areas are meeting the requirements of Article 4 (1c) for the 1st RBMP. Available at http://www.wfduk.org/tag_guidance/article_4/n2kareas_article4 (accessed on 25 April 2009)

Withrington D (2005) Water Framework directive (2000/60/EC): Interpreting the provisions applying to Protected Areas for habitats and species directly depending on water. J Water Law 16:197–200

Section 2
Biological Indicators for Freshwater Habitats

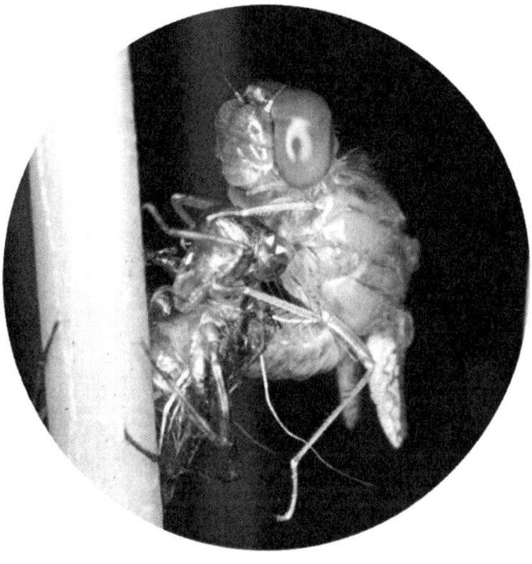

Chapter 4
Freshwater Mammals as Indicators of Habitat Condition

Michael Schneider

Introduction

In this chapter, I give a rather personal view of the feasibility of using freshwater mammals as indicators for the condition of sites or habitats. The conclusions drawn are open for discussion.

Mammals as Indicators

Mammals are not as often used as indicators as other groups of organisms, such as plants, lichens, invertebrates or birds (see Chapters 5–10). Freshwater habitats are no exception to this general rule. There are many reasons for that:

- Mammals are often relatively hard to find in the field, as they are often nocturnal and crepuscular, secretive and shy.
- Their biology and their ecological needs are often badly understood.
- They often roam over relatively large areas and thus would reflect ecological conditions of extended areas, not necessarily of the point where they are observed.
- Many mammals are affected by people directly through hunting or illegal killing. Therefore a decline may simply be an indication of human attitudes towards the species rather than the result of any decline in the abiotic or biotic condition of the environment.

M. Schneider
County Administration, Västerbotten County, SE-901 86, Umeå, Sweden
e-mail: Michael.Schneider@lansstyrelsen.se

C. Hurford et al. (eds.), *Conservation Monitoring in Freshwater Habitats: A Practical Guide and Case Studies*, DOI 10.1007/978-1-4020-9278-7_4,

- Many mammals have a long generation time and thus reflect changes in the environment relatively slowly at the population level. In many species, individuals are long-lived and survive for a while when conditions already have become adverse.
- Mammals are mobile and individuals may visit areas that are not optimal for the species.

For instance, Sweden has put considerable effort into the definition of "typical species" when developing a system for the monitoring of conservation status in areas and habitats included in the Natura 2000 framework. For a species to be designated typical for a habitat, it has to indicate favourable conservation status in this habitat, there have to be non-destructive ways of surveying the species, and it has to be easily identifiable. Not a single mammal species is included in the current (2008) list of typical species for freshwater habitats in Sweden (J. Grahn, 2009 personal communication).

Freshwater Mammals

Among the mammals within the current extent of the European Union, there are no truly aquatic freshwater species. There are, however, several species that are semi-aquatic and that in one way or the other are associated with or depend on freshwater habitats. There is also one seal species that inhabits brackish waters in the Baltic Sea as well as the freshwater lake Saimaa in Finland (see Chapter 27).

Here I define 16 species as being freshwater mammals (Table 4.1). All of them are dependent on freshwater habitats, but to different extent. Their possible value as indicators of condition differs between species and may for a given species even vary between areas.

Many of the species have been surveyed in a variety of areas throughout Europe. Commonly, these surveys were aiming at establishing the status of the species itself, not the condition of the habitat the species lives in.

The fact that freshwater mammals are not widely used as indicators for condition today does not mean that they cannot be used at all. However, mammals are more demanding than many other organisms. One has to know a lot about a species' ecology to be able to use it as an indicator. Good population estimates alone may not be very helpful.

An example to illustrate this is the case of the ringed seal in the Baltic Sea (Schneider *et al.* 2009). In this species, the observed changes in population size during the last 100 years indicated very different things. First, the rapid decline of the population was caused by over-hunting. Then, when hunting had been stopped, environmental contamination caused a further decline. Today, conservation efforts (no hunting, decrease in pollutants) have led to an increase of the population. In the future, climate change and a decreasing ice cover in the Baltic Sea will most probably cause a renewed decline of the ringed seal (see Chapter 27).

Table 4.1 Sixteen mammals that can be defined as being freshwater species. Most of them could have value as indicators, indicating either a positive or negative condition of the site or habitat

Common name	Scientific name	Potential indicator?	Indicating which condition?
Pyrenean desman	*Galemys pyrenaicus*	Yes	Positive
European water shrew	*Neomys fodiens*	No	–
Mediterranean water shrew	*Neomys anomalus*	No	–
Pond bat	*Myotis dasycneme*	Yes	Negative
Daubenton's bat	*Myotis daubentoni*	Yes	Negative
Long-fingered bat	*Myotis capaccinii*	Yes	Positive
Water vole	*Arvicola terrestris*	No	–
Southern water vole	*Arvicola sapidus*	Yes	Positive
Muskrat	*Ondathra zibethica*	Yes	Negative
Coypu	*Myocastor coypus*	Yes	Negative
European beaver	*Castor fiber*	No	–
American beaver	*Castor canadensis*	Yes	Negative
European mink	*Mustela lutreola*	Yes	Positive
American mink	*Mustela vison*	Yes	Negative
Otter	*Lutra lutra*	Yes	Positive
Ringed seal	*Phoca hispida*	Yes	Positive

Indicating What?

A recent assessment of the status of all native mammal species in Europe has compiled much information about the factors that affect and in some cases threaten mammal populations (IUCN 2007). Table 4.2 presents a list of specific threats for the different species. A healthy population of a species would indicate the absence of those adverse factors that have been listed as threats (see IUCN 2007 and references therein).

For six of the 16 freshwater species, no threats at all have been listed. These six species include Daubentons' bat, muskrat, coypu, European beaver, American beaver and American mink. Habitat loss and degradation, accidental mortality and pollution have been identified as general threats to most of the remaining species.

Alien Species

The presence of the alien species muskrat, coypu, American beaver and American mink is a negative indicator for the condition of freshwater habitats. Muskrats may create problems for water voles and may have a great impact upon the

Table 4.2 Specific factors threatening the survival of native freshwater mammals in Europe according to IUCN (2007). Species not considered to be threatened are not listed here. The alien species muskrat, coypu, American beaver and American mink were not included in IUCN's compilation

Specific threat	Pyrenean desman	European water shrew	Med. water shrew	Pond bat	Long-fingered bat	Water vole	Southern water vole	European mink	Otter	Ringed seal
Habitat loss or degradation	X	X	X	X	X	X	X	X	X	
Invasive alien species	X					X	X	X		
Harvesting								X	X	
Accidental mortality	X		X	X			X	X	X	X
Persecution	X									
Pollution	X	X	X		X	X		X	X	X
Changes in native species dynamics									X	
Intrinsic factors	X									
Human disturbance	X			X					X	X

vegetation in freshwater habitats (IUCN 2007; Pietsch 1982). Muskrats and coypu compete for food and dens with southern water voles (IUCN 2007). The American beaver apparently displaced the European Beaver in sites where both species were released in Finland (Tattersall 1999). The occurrence of American mink indicates problems for the European mink in areas where both species co-occur (IUCN 2007). American mink seem also to negatively affect water vole populations (Strachan *et al.* 1998) and populations of breeding water birds (Sundström and Olsson 2005).

European Beaver

Beavers have a very wide ecological amplitude and occur in a variety of waters (Fig. 4.1). European beavers have been re-introduced in many areas in Europe (Freye 1978). A disappearance of beavers most often indicates human intervention: excessive hunting, destruction of dams or road casualties. Therefore, beavers indicate human attitudes and activities rather than the condition of their habitat.

Fig. 4.1 A beaver pond in the agricultural landscape of northern Sweden. In spite of their ability to change ecosystems, European beavers are not good indicators for the condition of a given site. *Photo by Michael Schneider*

Pyrenean Desman

The Pyrenean desman is well adapted to an aquatic environment. It lives mostly in fast flowing, clean, well-oxygenated watercourses with structured banks. The species is threatened and has a restricted distribution (Juckwer 1990; IUCN 2007).

The species itself is threatened and seems to have high demands on its freshwater habitat. A thriving population of desmans should therefore indicate good condition for a site, although Queiroz (1999) mentions that the species can survive in some moderately polluted watercourses.

European Water Shrew

Analyses of the results of a nationwide, volunteer-based survey to determine the distribution and habitat occurrence of the water shrew in Great Britain showed that water shrews were ubiquitous and adaptable and that they are ecologically flexible (Carter and Churchfield 2006). Water shrews were observed in sites with a variety of substrate types, water depths and water widths. They were little influenced by bank height and incline and were apparently not disturbed by human activity. Neither bankside management nor the land use adjacent to a site influenced water shrew occurrence. Water shrews were found both in sites with and without aquatic vegetation (compare Fig. 4.2).

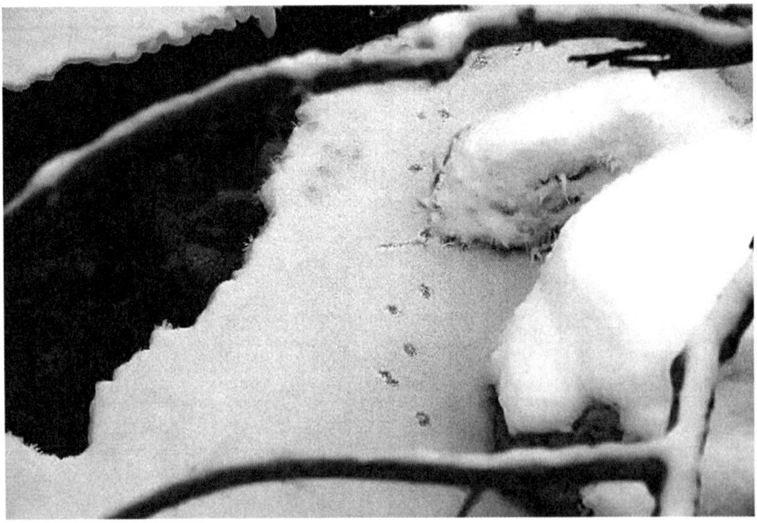

Fig. 4.2 Tracks of a European water shrew on snow-covered ice along a small Swedish forest river. The value of European water shrews as indicators of the condition of a site seems limited, and snow tracking is not a good method for surveying water shrews. *Photo by Michael Schneider*

Therefore, European water shrews do not seem to be good indicators of condition of a certain habitat, not in Great Britain at least. According to Spitzenberger (1990a, 1999a), water shrews need structural diversity of river banks and a sufficient prey base, the latter being indicative of low levels of pollution in the habitat.

Mediterranean Water Shrew

Where the two species co-occur, the Mediterranean water shrew is excluded from freshwater habitats by competition with the larger European water shrew (Spitzenberger 1999b). An abundant population of Mediterranean water shrews should therefore indicate the absence of European water shrews. As the species can utilise habitats of differing quality (Spitzenberger 1990b), its value as indicator of condition seems rather limited.

Pond Bat and Daubenton's Bat

The Pond bat and Daubenton's bat are two species which are favoured by the general increase in productivity that is obvious in many freshwater habitats, especially ponds and slow-flowing rivers (e.g. Meschede and Rudolph 2004). While the Daubenton's bat is one of the most common bat species in Europe, the pond bat is rare and threatened (IUCN 2007).

The life history of hibernating bats is complex. For the pond bat and Daubenton's bat, freshwater habitats comprise only a part of their home range, the part where they collect their food. Other important variables are summer roosts for both males and females and sites for over-wintering. Both foraging sites, summer roosts (cavities in trees, nest boxes, buildings for pond bats) and hibernacula (caves and cellars) are limiting factors if in short supply, and all these partial habitats have to be relatively close together. Even if food is over-abundant, bat populations will not increase in an area, if there are few roosts or if hibernacula are rare.

Therefore, an increase in foraging activity of pond bats or Daubenton's bats at a freshwater site could indicate an increase in productivity of the site, and thus a negative condition if the site is supposed to be nutrient-poor. A general population increase, on the other hand, could even depend on an increase in the abundance or quality of roost sites or hibernacula in the area.

Bat foraging activity can be surveyed with hand-held or automated bat detectors around water bodies. Surveys of roosts and hibernacula are complicated for the species. Summer roosts are mainly in hollow trees, woodpecker holes or nest boxes and therefore often hard to locate and to inspect. In hibernacula, the animals tend to creep into narrow crevices in the roof, walls or floor of the site and are therefore often hard to find (Meschede and Rudolph 2004).

Long-Fingered Bat

The long-fingered bat is a southern species and widespread but relatively rare in the Mediterranean (Guillén 1999). Similar to pond and Daubenton's bats, it mostly forages over wetlands and waterways. Summer roosts and hibernacula are mostly in caves. Nursery colonies contain usually several hundreds of animals. Movements between summer and winter sites are mostly within a distance of 50 km (IUCN 2007). Long-fingered bats may interact with Daubenton's bats and the species may competitively exclude each other (Guillén 1999). Hunting habitats of the species have been negatively influenced by agricultural pesticides in Greece (Dietz *et al.* 2007).

As summer roosts are relatively easy to find and monitor, their effect on the population dynamics of the species can be assessed. If the number of individuals hunting at a freshwater site decreases while roosts remain undisturbed, this may indicate a negative development of condition of the site due to a diminishing prey base, for instance caused by pollution.

Water Vole

The aquatic form of the water vole inhabits a variety of freshwater habitats in lowland and mountain areas, including rivers, streams, lakes and marshes. Steep riverbanks with lush grass and vegetation are preferred (compare Chapter 26). Declines in aquatic populations in parts of Western Europe have been attributed to habitat loss, water pollution, predation by introduced American mink and competition with muskrats (IUCN 2007; Jefferies 2003). As the species can utilise habitats of differing quality, its value as indicator of condition seems restricted.

Southern Water Vole

The southern water vole is distributed in south-western Europe and inhabits lakes, ponds and slow-moving rivers and streams with dense riparian vegetation (Saucy 1999). According to IUCN (2007), it is restricted to sites where banks and vegetation have not been significantly altered by human activity. Competition with coypu and muskrats and possibly the brown rat (*Rattus norvegicus*) for food and dens is also listed as a problem. As the species seems rather specialised regarding its habitat preferences and as concrete threats have been identified, a thriving population of voles should indicate favourable condition at a given site.

European Mink

European mink have specialised habitat requirements. They are semi-aquatic, inhabiting densely vegetated banks of lakeshores, rivers, streams and marshlands. They are nocturnal and have a varied diet. The European mink is a threatened species with a restricted range. Main threats include habitat loss, harvesting, accidental killing and pollution. American mink affect European mink negatively by competition and hybridisation (IUCN 2007 and references therein).

Due to its specialised habitat requirements and its restricted range, the European mink could be a good indicator of condition of the sites where it occurs. Healthy populations would indicate a positive condition, while decreasing populations would indicate a negative condition of sites.

Otter

Lots of work has been done on otters (Fig. 4.3). Mostly, this work focused on the otter as a threatened species and not on otters as indicators for the condition of their habitat. Starting from the threats that have been identified as affecting otters (IUCN 2007), this species may be indicative of a PCB-free environment, of sufficient fish populations, of low levels of persecution (including illegal hunting, road casualties,

Fig. 4.3 Otters have been surveyed in many areas to determine the species' conservation status. We are only beginning to think of the species as a possible indicator of the condition of a given site, however. *Photograph by Michael Schneider*

and by-catches) and of the structural complexity of river banks. Otters may also indicate low population levels of American mink (Jefferies 2003).

Spraint surveys can be used for retrieving presence-absence data for large areas at relatively low cost. However, to determine the size of a population and for counting family groups (which indicate successful breeding in an area) more labour-intensive and therefore costly methods have to be applied, such as snow tracking or surveys among anglers and the public.

Ringed Seal

Ringed seals are top consumers in the brackish waters of the Baltic Sea and in the freshwaters of Lake Saimaa. Annual aerial surveys are used in the Baltic to get an index of population size of ringed seals. Necropsy of dead animals is used to look at body conditions and health status of individuals (see Chapter 27).

Increased levels of PCB are believed to be responsible for a disease complex (pathological kidney alterations, uterine occlusions) found in ringed seals between 1965 and 1985 (Bergman *et al.* 2001). Since then, levels of contaminants have been decreasing, but still 30–40 % of adult females are sterile (Mattson and Helle 1995). This infertility in females is thought to be responsible for a reduced rate of population growth. Therefore, the health status of ringed seals and their rate of population increase can be used to indicate the condition of their habitat regarding the occurrence and the concentration of pollutants in the water.

Brief Summary

In this chapter, I define 16 mammals as being freshwater species (Table 4.1). Of these, four (European water shrew, Mediterranean water shrew, water vole, European beaver) do not seem to have any greater value as indicators for the condition of a given habitat or site. Six species (pond bat, Daubenton's bat, coypu, muskrat, American beaver, American mink) can be indicative of a negative condition. Six species (Pyrenean desman, long-fingered bat, southern water vole, European mink, otter, ringed seal) indicate a positive condition of a site.

As most of the species have a wide geographic distribution, their value as indicator may be different in different parts of their range and should be evaluated at a local or regional scale. It is also important to look at populations of the species rather than single individuals. Individuals can move over large distances and sometimes can be observed even in unfavourable habitats. When looking at populations, we should use long-term population dynamics to assess habitat or site condition and not put too much emphasis on single-season or 1-year observations.

Good population estimates alone may not be very helpful. You have to know the biology and ecology of your species well to understand what in fact it may indicate.

Acknowledgements I am grateful to Peter Reynolds for sharing his great experience of surveying otters and water voles and for updating my knowledge of the interactions of otters, American mink and water voles in Britain. Jonas Grahn gave insights into the Swedish system for monitoring within Natura 2000. Clive Hurford gave valuable comments on the manuscript.

References

Bergman A, Bergstrand A, Bignert A (2001) Renal lesions in Baltic grey seals (*Halichoerus grypus*) and ringed seals (*Phoca hispida botnica*). Ambio 30:397–409

Carter P, Churchfield S (2006) Distribution and habitat occurrence of water shrews in Great Britain. Environment Agency, Science Report SC010073/SR

Dietz C, von Helversien O, Nill D (2007) Handbuch der Fledermäuse Europas und Nordwestafrikas. Franck-Kosmos, Stuttgart (in German)

Freye H-A (1978) *Castor fiber* (Linnaeus, 1758) – Europäischer Biber. In: Niethammer J, Krapp F (eds) Handbuch der Säugetiere Europas, vol 1. Akademische Verlagsgesellschaft, Wiesbaden, pp 184–200 (in German)

Geiger H, Rudolph, B-U (2004) Wasserfledermaus *Myotis daubentonii*. In: Meschede A, Rudolph BU (eds) Fledermäuse in Bayern. Ulmer Verlag, Stuttgart, pp 127–138 (in German)

Guillén A (1999) *Myotis capaccinii* (Bonaparte, 1837). In: Mitchell-Jones *et al* (eds), pp 106–107

IUCN (2007) European mammal assessment. Available at http://ec.europa.eu/environment/nature/conservation/species/ema/ (accessed on 28 December 2008)

Jefferies DJ (ed) (2003) The water vole and mink survey of Britain 1996–1998 with a history of the long-term changes in the status of both species and their causes. The Vincent Wildlife Trust, Ledbury, UK

Juckwer E-A (1990) *Galemys pyrenaicus* (Geoffroy, 1811) – Pyrenäen-Desman. In: Niethammer J, Krapp F (eds) Handbuch der Säugetiere Europas, vol 3/1. Aula Verlag, Wiesbaden, pp 79–92 (in German)

Mattson M, Helle E (1995) Reproductive recovery and PCBs in Baltic seal populations. Eleventh biennial conference on the biology of marine mammals, 14–18 Dec 1995, Orlando, USA, p 74

Meschede A, Rudolph B-U (2004) Fledermäuse in Bayern, Ulmer Verlag, Stuttgart (in German)

Mitchell-Jones AJ, Amori G, Bogdanowicz W, Krystufek B, Reijnders PJH, Spitzenberger F, Stubbe M, Thissen JBM, Vohralik V, Zima J (eds) (1999) The atlas of European mammals. Academic Press, London

Pietsch M (1982) *Ondathra zibethicus* (Linnaeus, 1766) – Bisamratte, Bisam. In: Niethammer J, Krapp F (eds) Handbuch der Säugetiere Europas, vol 2/1. Akademische Verlagsgesellschaft, Wiesbaden, pp 177–192 (in German)

Queiroz AJ (1999) *Galemys pyrenaicus* (E. Geoffroy, 1811). In: Mitchell-Jones *et al* (eds), pp. 78–79

Saucy F (1999) *Arvicola sapidus* (Miller, 1908). In: Mitchell-Jones *et al* (eds) pp 220–221

Schneider M, Härkönen T, Hemmingsson M, Karlsson M (2009) Åtgärdsprogram för förvaltning av vikare *Phoca hispida botnica*. Unpublished manuscript (in Swedish with English summary)

Spitzenberger F (1990a) *Neomys fodiens* (Pennant, 1771) – Wasserspitzmaus. In: Niethammer J, Krapp F (eds) Handbuch der Säugetiere Europas, vol. 3/1. Aula Verlag, Wiesbaden pp 334–374 (in German)

Spitzenberger F (1990b) *Neomys anomalus* (Cabrera, 1907) – Sumpfspitzmaus. In: Niethammer J, Krapp F (eds) Handbuch der Säugetiere Europas, vol 3/1. Aula Verlag, Wiesbaden pp 317–333 (in German)

Spitzenberger F (1999a) *Neomys fodiens* (Pennant, 1771). In: Mitchell-Jones *et al* (eds) pp 60–61

Spitzenberger F (1999b) *Neomys anomalus* (Cabrera, 1907). In: Mitchell-Jones *et al* (eds) pp 58–59
Strachan C, Jefferies DJ, Barreto GR, Macdonald DW, Strachan R (1998) The rapid impact of resident American mink on water voles: Case studies from lowland England. In: Dunstone N, Gorman M (eds) Behaviour and ecology of riparian mammals. Cambridge University Press, Cambridge, pp 339–357
Sundström T, Olsson C (2005) *Västerbottens kustfågelfauna*. Länsstyrelsen Västerbotten, Meddelande 4-2005 (in Swedish)
Tattersall F (1999) *Castor canadensis* (Kuhl, 1820). In: Mitchell-Jones *et al* (eds), pp 198–199

Chapter 5
Waterbirds as Bioindicators of Environmental Conditions

Juan A. Amat and Andy J. Green

Introduction

One of the aims of monitoring is to provide information for ecological assessment, which can provide early warning of changes that could negatively affect species or ecosystems (Burger 2006). Since it is impractical to monitor all biological and physical components, a few of them can be used as indicators of wider conditions. Biological components chosen with this aim are called bioindicators (e.g. Matsinos and Wolf 2003).

Several aspects of the ecology of waterbirds make them useful as bioindicators. First, waterbirds have been shown to track environmental variations, at short (months) and long (years) temporal scales, and at both species and community levels (e.g. Nudds 1983; Amat *et al.* 1985; Guinet *et al.* 1998; Abraham and Sydeman 2004; Almaraz and Amat 2004; Rendón *et al.* 2008). Second, because many species are top predators and several contaminants often accumulate along the trophic chain, such species may be used as indicators of changes occurring at lower trophic levels (e.g. Matsinos and Wolf 2003; Burger and Eichhorst 2005). And third, either the waterbirds themselves or their prey are exploited by humans (e.g. hunting and fisheries), so that hunting bags of waterbirds may be indicative of productivity in nesting areas (Miller *et al.* 1988) or breeding parameters of birds may inform on fish stocks (Einoder 2009).

In this chapter we give some examples of the usefulness of using waterbirds as bioindicators. We explain why in other cases the use of waterbirds as indicators may be more limited, and we also identify how the design of studies can improve the utility of indicators.

J.A. Amat (✉) and A.J. Green
Department of Wetland Ecology, Estación Biológica de Doñana, Consejo Superior de Investigaciones Científicas, Calle Américo Vespucio s/n, 41092, Sevilla, Spain
e-mail: amat@ebd.csic.es

C. Hurford et al. (eds.), *Conservation Monitoring in Freshwater Habitats: A Practical Guide and Case Studies*, DOI 10.1007/978-1-4020-9278-7_5,

Waterbirds as Bioindicators of Environmental Conditions

Several studies show that waterbirds may be used as bioindicators of conditions encountered in wetlands, at both local and regional spatial scales. A classical example is the response of some waterbirds to eutrophication of wetlands. In the Mar Menor lagoon of south-eastern Spain the great crested grebe, *Podiceps cristatus*, increased in abundance as eutrophication increased as a result of nutrient inputs into the wetland, resulting from intensification of agricultural practices in the basin of the lagoon (Fig. 5.1). A parallel change in the number of grebes was not recorded in other sites of Spain, indicating that the increase in the Mar Menor lagoon did not result from external factors affecting the population at other spatial scales (Martínez Fernández *et al.* 2005).

Another example of the effects of agricultural changes in wetlands is provided by coots in southern Spain. There, the red-knobbed coot, *Fulica cristata*, declined in the second half of the 20th century and is now threatened with extinction. This decline was largely attributed to changes in the agricultural practices in the basins of wetlands, which accelerated siltation rates, and therefore shortened hydroperiods, affecting the quality of food plants of coots. Indeed, the assimilation efficiency of coots was negatively affected when the quality of their food plants was low, which usually occurs in early summer when water levels start to decline (Varo and Amat 2008). Therefore, the population dynamics of red-knobbed coots over long periods

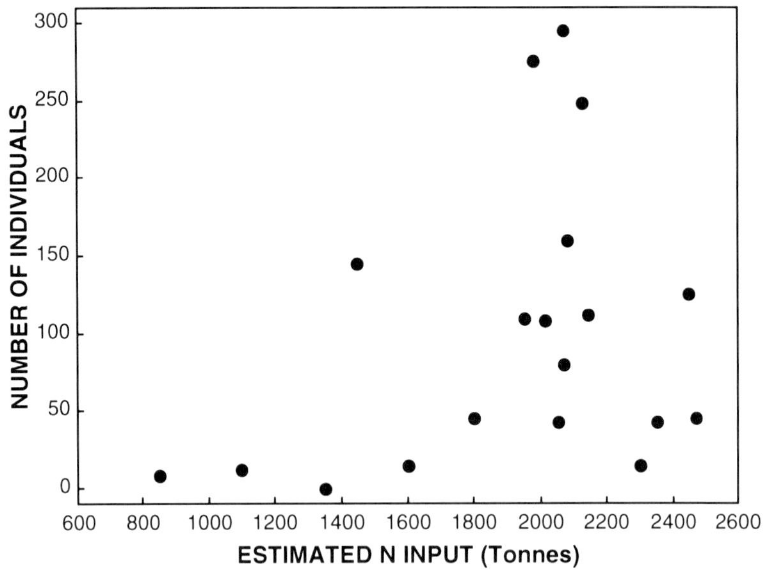

Fig. 5.1 Relationship between the number of great crested grebes *Podiceps cristatus* recorded at the Mar Menor lagoon (SE Spain) during January 1980–2001 and nutrient input estimates (tonnes of nitrogen) into the lagoon (modified from Martínez Fernández *et al.* 2005)

(e.g. decades) could be used as indicative of changes occurring in wetlands at slow rates (e.g. siltation processes).

Agricultural changes may also have effects on waterbird habitats at large spatial scales. As a result of feeding on agricultural crops in winter and on migration, some populations of snow goose, *Chen caerulescens*, have increased by 7% per year and these increasing numbers have had a strong negative long-term effect on intertidal salt-marsh vegetation at an Arctic coastal breeding site, located 5,000 km from wintering sites. Here, goose grubbing caused loss of vegetation and salinity of bare ground precluded re-establishment of vegetation (Abraham *et al.* 2005).

Changes in the nutrient budget of wetlands are not only due to the effects of human activities, and may be a result of the activities of the birds themselves, as shown by research on northern fulmars, *Fulmarus glacialis*, nesting on cliffs above a coastal plain with freshwater ponds (Michelutti *et al.* 2009). Fulmars behaved as biovectors that transported important quantities not only of nutrients, but also of contaminants, from the sea to the ponds. These ponds contained more chlorophyll, chironomids, and contaminants than those not affected by bird activity. These indicators of bird activity could be used to track population changes in other bird species for which chironomids are an important food source (Michelutti *et al.* 2009).

The monitoring of breeding colonies of waterbirds may provide information on the conditions of wetlands used for feeding. The main colony of greater flamingos, *Phoenicopterus roseus*, in southern Spain is located at Fuente de Piedra lake, but the birds mainly forage in the Guadalquivir marshes, located 130 km from the nesting site (Rendón-Martos *et al.* 2000; Amat *et al.* 2005). Here, colony size was affected by water levels in the foraging site (Fig. 5.2).

Limitations on the Use of Waterbirds as Bioindicators

As shown above, there may be important relationships between waterbirds and biotic and abiotic factors of wetlands, and the effects of the birds in these habitats may have important consequences on food webs. This justifies the incorporation of waterbirds into biomonitoring programs. Nevertheless, some have questioned the usefulness of waterbirds as bioindicators (Green and Figuerola 2003; Piatt *et al.* 2008).

The main criticisms come from the lack of relationships between the diversity of waterbirds and that of other organisms. Community concordance measures the degree to which patterns in community structure in a set of sites are similar between different taxonomic groups (Paszkowski and Tonn 2000). Concordant patterns have been found among guilds of waterbirds, and even between waterbirds and fish in several lakes (Paszkowski and Tonn 2000, 2006), indicating that monitoring the status of one group may provide a useful bioindicator of the status of other groups (Paszkowski and Tonn 2006).

However, such patterns may not be so evident in other cases. For instance, the similarity among a set of lakes in southern Spain in their waterbird communities is

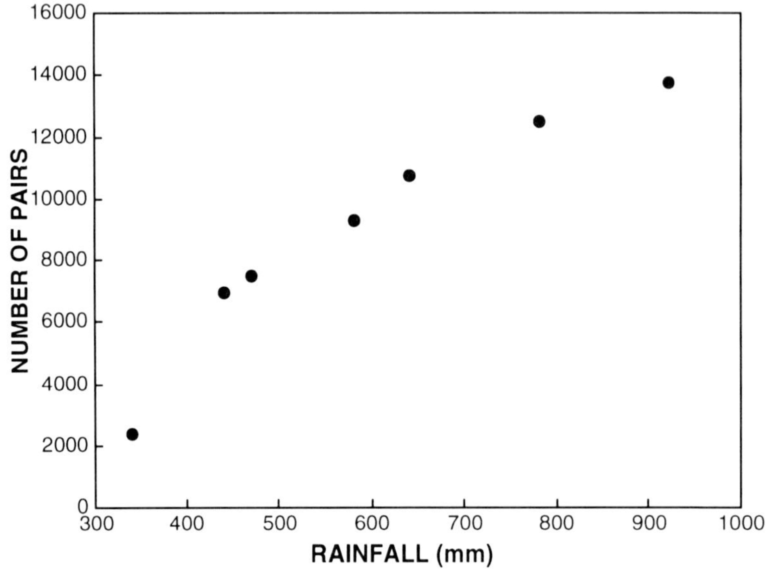

Fig. 5.2 Relationship between the size of a breeding colony of greater flamingos *Phoenicopterus roseus* at Fuente de Piedra lake and rainfall recorded during the preceding months at the marshes of the Guadalquivir in southern Spain (from Rendón-Martos 1996). The marshes are located 130 km from Fuente de Piedra, and are the main foraging site of flamingos breeding at the lake

very different to the similarities in their zooplankton or submerged macrophyte communities (Amat *et al.* 1985). Also in these lakes, the diversity of waterbird guilds (ducks, shorebirds) differs according to water levels (Amat 1984). In the case of highly dynamic wetlands, the responses to environmental variations may vary according to type of organisms, since different types of organisms may not perceive environmental variations in the same way. Under these circumstances, monitoring one group may not be a useful bioindicator of the status of another group.

Another difficulty of using waterbirds as bioindicators is related to their high mobility throughout the year. Migratory populations are subject to changes occurring not just in an area, but right across the migratory flyway. Even on a daily scale, their high mobility complicates their use as indicators. Although there may be a positive relationship between the size of wetlands and the number of waterbirds using them (Amat 1984; Nudds 1992; Weller 1999, Fig. 5.3), there are also cases in which such a relationship is not found, which may be explained in part by the differential use of wetlands throughout the day by waterbirds. Thus, many dabbling duck species use some wetlands for resting during the day, but forage in different wetlands during the night (Tamisier and Dehorter 1999). As waterbird counts are usually conducted during the day, trying to establish a relationship between biotic and abiotic factors of wetlands that are used as roosting sites and the carrying capacity of such wetlands for ducks may be misleading (Yésou 1983).

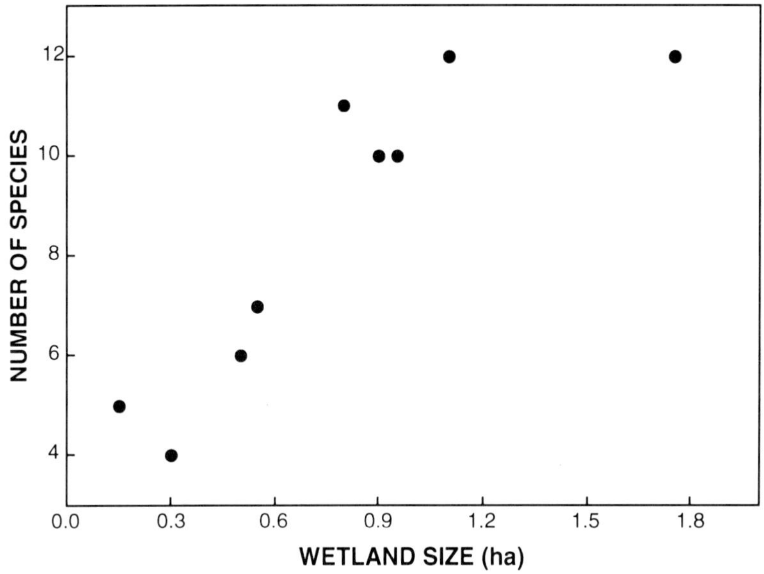

Fig. 5.3 Relationship between species richness of ducks and wetland size in Manitoba, Canada (modified from Nudds 1992)

Therefore, when considering the use of waterbirds as bioindicators, the identification of biologically meaningful parameters is vital. Also, because wetlands are highly dynamic ecosystems, it may be difficult to "capture" this variability by biomonitoring a single group of waterbirds. For a monitoring program to be successful, specific quantitative objectives should be defined, the objectives should be expressed as null hypotheses, and the sampling and analysis plan should be designed to test these hypotheses (Segar *et al.* 1985).

Which Indicators Are More Relevant?

There are species that indicate areas of high diversity, and others that measure environmental changes (Caro and O'Doherty 1999). The relevance of indicator species should depend on the aims of the monitoring program.

In the case of species indicating diversity hotspots, the monitoring of such species may be enough to assess the conservation status of the entire area. For instance, the red-knobbed coot is found in Morocco in wetlands with a high diversity of submerged macrophytes (Green *et al.* 2002). Here conserving the aquatic plants would ensure the conservation of the coot, but the monitoring of coots would be an efficient surrogate for the more difficult task of monitoring the vegetation. In the case of species that measure environmental change, some population parameters (e.g. nesting success, dynamics), may be used as indicators (e.g. Fig. 5.2).

Because the natural environment affects many physiological processes, waterbird physiology can provide information to detect stressors and to predict possible negative effects on populations. For instance, interannual variations in the body condition of flamingo chicks are linked to variations in several blood parameters, which reflect the feeding conditions encountered by adults (Amat *et al.* 2007). Long-term temporal changes in diets can also be examined with the use of stable isotopes in both tissues of museum specimens and live individuals (Chamberlain *et al.* 2005; Becker and Beissinger 2006). The heat-shock protein response may have also applications in biomonitoring, because of their responsiveness to stressors (Feder and Hofmann 1999).

Traditionally, the most basic objective of biomonitoring has been to detect trends. Although in some countries there may be databases that cover long periods of time, in some other countries such information may be scarce. Errors in trend analysis may be more likely with limited databases and can have serious consequences. Hence, the trends detected by a monitoring programme should be evaluated with power analysis (Lougheed *et al.* 1999).

Conclusions

Birds are popular subjects for research and monitoring, and long-term datasets of waterbird counts often provide a useful resource as indicators of ecological change. However, different waterbird species undergo population fluctuations for different reasons, and a thorough knowledge of the ecology of a given species is required if trends are to be interpreted correctly. Waterbirds do not merely respond to environmental change, they can also be the cause of change as their populations increase, due to overgrazing or because they act as vectors of nutrients and contaminants. In some cases birds may not respond to wetland characteristics in the same way as other groups of organisms, in which case birds may not be considered as surrogates of other organisms in biomonitoring programs. In other cases, however, birds can be reliable indicators of nutrient status, fish stocks or the abundance of aquatic plants. As the difficulties inherent in monitoring some groups of organisms (e.g. aquatic vegetation) might be best avoided if a reliable indicator is available, in these last cases birds may be considered as relatively easily measurable surrogates. When using waterbirds as indicators, clear objectives for the monitoring programme are essential.

References

Abraham CL, Sydeman WJ (2004) Ocean climate, euphausids and auklet nesting: Inter-annual trends and variation in phenology, diet and growth of a planktivorous seabird. Marine Ecol Prog Ser 274:235–250

Abraham KF, Jefferies RL, Alisauskas RT (2005) The dynamics of landscape change and snow geese in mid-continent North America. Global Change Biol 11:841–855

Almaraz P, Amat JA (2004) Multi-annual spatial and numeric dynamics of the white-headed duck *Oxyura leucocephala* in southern Europe: Seasonality, density-dependence and climatic variability. J Anim Ecol 73:1013–1023

Amat JA (1984) Las poblaciones de aves acuáticas en las lagunas andaluzas: Composición y diversidad durante un ciclo anual. Ardeola 31:61–79 (in Spanish)

Amat JA, Díaz Paniagua C, Herrera CM, Jordano P, Obeso JR, Soriguer RC (1985) Criterios de Valoración de Zonas Húmedas de Importancia Nacional y Regional en Función de las Aves Acuáticas. Instituto Nacional para la Conservación de la Naturaleza, Madrid, p 79 (in Spanish)

Amat JA, Rendón MA, Rendón-Martos M, Garrido A, Ramírez JM (2005) Ranging behaviour of greater flamingos during the breeding and post-breeding periods: Linking connectivity to biological processes. Biol Conserv 125:183–192

Amat JA, Hortas F, Arroyo GM, Rendón MA, Ramírez JM, Rendón-Martos M, Pérez-Hurtado A, Garrido A (2007) Interannual variations in feeding frequencies and food quality of greater flamingo chicks (*Phoenicopterus roseus*): Evidence from plasma chemistry and effects on body condition. Compar Biochem Physiol A 147:569–576

Becker BH, Beissinger SR (2006) Centennial decline in the trophic level of an endangered seabird after fisheries decline. Conserv Biol 20:470–479

Burger J (2006) Bioindicators: Types, development, and use in ecological assessment and research. Environ Bioindicators 1:22–39

Burger J, Eichhorst B (2005) Heavy metals and selenium in grebe eggs from Agassiz National Wildlife Refuge in northern Minnesota. Environ Monitor Assess 107:285–295

Caro TM, O'Doherty G (1999) On the use of surrogate species in conservation biology. Conserv Biol 13:805–814

Chamberlain CP, Waldbauer JR, Fox-Dobbs K, Newsome SD, Koch PL, Smith DR, Church ME, Chamberlain SD, Sorenson KJ, Risebrough R (2005) Pleistocene to recent dietary shifts in California condors. Proc Natl Acad Sci USA 102:16707–16711

Einoder LD (2009) A review of the use of seabirds as indicators in fisheries and ecosystem management. Fish Res 95:6–13

Feder ME, Hofmann GE (1999) Heat-shock proteins, molecular chaperones, and the stress response: Evolutionary and ecological physiology. Ann Rev Physiol 61:243–282

Green AJ, Figuerola J (2003) Aves acuáticas como bioindicadores en los humedales. In: Paracuellos M (ed) Ecología, Manejo y Conservación de los Humedales. Instituto de Estudios Almerienses, Almería, pp 47–60 (in Spanish)

Green AJ, El Hamzaoui M, El Agbani MA, Franchimont J (2002) The conservation status of Moroccan wetlands with particular reference to waterbirds and to changes since 1978. Biol Conserv 104:71–82

Guinet C, Chastel O, Koudil M, Durbec JP, Jouventin P (1998) Effects of warm sea-surface temperature anomalies on the blue petrel at the Kerguelen Islands. Proc Roy Soc Lond B 265:1001–1006

Lougheed LW, Breault T, Lank DB (1999) Estimating statistical power to evaluate ongoing waterfowl population monitoring. J Wildlife Manage 63:1359–1369

Martínez Fernández J, Esteve Selma MA, Robledano Aymerich F, Pardo Sáez MT, Carreño Fructuoso MF (2005) Aquatic birds as bioindicators of trophic changes and ecosystem deterioration in the Mar Menor lagoon (SE Spain). Hydrobiologia 550:221–235

Matsinos YG, Wolf WF (2003) An individual-oriented model for ecological risk assessment of wading birds. Ecol Model 170:471–478

Michelutti N, Keatly BE, Brimble S, Blais JM, Liu H, Douglas MSV, Mallory ML, Macdonald RW, Smol PL (2009) Seabird-driven shifts in Arctic pond ecosystems. Proc R Soc B 276: 591–596

Miller MR, Beam J, Connelly DP (1988) Dabbling duck harvest dynamics in the Central Valley of California – Implications for recruitment. In: Weller MW (ed) Waterfowl in winter. University of Minnesota Press, Minneapolis, MN, pp 553–569

Nudds TD (1983) Niche dynamics and organization of waterfowl guilds in variable environments. Ecology 64:319–330

Nudds TD (1992) Patterns in breeding waterfowl communities. In: Batt BDJ, Afton AD, Anderson MG, Ankney CD, Johnson DH, Kadlec JA, Krapu GL (eds) Ecology and management of breeding waterfowl. University of Minnesota Press, Minneapolis, MN, pp 540–567

Paszkowski CA, Tonn WM (2000) Community concordance between fish and aquatic birds of lakes in northern Alberta, Canada: The relative importance of environmental and biotic factors. Freshwater Biol 43:421–437

Paszkowski CA, Tonn WM (2006) Foraging guilds of aquatic birds on productive boreal lakes: Environmental relations and concordance patterns. Hydrobiologia 567:19–30

Piatt JF, Harding AMA, Schultz M, Speckman SG, van Pelt TI, Drew GS, Kettle AB (2008) Seabirds as indicators of marine food supplies: Cairns revisited. Marine Ecol Prog Ser 352:221–234

Rendón MA, Green AJ, Aguilera E, Almaraz P (2008) Status, distribution and long-term changes in the waterbird community wintering in Doñana, south-west Spain. Biol Conserv 141:1371–1388

Rendón-Martos M (1996) La laguna de Fuente de Piedra en la Dinámica de la Población de Flamencos (*Phoenicopterus ruber*) del Mediterráneo Occidental. PhD Thesis, University of Málaga, Málaga p 386 (in Spanish)

Rendón-Martos M, Vargas JM, Rendón MA, Garrido A, Ramírez JM (2000) Nocturnal movements of breeding greater flamingos in Spain. Waterbirds 23(Special Publ. 1):9–19

Segar D, Phillips D, Stamman E (1985) Design of bioindicator based programs to monitor ocean status and trends. Oceans 17:926–931

Tamisier A, Dehorter O (1999) Camargue: Canards et Foulques. Centre Ornithologique du Gard, Nimes: p 369 (in French)

Varo N, Amat JA (2008) Differences in food assimilation between two coot species assessed with stable isotopes and particle size in faeces: Linking physiology and conservation. Compar Biochem Physiol A 149:217–223

Weller MW (1999) Wetland birds. Habitat resources and conservation implications. Cambridge: Cambridge University Press, p 271

Yésou P (1983) Anatidés et zones humides de France métropolitaine. Bulletin mensuel de l'Office National de Chasse, Numéro special, scientifique et technique, 1–315 (in French)

Chapter 6
Monitoring Fish Populations in River SACs

Ian G. Cowx, Jonathan P. Harvey, Richard A. Noble, and Andrew D. Nunn

Introduction

The EC Habitats Directive (92/43/EEC) on Conservation of Natural Habitats and of Wild Fauna and Flora requires European Member States to take measures to maintain or restore certain habitats and species that ensure their favourable conservation status across the European Community. Managing SAC rivers designated for fish species requires monitoring programmes to assess the status for the fish populations relative to their overall conservation objectives and management prescriptions, a process known as condition assessment in the UK. Condition assessment establishes the status of the target species and provides a broad indication of population trends. Monitoring programmes must be able to detect changes in temporal and spatial information relating to the fish species distribution and abundance if management is to be effective.

This paper reviews the general requirements for establishing strategies for assessing the condition of fish populations to help nature conservation agencies prepare monitoring protocols for fish species in SAC rivers. The need for pragmatic, cost-effective strategies that are likely to provide the necessary information for assessing species status is recognised.

Monitoring Protocols for SAC Rivers

Important factors when designing a monitoring protocol for assessment of the condition status of a fish species include: representative coverage of the distribution range of the fish species in the target river; appropriate sampling methodology; accurate recording of fish data; appropriate timing of surveys to ensure that the assessment

---

I.G. Cowx (✉), J.P. Harvey, R.A. Noble, and A.D. Nunn
Hull International Fisheries Institute, University of Hull, Hull, HU6 7RX, UK
e-mail: i.g.cowx@hull.ac.uk

C. Hurford et al. (eds.), *Conservation Monitoring in Freshwater Habitats: A Practical Guide and Case Studies*, DOI 10.1007/978-1-4020-9278-7_6,

accurately reflects the status of the fish populations; and collection of appropriate environmental data to help elucidate the cause of any change in population status. Where possible, the survey design should also be geared towards modifying existing monitoring programmes to minimise additional resource implications.

Sampling Procedure

Most fish monitoring effort in Europe focuses on juvenile salmonids or coarse fishes using electric fishing in shallower reaches of rivers as the primary sampling method (Cowx 1995; Roset *et al.* 2007). This is because the results of electric fishing surveys can provide a good measure of the distribution and abundance of species, identification of population trends between years, an assessment of the demographic structure, identification of adverse environmental impacts and, in some cases, detection of recruitment failures (Cowx and Lamarque 1990). Generally, sampling of this nature involves single catch or multiple catch depletion methods in a fixed area of river. However, these existing methodologies can be ineffective for monitoring fish populations in large lowland rivers, adults of diadromous species, and, unless specifically targeted, small-sized species such as bullhead or spined loach that live in cryptic habitats. For example, sampling of bullhead is usually qualitative using presence/absence or categorical data (e.g. 0–4, 5–10, 10–50, >50 bullhead caught per 100 m^2 of river) because of difficulties capturing individual fish. Also, existing survey protocols do not effectively sample the appropriate habitat for lamprey ammocoetes, and specialist gears may need to be developed (Nunn *et al.* 2008; Cowx *et al.* 2009; Chapter 19 this volume). To achieve effective monitoring requires careful planning to ensure all life stages and habitats are surveyed. This may require multiple methods to sample different life stages, such as in salmon where the adults are monitored by fish counters, trapping, and angler and commercial catch statistics, and juveniles by electric fishing in nursery areas. It is beyond the scope of this chapter to describe the survey protocols in detail, and the reader is referred to Cowx and Fraser (2003), Cowx and Harvey (2003), Harvey and Cowx (2003) and Hillman *et al.* (2003) for details of the rationale and methods for salmon, bullhead, lamprey and shad, respectively. A summary of the various sampling procedures is, however, given in Table 6.1 for guidance.

Distribution Range

Site selection is fundamental for providing an overall condition assessment of target species in the SAC river. Location of sampling sites needs to take account of the extant and historical distribution range and abundance of discrete target fish populations to avoid unnecessary sampling in areas where the species has never existed, e.g. upstream distribution constrained by impassable natural barriers. Previous survey data should be able to identify approximately where the upper and lower limits of extant species distribution are within the river, and historical data and reports should allow assessment of any reduction in distribution range. Where information is limited, a comprehensive sampling programme should provide assessment of the current population distribution range as a baseline for detecting change in the future. Figure 6.1 illustrates an extensive sampling programme implemented to assess the distribution of the endangered endemic fish species *Anaecypris hispanica* in the Portuguese sector of the River Guadiana (Collares-Pereira *et al.* 2000).

The species was originally distributed across the catchment but is now constrained to a few tributaries because of habitat degradation, overabstraction of water and introduction of non-native species (Collares-Pereira *et al.* 2000).

Table 6.1 Recommended sampling methods for assessing the conservation status of fish species in SAC rivers (from Cowx *et al.* 2009)

Species	Method	Size of sampling area
Lamprey		
Adults	Direct observation of spawning habitat, counters, impingement at power station intakes	
Ammocoetes in optimal habitat	Generator driven electric fishing in quadrat area; 3-catch depletion	Three, 1-m^2 quadrats per 100 m of river length
Ammocoetes in sub-optimal habitat	Back-pack or generator driven electric fishing; single anode, qualitative assessment or catch per unit river length	50 m of river bank
Bullhead		
Adults and juveniles	Back-pack or generator driven electric fishing; single or multiple anodes depending on width of river; 3-catch depletion or calibrated gear	Minimum 50 m or 10 times river width where only occasional fish caught; 10–20 m of river length where 0.5 bullhead m^{-2}
Salmon		
Adults	Commercial net catches, rod catches, traps and counters	
Juveniles	Back-pack or generator driven electric fishing; single or multiple anodes depending on width or river; 3-catch depletion or calibrated gear	Minimum 50 m or 10 times river width
Shad		
Adults	Catches in commercial nets and impingement at power station intakes	
Eggs	Kick sampling	Across dimensions of spawning habitat
Juveniles	Micro-mesh seine netting	Upper estuary/lower river

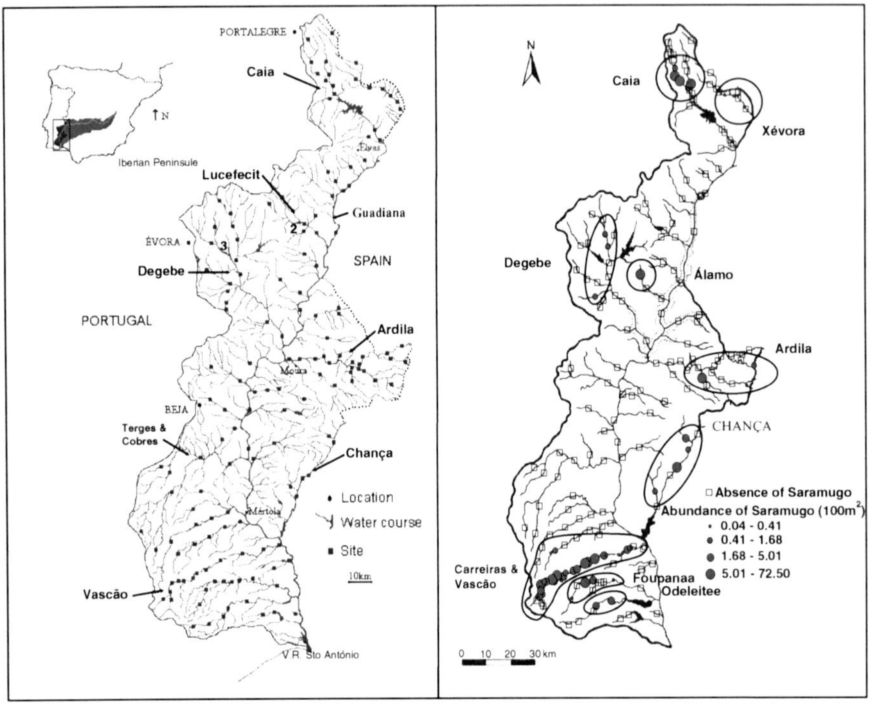

Fig. 6.1 (**a**) Distribution of survey sites in the lower Guadiana River catchment to assess conservation status of *Anaecypris hispanica*. (**b**) Relative abundance ind. 100 m^{-2} of *Anaecypris hispanica* in Portuguese tributaries of the River Guadiana where the species still persists

This example highlights the need to collect information on anthropogenic pressures likely to be responsible for any contraction of the distribution range of the target species whilst undertaking such surveys. Specific environmental information using established measurement procedures, such as those outlined by Bain and Stevenson (1999), or for the River Habitat Survey (RHS) (Environment Agency 2003) and HABSCORE (Milner *et al.* 1998) protocols, should be collected on each sampling occasion. This habitat assessment should include factors that may hinder the river's ability to meet favourable condition, such as migration barriers, channelisation, flow regulation and land-use changes. Suggestions on what constitutes favourable condition in relation to water quality, river morphology, flow, substrate and environmental disturbance are available on:www.englishnature.org.uk/lifeinukrivers/index.htm.

When designing such a survey programme, it is essential that the sampling sites selected cover all the habitats used by the various freshwater life stages of the target species, including reaches outside the SAC boundaries. The latter is important because it identifies factors outside the SAC that may potentially affect the status of the population.

Number of Sites and Frequency of Sampling

When designing a survey programme to assess condition status, it is important to determine the accuracy and precision of the data that must be achieved with respect to individual populations (Southwood 1978; Bohlin *et al.* 1990). Condition assessment must be able to measure stock parameters such as abundance and the magnitude of change in relation to the conservation objectives. This means that sufficient sites must be sampled at appropriate time intervals to provide reliable measures of both spatial and temporal changes in abundance of the species.

Detailed methodologies for fish survey design, which determine the number of sites that must be sampled to gain an acceptable level of precision, are provided by Southwood (1978) and Bohlin *et al.* (1990) and the guidelines of the LIFE Rivers projects (Cowx and Fraser 2003; Cowx and Harvey 2003; Harvey and Cowx 2003; Hillman *et al.* 2003). In essence, regular surveillance monitoring programmes need to be able to detect changes in population characteristics over a number of years or between sites.

To comply with the Habitats Directive, the nature conservation agencies are required to report the status of fish populations in SAC rivers every 6 years. It is entirely possible that populations could suffer acute problems within this 6-year period that would remain undetected for some considerable time, thus limiting the possibility of remedial actions. It is therefore recommended that full assessment surveys are carried out every 6 years, but a smaller number of 'index' sites, selected to represent the populations as a whole, are sampled every 2 years. If problems are found with the fish population status in these biennial surveys this should trigger a full survey and investigative work to identify remedial action.

The timing of surveys is also important to ensure that a representative sample of the population is assessed that can be related to condition status. In general, the earliest that monitoring surveys for juvenile life stages should be carried out in any 1 year is mid to late August, and preferably in September or October. This will allow the young-of-the-year to grow to a size where they can be caught more easily and assessed more accurately. Note, adult stocks are usually monitored during the migratory period of their life history to benefit from indirect methods of monitoring such as counters and catch statistics.

Fish Data Collection

All fish should be measured as fork length to the nearest millimetre. If large numbers of fish are collected, a sub-sample of at least 100 randomly selected individuals should be measured and the remainder counted. This information is important because analysis of age structure of the sampled population, a key attribute for the condition assessment to prove annual recruitment, will rely on assessment of size distribution. After completing measurements, all fish captured should be released at the site they were caught.

Assessment of Condition

Condition assessment of the species in the specific SAC is determined from a combination of attributes that reflect the overall population status. These generally include: no contraction in species distribution range; achieving threshold criteria for species abundance (density per unit area), demographic population structure showing evidence of recruitment and self sustaining populations, negligible deterioration in habitat quality. Note, for salmon, these should be supplemented by spawning target reference points (Milner *et al.* 2000; Cowx and Fraser 2003). Attributes based on abundance and demographic structure are complementary to those recommended for fish assessment under the EC Water Framework Directive [WFD] (EC Directive 2000/60/EC; see also Schmutz *et al.* 2007).

Abundance classification enables the density of the target species to be linked to the relative condition of fish populations. Because abundance is highly variable within a river catchment and the majority of sites can have few or no fish, attainment of favourable condition is determined from the average density for all the sampling sites (see Table 6.2 for threshold densities used in the UK). Threshold densities may vary between different zones of the river because of the known preference of some species to certain habitat. For example, bullhead densities tend to be much higher in lowland rivers than in upland areas (Noble *et al.* 2007), while lamprey ammocoete abundance is greater in localities with extensive silt loading compared with marginal silted habitat adjacent to gravel areas (Cowx and Harvey 2003; Harvey and Cowx 2003).

Abundance levels less than the targets proposed in Table 6.2 indicate unfavourable condition. It is important that a representative number of sites are surveyed to account for natural spatial variation in population size.

A second key attribute is demographic structure of the population. This establishes the contribution of juveniles and adults to the population and demonstrates recruitment. As with abundance, population demographic structure varies between species and locality within a river catchment. Criteria to establish favourable condition (Table 6.3) were based on previous studies that showed, where locally abundant, young-of-the-year fish or juveniles contribute discrete modes within the demographic structure of the population (see Fig. 6.2 for bullhead), or in the case of lamprey ammocoetes, several year groups (Fig. 6.3). In the examples (Figs 6.2 and 6.3), discrete modes are found representing year groups or cohorts. Evidence of good recruitment in most, but not all, years (Table 6.3) is required to achieve favourable status. For salmon and shad, demographic criteria are based on adult population structure and recruitment success, respectively (Table 6.3; Cowx and Fraser 2003; Hillman *et al.* 2003). Care is needed when interpreting the output because of natural annual variability of population recruitment in river systems. Recruitment failure in one or 2 years over the 6-year reporting period for the Habitats Directive may not necessarily constitute unfavourable condition as long as good recruitment is observed in the intervening years.

Table 6.2 Minimum abundance criteria proposed for favourable conservation status for fish species in UK SAC rivers

Species	Habitat	Abundance assessment criteria
Lamprey	Optimal habitat	
	River/brook lamprey ammocoetes	>10 individuals m^{-2}
	River/brook lamprey ammocoetes – chalk streams	>5 individuals m^{-2}
	Sea lamprey ammocoetes	0.2 individuals m^{-2}
	Catchment perspective	>2 individuals m^{-2}
	River/brook lamprey ammocoetes	
	Sea lamprey ammocoetes	0.1 individuals m^{-2}
Bullhead	Upland streams	>0.2 individuals m^{-2}
	Lowland rivers	>0.5 individuals m^{-2}
Salmon		Deviation from reference juvenile densities based on HABSCORE predictions (Milner *et al.* 1998)

Table 6.3 Population demographic structure criteria proposed for favourable conservation status for fish species in SAC rivers in the UK

Species	Assessment criteria
Bullhead	>40% of the individuals in 0+ age class
River and brook lamprey ammocoetes	Where abundant, >two age classes in population
Salmon	Deviation for 3–5 years from long-term trends in run timing and age composition of catch
	Adult catch must exceed annual spawning escapement equivalent in any 4 years over a 5-year period
Shad	Presence of eggs and/or 0+ individuals on spawning/nursery grounds 4 out of every 6 years

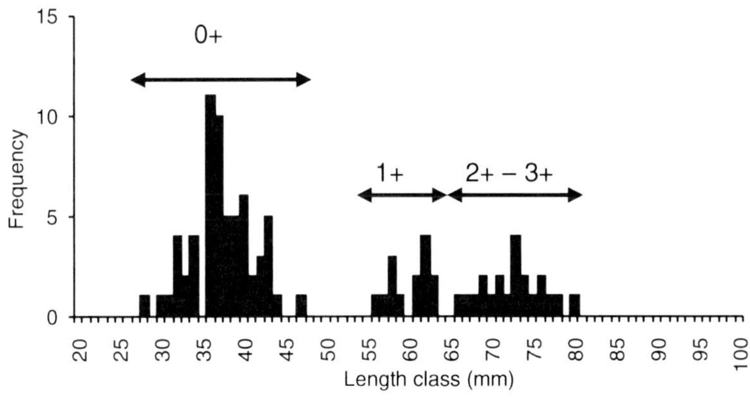

Fig. 6.2 Length frequency distribution of bullheads at Cyrtau on the Afon Groes (Afon Teifi catchment). Age classes of fish are indicated (n = 97)

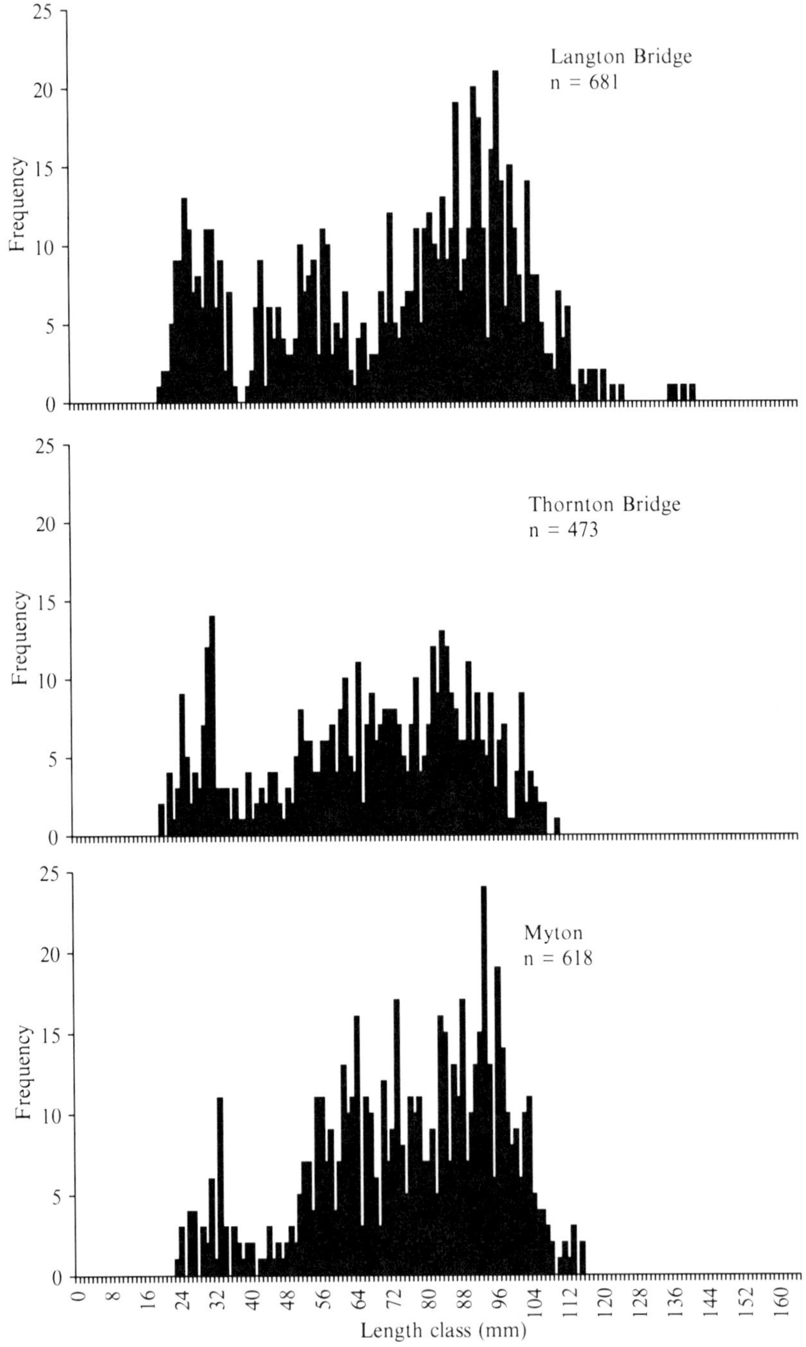

Fig. 6.3 Length frequency distributions of *Lampetra* ammocoetes captured from the River Swale, Yorkshire

Further assessment of the status of fish populations can be derived from mapping their distribution in individual rivers. Provided that abundance and demographic criteria are met, favourable condition can be inferred if there is no decline in distribution range from the previous surveys. This distribution pattern can also be linked to criteria for favourable habitat condition to establish the impact of changes in habitat and water quality, for example.

Reporting on the conservation status of specific fish species in river SACs is now a statutory requirement for Member States in the European Union. The protocols and methodology developed as part of the LIFE in UK Rivers Project, and described here, could be applied as a generic framework for condition assessment in the UK and other EU Member States. The framework can be applied to all fish species listed in annexes of the Habitats Directive. The output of conservation monitoring should not be treated in isolation, and any management intervention should be based on a catchment-scale approach (Collares-Pereira and Cowx 2004), because factors upstream and downstream of the designated area may contribute to the pressures on the fish stocks.

Acknowledgements This paper is based on contributions to the EU LIFE in UK Rivers Project and projects to assess condition status in UK rivers funded by the Environment Agency and Countryside Commission for Wales.

References

Bain MB, Stevenson NJ (1999) Aquatic habitat assessment: common methods. American Fisheries Society Special Publication, Bethesda, MD

Bohlin T, Heggberget TG, Strange C (1990) Estimation of population parameters using electric fishing: aspects of the sampling design with emphasis on salmonids in streams. In: Cowx IG, Lamarque P (eds) Fishing with electricity. Fishing News Books, Oxford, pp 156–173

Collares-Pereira MJ, Cowx IG (2004) The role of catchment scale environmental management in freshwater fish conservation. Fish Manage Ecol 11:303–312

Collares-Pereira MJ, Cowx IG, Ribeiro F, Rodrigues L, Rogado JA (2000) Threats imposed by water development schemes on the conservation of endangered species in the Guadiana River, Portugal. Fish Manage Ecol 7:167–178

Cowx IG (1995) Fish stock assessment – biological basis for sound ecological management. In: Harper D, Ferguson A (eds) Biological basis for river management. John Wiley, Chichester, pp 375–388

Cowx IG, Lamarque P (eds) (1990) Fishing with electricity. Oxford, Fishing News Books 245 pp

Cowx IG, Harvey JP, Noble RA, Nunn AD (2009) Establishing survey and monitoring protocols for the assessment of conservation status of fish populations in river Special Areas of Conservation in the UK. Aquat Conserv Marine Freshwater Ecosyst 19:96–103

Cowx IG, Fraser D (2003) A standardised survey and monitoring protocol for the assessment of salmon populations within SAC rivers. Conserving Natura 2000 Rivers Monitoring Series No. 3, English Nature, Peterborough

Cowx IG, Harvey JP (2003) A standardised survey and monitoring protocol for the assessment of bullhead populations within SAC rivers. Conserving Natura 2000 Rivers Monitoring Series No. 3, English Nature, Peterborough

Environment Agency (2003) River habitat survey in britain and ireland. Field survey guidance manual: 2003 version. Environment Agency, Bristol

Harvey JP, Cowx IG (2003) A standardised survey and monitoring protocol for the assessment of lamprey populations within SAC rivers. Conserving Natura 2000 Rivers Monitoring Series No. 3, English Nature, Peterborough
Hillman R, Cowx IG, Harvey JP (2003) A standardised survey and monitoring protocol for the assessment of shad populations within SAC rivers. Conserving Natura 2000 Rivers Monitoring Series No. 3, English Nature, Peterborough
Milner NJ, Wyatt RJ, Broad K (1998) HABSCORE – Applications and future development of related habitat models. Aquat Conserv Marine Freshwater Ecosyst 8:633–644
Milner NJ, Davidson IC, Wyatt RJ, Aprahamian M (2000) The use of spawning targets for fishery management in England and Wales. In: Cowx IG (ed) Management and ecology of river fisheries. Fishing News Books, Oxford, pp 361–372
Noble RA, Starkie AS, Cowx IG (2007) Development of fish-based methods for the assessment of ecological status in English and Welsh rivers. Fish Manage Ecol 14:495–508
Nunn AD, Harvey JP, Noble RAA, Cowx IG (2008) Condition assessment of lamprey populations in the Yorkshire Ouse catchment, north-east England, and the potential influence of physical migration barriers. Aquat Conserv Marine Freshwater Ecosyst 18:175–189
Roset N, Grenouillet G, Goffaux D, Pont D, Kestemont P (2007) A review of existing fish assemblage indicators and methodologies. Fish Manage Ecol 14:393–405
Schmutz S, Cowx IG, Haidvogl G, Pont D (2007) Fish-based methods for assessing European running waters: a synthesis. Fish Manage Ecol 14:369–380
Southwood TRE (1978) Ecological methods. Chapman and Hall, London

Chapter 7
Assessment of Aquatic Invertebrates

Richard Chadd

Introduction

Invertebrates represent a useful 'indicator community' for assessing the condition of aquatic habitats, both in freshwaters and in transitional coastal zones. This is because their taxonomy, status and ecology, including phenology as well as associations with habitat, are generally well known. Additionally, their responses to various stressors have been extensively studied (Hynes 1960; Mellanby 1967; Hart and Fuller 1974; Hellawell 1986; Rosenberg and Resh 1993; Wood *et al.* 2007; Lancaster and Briers 2008).

Broad Indices

Although the information derived from surveys and experiments applies across all taxonomic levels, the use of small suites of 'indicator species' to determine stresses on the aquatic environment can be distorted by a range of interacting factors. This is particularly true in lotic environments. For example, reducing concentrations of dissolved oxygen are both caused and exacerbated by reducing flow. Many indexing systems, designed to respond to stresses, therefore operate much better at the community or population level. This principle integrates the distorting factors to derive useable information. Such systems include the Lincoln Quality Index (LQI) (Extence *et al.* 1987) and the Lotic-invertebrate Index for Flow Evaluation (LIFE) (Extence *et al.* 1999). Other such indexing methodologies are outlined in Table 7.1. In such circumstances, sampling methodologies and strategies in freshwaters are necessarily indiscriminate and generalised.

R. Chadd
Environment Agency of England and Wales, Anglian Region (Northern Area), Stepping Stone Walk, Winfrey Avenue, Spalding, Lincolnshire, PE11 1DA, UK
e-mail: richard.chadd@environment-agency.gov.uk

C. Hurford et al. (eds.), *Conservation Monitoring in Freshwater Habitats: A Practical Guide and Case Studies*, DOI 10.1007/978-1-4020-9278-7_7,

Table 7.1 Selected indexing systems and computer packages used to derive information from aquatic macroinvertebrate data in the United Kingdom

Index	Indicator/model type and scope of use	Advantages	Disadvantages
Biological Monitoring Working Party Score (BMWP) Average Score Per Taxon (ASPT) = BMWP/Number of taxa Reference: Chesters (1980)	Organic pollution Designed for rivers, but can apply in all waters	Gives a broad indication of water quality Can respond to inorganic pollutants	Does not include an abundance – element for individual taxa, so not sensitive to low-level enrichment Operates at family level only, so scores apply to least sensitive species BMWP alone influenced by habitat, sampling effort (additive) and season. ASPT addresses this to some extent Some families mis-classified – process of classification was subjective (committee-based)
Walley Hawkes Paisley Trigg Score (WHPT) Reference: Walley and Hawkes (1996), Walley and Hawkes (1997), and Paisley *et al.* (2007)	Organic pollution (revision of BMWP) Designed for rivers, but can apply in all waters	Incorporates abundance Based on actual data (10,000 samples), so more objective	Operates at family level only (but may be future development using species-level data) Additive (but some allowance for habitat)
Lincoln Quality Index (LQI) Reference: Extence *et al* (1987)	Organic pollution (combination of BMWP and ASPT) Designed for rivers, but can apply in all waters	Allowance for habitat richness and type Combination allows for additivity problem of BMWP	No abundance element Family level only Mis-classification of some families in BMWP
Lotic-invertebrate Index for Flow Evaluation (LIFE) Reference: Extence *et al.* (1999)	Evaluation of the effect of flow variation (drought, spate, abstraction – pressure, etc.) Only for use in lotic environments	Based on data (objective) Sensitive to abundance of individual taxa Statistically significant empirical response to all flow parameters Linear response to habitat degradation Operates at family and species – level	Generally, cannot be applied to deep, silty rivers Output may be compromised by poor water quality Some mis-classification of taxa and some taxon omissions (minor revisions planned)

System for Evaluating Rivers for Conservation (SERCON) Reference: Boon *et al.* (1996)	Conservation evaluation Can be used in all river habitats	Multiple component approach (typicality, habitat, etc.) in deriving output	Some elements (e.g. presence of alien species) take no account of relative risk to conservation value Only applicable to river habitats
Community Conservation Index (CCI) Reference: Chadd & Extence (2004)	Conservation evaluation Applicable to all inland aquatic habitats in Britain and Ireland	Weighted for taxon-richness, as well as rarity Can be adjusted to allow for local rarity of species	Only directly applicable to Britain and Ireland (but capable of international adaptation, following certain rules) Some components of reference dataset subset requiring review
Species Quality Score (SQS) Reference: Foster and Eyre (1992)	Conservation evaluation Use in stillwater habitats in Britain and Ireland	Good evaluation of wetland habitats	Uses water beetles only (useful Order for such purposes – see text) Rather old – component elements probably out of date
Predictive System for Multimetrics (PSYM) Reference: Environment Agency and Ponds Conservation Trust (2002)	Multimetric model for ecological quality assessment Applicable in ponds, lakes and canals	Multimetric model, so multiple components in output (including plant community)	Operates at family level only for invertebrates Value judgement -does not allow for rarity of taxa
Shannon-Weaver Diversity Index Reference: Shannon and Weaver (1963)	Evaluation of diversity Universally applicable	Operates across a broad range of taxa Describes community diversity	Only describes numerical balance between taxa. Does not allow for ecological needs of constituents of the community
River InVertebrate Prediction and Classification System (RIVPACS) Reference: Wright *et al.* (2000)	Predictive model to assess observed versus expected ecology Use in rivers in Britain only	Allows comparison against reference state Can apply against different indices (quality or flow)	Some component of reference dataset suspect Tends to under-predict in calcareous environments Only applicable to river environments. Can only predict at family-level, where abundance – element needed
River Invertebrate Classification Tool (RICT) Reference: Davy-Bowker *et al.* (2008)	Predictive model (revision and development of RIVPACS) Use in rivers in Britain only	As for RIVPACS	Only applicable to river environments NOTE: Most of the known deficiencies of RIVPACS have been addressed, or are planned to be addressed in the near future, in development of the new model

(continued)

Table 7.1 (continued)

Index	Indicator/model type and scope of use	Advantages	Disadvantages
River Pollution Bayesian Belief Network (RPBBN) Walley *et al.* (1992)	Automated probabilistic reasoning tool (Artificial Intelligence) Limited to British rivers, but could have universal application	Automates the process of matching stresses to data, using an extensive database	'Black-box' system – must be used alongside some knowledge of ecological processes, to inform judgement of output Only applicable to river environments
River Pollution Diagnostic System (RPDS). O'Connor and Walley (2002)	Automated Pattern Recognition (Artificial Intelligence) Limited to British rivers, but could have universal application	Stresses (Stafford University are not big on hair-based datasets!)	Depends on accuracy of stress database (derived from expert judgement)

Directed Investigations

When the required information relates more to the conservation value of a habitat, the strategy and methods employed may be more intensive and directed towards key groups. This does not, however, preclude more generalised approaches.

The Community Conservation Index (CCI) (Chadd and Extence 2004) seeks to place a 'conservation value' on whole communities of aquatic invertebrates, both within and between ecosystems, and therefore uses as many taxa as can practicably be identified. Diversity indices, such as Shannon-Weaver (Shannon and Weaver 1963) rely on datasets which include a substantial range of taxa combined with relative abundance. The latter only describes the numerical balance between species, however, not the ecological needs of the constituent taxa.

Thus, one is able to derive information on stressors in the aquatic environment, both by index approaches and by pattern recognition. It is also possible to place value judgements on the importance, at a site-scale or national scale, of aquatic habitats. The methods employed depend to a great extent on the questions to be asked.

Sampling Tools

The usual 'tool of choice' for assessing aquatic invertebrates is the standard Freshwater Biology Association pattern pond-net (Fig. 7.1). This has a square frame of around 250 × 200 mm aperture, fitted with a net of 0.9 mm mesh-size. It can be deployed in flowing waters by placing the frame on the river bed and disturbing the upstream substratum with a foot, allowing benthic invertebrates to be washed into the net – known as 'kick-sampling' (Armitage *et al.* 1983). The sample is then augmented by sweeping the net in still, marginal areas as well as hand-searching. In the absence of flow (and in marginal areas) kick-sampling may still be employed, but the net is moved by the sampler into the column of water filled by disturbed invertebrates, rather than relying on a flow to deposit them in the net-bag. Contrary to popular belief, this approach covers all available habitats in lotic systems – marginal and mid-water, still and flowing areas – in proportion to their occurrence.

Generally, this method is employed in environments of less than 1 m in depth, so as to be able to stand in the water, but also to facilitate the hand-search. Marginal 'sweeping' from the bank of a river, lake, pond or drain can be used in deeper waters, as discussed below. Sampling effort (and therefore comparability between sites) is limited by time. Three minutes of active kicking, sweeping and kick/sweeping is followed by 1 min of hand-searching for sedentary, firmly-stuck and highly active taxa. This collects around 70 % of available invertebrate families, and results in a sample of a size which can be sorted in a relatively short time.

Such an approach provides data which can be analysed to indicate the stresses operating in any given watercourse, relating to flow, pollution or other stressors, and is the basis of the monitoring strategy employed by regulatory bodies.

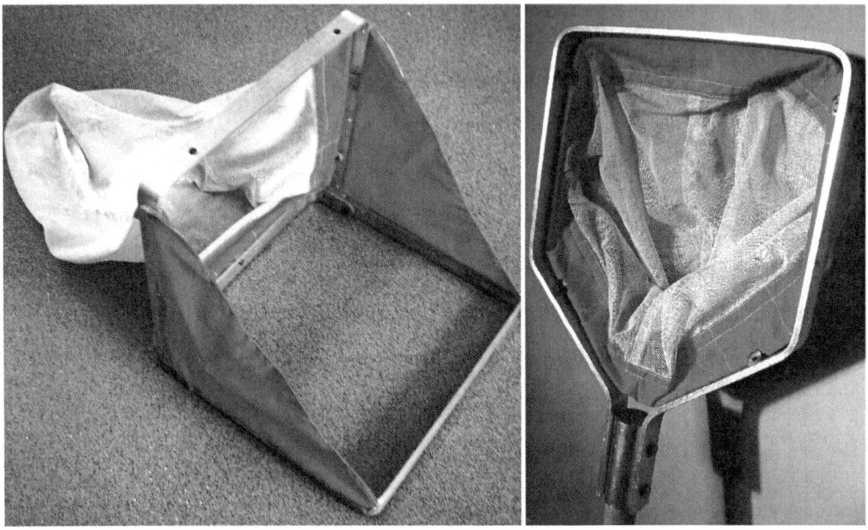

Fig. 7.1 On the left is a Surber Sampler, used for quantitative assessment of invertebrate assemblages in flowing waters, while on the right is a standard FBA pattern pond net: the most frequently-used tool for assessment of freshwater invertebrates. *Photographs by Richard Chadd*

Family or Species Level?

Although most index methods can operate at family level, analysis at species-level allows a much greater accuracy of output and is the ideal approach. Clearly, this requires an extra level of expertise and an initial time-investment to get less experienced workers 'up to speed'. The time invested rapidly reduces, however, as less experienced workers learn to recognise common or distinctive species instantly or with the minimum of effort. Within a few weeks or months, the time invested in analysis to species level barely differs from analysis to family level. Those taxa requiring more substantial effort, such as dissection, to arrive at a species' identification, can be left at the lowest practical level of resolution – genus or species-group – if time is an issue.

Sampling Effort

Use of pond nets for conservation assessment is not generally limited by sampling time, since sampling effort may be an irrelevance – one is attempting to find as many species as possible. On the other hand, sampling with enough vigour and enthusiasm to produce a mass of material enough to fill a bath is not sensible. It may result in damage to the habitat and its associated communities, as well as being too much to analyse in timescales which make the data useful!

The method employed in still waters utilises the 'sweep' approach used in still or sluggish areas of lotic environments. The strategy to be used in any given wetland involves sampling across the range of habitats present. For example, in a complex environment of ponds and ditches, one might pick a series of sites covering a range of habitat structures – a weedy ditch, a weedless ditch, a shallow, temporary pond, a deep, permanent pond, a series of ponds and ditches differing in the plant communities they support, and so on. One is never likely to cover all circumstances unless the site is repeatedly visited over a series of months, or even years. Nevertheless, a single day's targeted effort could produce a series of samples which can be analysed in a week. This can then give a useful 'snapshot' of the overall value of the wetland and its component habitats.

Specialised or General Search?

As mentioned above, conservation assessment can involve a less generalist approach. Direct searching of detritus, leaf axils, plant stems or the surfaces of stones or logs may be employed where particular species or orders are of interest. Active disturbance of the intensively searched substratum may be involved. For example, Desmoulin's Whorl Snail *Vertigo moulinsiana*, a rare and protected species in Europe, is surveyed by bending the stems of wetland plants, such as Reed sweet-grass (*Glyceria maxima*), over a tray and shaking them vigorously (Killeen and Moorkens 2003). The tiny Sphagnum Bug *Hebrus ruficeps* – less than 2 mm long – is found by tearing-out handfuls of *Sphagnum* spp., lightly wringing these out and placing them on a tray in direct sunlight. As the moss clumps warm up and dry out, the bugs run out over the tray and can be caught with ease (Southwood and Leston 1959).

Even when using a generalist approach, the intention may be to derive species lists of a limited range of taxa. Many 'still-water' habitats are assessed using only the resident community of water beetles. This is because there are sufficient numbers of species (around 400 in Britain) and their ecology, taxonomy and status have been studied for many years. These features are, therefore, very well understood. A sweep or kick/sweep sample, perhaps augmented by active searching, is taken, and the beetles recorded in the field or extracted and retained for identification. Certain conservation assessment approaches for wetlands have been designed with water beetles in mind, such as the Species Quality Score (SQS) methodology of Foster and Eyre (1992).

Quality or Quantity?

Pond-net and direct searching methodologies generally produce qualitative or semi-quantitative datasets, with the latter usually expressed as a relative abundance of taxa on a subjective scale (DAFOR – Dominant, Abundant, Frequent, Occasional, Rare) or a logarithmic one (1–9, 10–99, 100–999, etc.). The former approach is understandably subject to observer variation, as a product of its subjectivity, whereas the latter is more objective. Both approaches are, however, vastly more useful than production of simple presence/absence data. This is because the required information, whether it be identification of stresses or judgement of conservation value, is enhanced by an appreciation of the population as well as the community. As a rule, a semi-quantitative approach is sufficient to derive the appropriate conclusions. Nevertheless, fully quantitative methodologies can be employed where empirical comparison of habitats at various scales, but usually smaller ones, is the desired outcome.

In flowing waters, a Surber Sampler may be used for fully quantitative sampling (Fig. 7.1). The frame folds out so that a net (0.9 mm mesh) is held at 90° to a frame of approximately 0.33 m^2. This is placed randomly on the river bed, with the aperture of the net aligned downstream of the frame and the substratum disturbed to a standard depth (usually 0.1 m). Thus, the invertebrates in a standard volume of substratum can be collected and compared. Alternatively, standard 'chunks' of river or lake/pond substratum can be taken by using grabs – steel boxes with spring-loaded jaws which 'bite' the substratum to a standardised depth – or cylinder or box corers. All of these methodologies provide fully quantitative data, but are limited in deployment. Surber Samplers may only be used in flowing waters and are unlikely to be suitable in assessing invertebrate communities associated with plants or marginal deadwaters. All quantitative methodologies are very limited in coverage. They are only likely to derive taxonomic information for a small section of the available invertebrate community (coring at depth is difficult). They can be useful, however, in certain circumstances where pond nets are unsuitable. Deployment of Eckman Grabs at Rutland Water in the early 1990s was used to demonstrate damage to the deep benthos, caused by dosing of the reservoir with ferric sulphate to limit phosphate availability for toxic algae.

When to Sample?

The only question which remains in a sampling strategy is the optimum time of year to undertake it. Freshwaters, especially lentic habitats, are generally stable environments, especially in relation to temperature, so seasonality is less of an issue than in terrestrial habitats. One can still derive perfectly usable information in the depths of winter, as many species overwinter within the benthos. Some taxa (e.g. certain water beetles) may be out of reach of a generalist approach in winter, however, as they hibernate in nearby terrestrial zones or move to deeper waters. The ideal strategy is to sample any one habitat seasonally, so that the likelihood of obtaining an appropriate life stage for ready identification is maximised and seasonal limitations are minimised. If, however, the time to dedicate to a site is limited, the best option is to sample in late summer to early autumn (August to October). This maximises stresses related to water resources, so that problems are more readily identified and, more importantly, is the time of year when most freshwater invertebrates are present in larval or adult stages and the maximum dataset can be derived. Clearly, the more one limits the sampling seasons, the more species are likely to be missed. This may be a necessary trade-off against practicality and resources.

Analysis of Results

Analysis of a general invertebrate sample from an aquatic ecosystem for conservation value is facilitated by the Community Conservation Index (CCI) of Chadd and Extence (2004). This places a score of 1–10 (Conservation Score or CS) on virtually all members of the British aquatic macroinvertebrate fauna in freshwater and brackish ecosystems, where very common species have a CS of 1 and Endangered (RDB1) species have a CS of 10. The scores are summed, and then divided by the number of scoring species in the dataset to derive an average. This is then enhanced using a multiplier (Community Score or CoS) based on BMWP score (Chesters 1980), which is strongly influenced by habitat as well as water quality, and acts as a useful surrogate for measures of species richness or community diversity. The CoS may also be based on the rarest taxon in the dataset. Whichever CoS is the higher is the one which is used in multiplication. Thus, samples supporting a large number of common species or a sample supporting a limited fauna, but including rare species, are accentuated. This method of analysis can be used in any freshwater or brackish ecosystem in still or flowing waters.

Several computer-based models exist, to assist in interpretation of aquatic macroinvertebrate datasets. These include the River InVertebrate Prediction And Classification System (RIVPACS) (Wright *et al.* 2000), River Invertebrate Classification Tool (RICT) (Davy-Bowker *et al.* 2008), River Pollution Bayesian Belief Network (RPBBN) (Walley *et al.* 1992) and River Pollution Diagnostic System (RPDS) (O'Connor and Walley 2002). Both of the latter are the work of Staffordshire University, sponsored by the Environment Agency of England & Wales, and comprise diagnostic tools based on Artificial Intelligence (AI). Essentially, the AI systems seek to derive information from the datasets using known stresses and pattern recognition or probabilistic reasoning. RIVPACS and RICT (the latter is an update, derived from the RIVPACS model) seek to provide a 'reference state' for river sites, using environmental information compared against reference datasets. The observed data is then compared against reference-state. The further the departure from reference-state, the greater the stresses imposed on the fauna.

Outlook

With projected changes in climate leading to modelled stresses on water resources, relating to both flood and drought, the pressure on aquatic invertebrate communities and populations will inevitably increase in the coming decades. Added to this is the impact of habitat modification and invasion by exotic species, plus a range of potential changes relating to water chemistry. Assessment of stress on these communities, and management actions for the more important ones, is essential. In this way, we may be able to conserve as much of the native fauna – species, populations and community structures – as possible.

References

Armitage PD, Moss D, Wright JF, Furze MT (1983) The performance of a new biological water quality score system based on macro-invertebrates over a wide range of unpolluted running-water sites. Water Res 17:333–347

Boon PJ, Holmes NTH, Maitland PS, Rowell TA (1996) SERCON: System for evaluating rivers for conservation: Version 1 manual. Scottish Natural Heritage Research, Survey and Monitoring Report No. 61

Chadd R, Extence C (2004) The conservation of freshwater macroinvertebrate populations: A community-based classification scheme. Aquat Conserv Marine Freshwater Ecosyst 14:597–624

Chesters RK (1980) Biological monitoring working party. The 1978 national testing exercise. Department of the Environment, Water Data Unit, Technical Memorandum 19

Davy-Bowker J, Clarke R, Corbin T, Vincent H, Pretty J, Hawczak A, Blackburn J, Murphy J, Jones I (2008) River Invertebrate classification Tool. Final Report Project WFD72c. Edinburgh, SNIFFER

Environment Agency & Ponds Conservation Trust (2002) A guide to monitoring the ecological quality of ponds and canals using PSYM. Ponds Conservation Trust Policy and Research, Oxford

Extence CA, Bates AJ, Forbes WJ, Barham PJ (1987) Biologically based water quality management. Environ Pollut 45:221–236

Extence CA, Balbi DM, Chadd RP (1999) River flow indexing using British benthic macroinvertebrates: A framework for setting hydroecological objectives. Regulated Rivers Res Manage 15:543–574

Foster GN, Eyre MD (1992) Classification and ranking of water beetle communities. Joint Nature Conservation Committee, Peterborough

Hart CW Jr, Fuller SLH (eds) (1974) Pollution ecology of freshwater invertebrates. Academic, New York and London

Hellawell JM (1986) Biological indicators of freshwater pollution and environmental management. Elsevier, London and New York

Hynes HBN (1960) The biology of polluted waters. Liverpool University Press, Liverpool

Killeen IJ, Moorkens EA (2003) Monitoring Desmoulin's Whorl Snail Vertigo moulinsiana. Conserving Natura 2000 Rivers Monitoring Series No. 6. English Nature, Peterborough

Lancaster J, Briers RA (eds) (2008) Aquatic insects: Challenges to populations. CAB International, Wallingford

Mellanby K (1967) Pesticides and pollution. The New Naturalist, vol 50. Collins, London

O'Connor MA, Walley WJ (2002) River pollution diagnostic system (RPDS) – Computer-based analysis and visualisation for bio-monitoring data. 2nd world water congress: Environmental monitoring, contaminants and pathogens. Water Sci Technol 46:17–23

Paisley MF, Trigg DJ, Walley WJ (2007) Revision and testing of BMWP scores. Final report SNIFFER project WFD72a. Scotland and Northern Ireland Forum for Environmental Research (SNIFFER), Edinburgh

Rosenberg DM, Resh VH (eds) (1993) Freshwater biomonitoring and benthic macroinvertebrates. Chapman and Hall, New York and London

Shannon C, Weaver W (1963) The mathematical theory of communications. University of Illinois Press, Urbana

Southwood TRE, Leston D (1959) Land and water bugs of the British Isles. Frederick Warne and Co., London and New York

Walley WJ, Hawkes HA, Boyd M (1992) Application of Bayesian inference to river water quality surveillance. In: Grierson DE, Rzevski G, Adey RA (eds) Proceedings of the 7th international conference on applications of artificial intelligence in engineering. Elsevier, Amsterdam, pp 1030–1047

Walley WJ, Hawkes HA (1996) A computer-based reappraisal of the Biological Monitoring Working Party scores using data from the 1990 river quality survey of England and Wales. Water Res 30:2086–2094

Walley WJ, Hawkes HA (1997) A computer-based development of the Biological Monitoring Working Party score system incorporating abundance rating, site type and indicator value. Water Res 31:201–210

Wood PJ, Hannah DM, Sadler JP (eds) (2007) Hydroecology and ecohydrology: Past, present and future. Wiley, Chichester

Wright JF, Sutcliffe DW, Furse MT (eds) (2000) Assessing the biological quality of fresh waters: RIVPACS and other techniques. Freshwater Biological Association, Ambleside

Chapter 8
Riverine Plants as Biological Indicators

Richard Lansdown and Sam Bosanquet

Introduction

There are no standardised methods for the use of riverine plants in monitoring in Europe, in the sense of recording against pre-defined targets. There are currently three main purposes for which data on riverine plants are collected:

- Species-specific surveys of rare taxa
- Data collection to answer a particular question
- Standard survey for the purpose of condition assessment or conservation

Of these, the second and third employ plants as biological indicators and are considered in detail here. Species-specific surveys almost always involve a specialist who will search for and collect data on the target species to their own protocol and rarely employ any form of standardised survey (e.g. Giavarini 2000; Holyoak 2001), although the countryside agencies sometimes provide exemplar surveys by specialists (e.g. Bosanquet 2008) as an example of how to collect or present data. Data collection to answer particular questions will usually either involve simple listing of species present or will employ one of the three standard survey methods, which are also the basis for conservation and condition assessment:

- JNCC Macrophyte Survey (MS) (Holmes 1983; Holmes *et al.* 1999a)
- Mean Trophic Rank (MTR) (Dawson *et al.* 1999; Holmes *et al.* 1999b) (Chapter 12) and its successor LEAFPACS (Willby 2005)
- River Corridor Survey (RCS) (NRA 1992)

R. Lansdown (✉)
45 The Bridle, Stroud, Glos, GL5 4SQ
e-mail: rlansdown@ardeola.demon.co.uk

S. Bosanquet
Cnwc y Llwyn, Brechfa, Carmarthen, Carms, SA32 7QR

C. Hurford et al. (eds.), *Conservation Monitoring in Freshwater Habitats: A Practical Guide and Case Studies*, DOI 10.1007/978-1-4020-9278-7_8,

All of these except RCS are based on recording what are termed "macrophytes" which are usually considered to conform to the following definition "larger alga or higher aquatic plant (including bryophytes), observable to the naked eye and nearly always identifiable when observed" (Holmes *et al.* 1999a). RCS differs in that it is based on preparation of a simplistic map of the river corridor, major structures and vegetation, supported by a text describing these in some detail. There is no interpretation and the method can be modified to focus on a particular aspect without compromising the quality of data. RCS is not considered further here.

MS, MTR and LEAFPACS are essentially similar in that they involve what may be termed "comprehensive" recording, where the surveyor is expected to be aware of all macrophytes present in a standard length of river, even if only a subset of these is recorded, as opposed to "indicative" recording where the surveyor would seek one or more indicator species. However, the influence of factors such as dispersal and colonisation on these data is rarely, if ever, acknowledged. All three methods require surveyors to focus on taxa on a standard list. They differ mainly in that MS is applied to a 500 m section (originally two abutted 500 m sections) and requires separation of plants into "bank" and "channel" populations, whereas the other two are applied to 100 m sections and make no attempt to differentiate marginal and channel species, although they do exclude bank species. All three of these methods rely at least in part on standardised interpretation methods ("typification" or scoring) to derive a characterisation of the vegetation or assessment of the trophic status of the section surveyed. MTR scores were derived on a somewhat subjective basis, from a combination of judgement and the literature. LEAFPACS scores were derived from statistical analysis of a large data-set collected on 100, 500 or 1,000 m long sections of rivers (N. J. Willby pers. comm. 2007).

Plants may also be used to indicate a particular situation. For example, all British and Irish vascular plants and bryophytes have been assigned Ellenberg indicator values (Hill *et al.* 2004, 2007) that score their tolerance of five conditions (light, moisture, reaction/pH, nutrient levels and salt). In theory, these values could be used to assess the results of MS, MTR or LEAFPACS to work out average (mean and mode) nutrient values or light scores for a section of river. This approach would work much better with comprehensive lists than with indicator species, but there is no evidence to suggest that this comparison has been made. One minor flaw of the Ellenberg values is that many species have values derived from extrapolations based on associated taxa, rather than values gained through direct observation; although this probably has little impact on average scores when a whole suite of species is assessed. It is possible that Ellenberg values could show trends in the floristic composition within a river channel, assuming of course that the species lists being analysed are genuinely comprehensive.

MS uses a classification to compare different rivers and parts of rivers for conservation assessment. The other survey methods all use records of plants to indicate changes in the chemical composition of rivers, as a means of monitoring pollution

Pollution Monitoring

It is almost impossible to reach definitive conclusions about the factors dictating the distribution of aquatic plant species from the literature, because differences in data collection methods combined with differences in the physical and chemical characteristics of the study sites preclude direct comparison. At the same time there has been very little ex-situ testing of the tolerances of aquatic plants to chemicals to support conclusions reached from field studies. It seems that for almost every conclusion reached, the opposite will have been demonstrated, to the extent that when Benoit Demars (pers. comm. 2009) has subjected data from published works to alternative analyses, the same data have produced different results. The suggestion that standard-length based river surveys may not result in reliably replicable data (Lansdown 2007, Chapter 14) may, to some degree, explain poor comparability of results. Therefore, the following review has, wherever possible, excluded conclusions reached using data collected from standard-length surveys.

It is generally assumed that nutrients dictate riverine plant distribution (see Box 8.1), however, Demars and Thiébault (2008) carried out detailed and rigorous analysis of the data employed to derive UK indices and concluded that "the lack of strong correlations between inorganic N and P and aquatic plant composition and abundance seriously question the ability of the current macrophyte indices or vegetation-based methods for ecological diagnosis". In spite of the fact that there is reasonable evidence to suggest that flow and nutrients influence the distribution of plants (e.g. Watson 1919; Clarke and Wharton 2001), a variety of other parameters have been described as influencing their distribution and abundance in rivers, and there is disagreement over the influence of almost every parameter.

Plant ecology information is relatively compendious, although much is focussed on the ecology of single species rather than phytosociological units, and there have

Box 8.1 Aquatic vascular plants as indicators of water pollution

Clarke and Wharton (2001) provide an excellent review of studies of eutrophication and aquatic plants and suggest that their use for bio-indication or biomonitoring has the following advantages:

- Macrophytes are stationary and so absence is easily ascertained.
- They are by definition visible to the naked eye.
- Monitoring is rapid and requires little or no subsequent laboratory identification.
- There are relatively few species within any one region.
- Many are rooted and thus reflect seasonal or disturbance factors.

They go on to describe disadvantages of using macrophytes as monitoring tools, including poor information on the ecology of many species.

been numerous studies of the effects of different environmental variables on riverine vascular plants. Unfortunately, many of these are contradictory, making it difficult to draw conclusions as to which species or species groups genuinely give useful information. Most of the emphasis in river monitoring has been based on vascular plants, with only minor emphasis on lower plants. However Gilbert (1996) and Gilbert and Giavarini (1997) suggest that substrate, sediment deposition and algae are the most important factors influencing the distribution of lichens. Logic suggests that, apart from totally submerged populations, the distribution of vascular plants will be influenced more by the sediment than the water column, although Clarke and Wharton (2001) found only a weak relationship between sediment nutrients and plant distribution. The only plants which are critically dependent upon the water column are algae and bryophytes, as these lack a vascular system and therefore take in water from their surroundings, making them theoretically better able to respond rapidly to changes in water conditions. However, there is apparently a dynamic equilibrium of nutrients at the sediment-water column interface so bryophytes would also at least partly reflect sediment chemistry.

In the 1990s, several studies in continental Europe looked at the effects of environmental pollution on bryophytes, either by transplanting them from clean to polluted rivers, or by comparing a series of rivers in an area. Martinez-Abaigar *et al.* (1993) transplanted three species from clean upper reaches of the Rio Iregna in Spain into its highly sewage-enriched lower reaches and found *Jungermannia exsertifolia* to be very sensitive to pollution events, whilst *Fontinalis antipyretica* was less sensitive. *F. antipyretica* is also tolerant of heavy metal pollution, accumulating metal deposits on its cell walls (Sérgio *et al.* 2000), something also found in *Rhynchostegium riparioides*. These studies suggest that although some bryophytes may be good indicators of river quality, others have broader tolerances and tell one relatively little. Vanderpoorten *et al.* (1999) studied lowland reaches of the river Rhine. They showed that *Chiloscyphus polyanthos*, *Pellia endiviifolia* and *Amblystegium tenax* are sensitive to increases in inorganic nitrate and phosphates, whilst *Amblystegium fluviatile* and *Fissidens crassipes* are relatively tolerant and *F. antipyretica* and *R. riparioides* appear unaffected. It appears likely that the leafy liverwort *C. polyanthos* is the most useful indicator bryophyte on British rivers. The *Amblystegium* species are notoriously difficult to distinguish, as are the suite of aquatic *Fissidens*, whilst *C. polyanthos* is sufficiently widespread and tolerant of both light and shade to be used in most of the west of the country.

Jungermannia species may also be useful for monitoring in upland areas, judging by Martinez-Abaigar *et al.* (1993). Other upland indicators were described by Vanderpoorten and Klein (1999) for the Upper Rhine: *Brachythecium plumosum*, *Hyocomium armoricum* and *Marsupella emarginata* declined through sewage effluent pollution; these three species and *Scapania undulata* were replaced by *Cratoneuron filicinum* and *Hygrohypnum luridum* after large-scale physical disturbance; and *Chiloscyphus polyanthos*, *Rhynchostegium riparioides* and *Thamnobryum alopecurum* were particularly sensitive to acidification. Ellenberg values for nutrient tolerance back up these studies: *C. polyanthos* and *A. tenax* have a nutrient value of N = 4, whilst *F. antipyretica* has N = 5 and *R. riparioides* has N = 6. *Scapania undulata* has an even lower value (N = 2), whilst *Jungermannia* species, including *J. exsertifolia*, have N = 2 or 3. It seems possible that leafy liverworts, which are readily identifiable for general ecologists as a group, if not always to individual species level, might provide a realistic set of indicators of the nutrient levels of rivers. Assessment of acidification is more difficult because the acid-sensitive *Rhynchostegium riparioides* and *Thamnobryum alopecurum* (Vanderpoorten and Klein 1999) have two similar-looking counterparts on naturally acidic rivers, *R. alopecuroides* and *Isothecium holtii*.

It is typical of studies of the chemical determinants of aquatic plant abundance and distribution, that each study shows differences from results of other studies of the same parameters. Carbiener *et al.* (1990) carried out long-term studies on the River Ill in Alsace. They showed that the headwaters were groundwater fed and oligotrophic and supported a particular plant community. Downstream, discharge from a trout farm led to raised trophic levels and consequent development of a different plant community, but this was restored to the upstream community when groundwater again entered the channel downstream. However B. Demars (pers. comm. 2009) re-analysed the data employed and concluded that "pCO_2 is the strongest factor influencing plant distribution, although NH_4, PO_4 and temperature were all interacting significantly". The conclusions of Carbiener *et al.* (1990) agree with those of Kohler (1971) and Kohler and Zeltner (1981) (both cited in Carbiener *et al.* 1990) but at lower concentrations, which they suggest may be due to the presence of peat bogs in the catchments studied by Kohler. However, they present no evidence to support this suggestion.

Thiébaut and Muller (1995) suggest that eutrophication of acidic streams results in replacement of oligotrophic taxa, such as *Potamogeton polygonifolius*, by taxa tolerant of increased trophy, such as *Callitriche obtusangula*, *Elodea canadensis* and *E. nuttallii*. However B. Demars (pers. comm. 2009) re-analysed the data employed and concluded that alkalinity/pH were the dominant factors rather than nutrients. In another contrast, James *et al.* (2006) consider that relative growth rate, independent from response to nutrient enrichment, may explain why *Elodea canadensis* was displaced by *E. nuttallii* and then *Lagarosiphon major* in British water bodies.

Carbiener *et al.* (1990, 1995) state that NO_3 has no effect on aquatic plant distribution. Clarke and Wharton (2001), in a comparison of data from different river systems, showed that "Macrophyte species showed a broad tolerance to all sediment variables and it was difficult to separate the influence of sediment nutrients from other sediment parameters or differences between rivers" and go on to suggest that catchment geology, channel shading and sediment total nitrogen concentrations were important in explaining the variation in macrophyte species distribution.

In most cases, for each "conclusive" finding of a trend, there is an equally conclusive finding of the opposite. Thus, Zhu *et al.* (2008) suggest that P does not affect species diversity or biomass, but Carbiener *et al.* (1990, 1995) state that it is the most important determinant in riverine plant distribution. Alternatively, Richard and Ivey (2004) (studying *Sagittaria lancifolia*) and Mony *et al.* (2007) (studying *Ranunculus peltatus*) suggest that P affects leaf morphology and production of sexual organs respectively, but do not suggest any effect of P on abundance or survival, while Mony *et al.* (2007) suggest that the nature of the influence of P is dependent upon season.

It is very likely that the influences of chemicals and physical conditions on plant abundance and distribution are not simple, not only may one chemical influence the effects of another, but the influence of one chemical on one species may influence another species. For example Irfanullah and Moss (2004) suggest that recovery of *Elodea nuttallii* is impeded mainly by growth of epiphytic filamentous algae in response to high nutrient levels, even though Thiébaut and Muller (1995) showed that *E. nuttallii* colonised when nutrient levels increased. Dormann (2007) showed that in monoculture *Calliergonella cuspidata* increased biomass by 60% as a result of fertiliser application. However in mixed cultures with *Plagiomnium undulatum* and a combination of *Plagiomnium undulatum* and *Rhytidiadelphus squarrosus*, the increase was reduced or eliminated. Thus, there is strong evidence that the presence of the other taxa affected the response of *C. cuspidata* to fertiliser. Further to this, Edvarsen and Økland (2006) say that the influence of the trophic state in a pond follows a Gaussian distribution, where low trophic state results in low species-richness; intermediate trophic state allows an increase in species-richness; and high trophic state again results in low species-richness. This agrees with the findings of Tylová *et al.* (2008) who found that *Glyceria maxima* showed increased growth with low levels of N availability, but suppression at very high levels (probably exceeding those found in natural systems). This could also be the reason that Takaki (1976 and 1977, cited in Glime and Saxena 1991) found that bryophytes were absent from the upper reaches of an unpolluted river, but started to appear downstream with pollution from villages, industries or mines.

Physical determinants of aquatic plant distribution are similarly subject to claim and counter-claim. For example, Carbiener *et al.* (1990, 1995) state that current velocity has no effect on aquatic plant distribution (an unlikely conclusion) but O'Hare *et al.* (2007) suggest that drag (which must be related to velocity) affects the distribution of aquatic plants, while Butcher (1927) says that the nature of the rooting system has more influence over the potential of plants to be removed by scour, and consequently their distribution. Boeger and Poulson (2003) suggest that the morphology of *Veronica anagallis-aquatica* is affected by high flow velocity, but at no point suggest that higher velocity leads to death or uprooting. Šraj-Krži *et al.* (2007) state that *Mentha aquatica*, *Myosotis scorpioides*, *Persicaria amphibia*, *Rorippa amphibia* and *Ranunculus trichophyllus* (all taxa which have MTR scores) are most typical of intermittent flow due to their morphological and functional plasticity.

The wide variation in claims, theories and conclusions regarding the factors influencing the distribution and abundance of plants in rivers is very difficult to interpret for the purposes of bio-indication. The findings of Demars and Edwards (2009) provide what might be considered the most honest interpretation of plant data "The distribution of aquatic plants results not only from interactions of environmental pressures but also from dispersal barriers and biotic interactions. While some species are more restricted to certain environmental conditions, many are indifferent, and no species can safely be associated with any particular environmental variable. Hence the idea of building indices (ultimately) based on species composition to indicate reliably inorganic N or P enrichment does not seem reasonable". The situation might be different in controlled experiments but this conclusion may be reasonable for studies of plants *in situ*.

Conclusions and Ways Forward

Concerns over potential replication of standard-length surveys, coupled with the demonstrable uncertainty over the reliability of named plants as pollution indicators, casts doubt over the use of aquatic plants as indicators based on current information. However, it is possible that some taxa do respond to changes in water chemistry. Therefore, with rigorous scientific research, it may be possible to identify a suite of species that are so critically associated with the presence or absence of a chemical that they can be used to indicate chemical changes in water.

It is clear that, whilst some vascular plants may show strong chemical associations, the best indicators are likely to be found among algae and bryophytes. The taxonomy and ecology of freshwater algae are still so poorly known that it will be many years before suitable taxa that can be readily identified by non-specialists are found to be reliable indicators. Therefore, perhaps the best indicators should be sought among bryophytes.

As suggested above, leafy liverworts might form a useful and realistically identifiable group to indicate the nutrient status of a river. Ideally, individual species would be identified and river specialists would learn to recognise uncommon plants, such as *Porella pinnata* and *Jungermannia hyalina*, but records at genus or even broad 'leafy liverwort' level might possibly suffice. An indication of abundance on short stretches of river would be practical if this broad level was being used. This would be important because 100 or 500 m long reaches are unlikely to pick up potentially fine-scale variation in the occurrence of liverworts. Local absence of the rocky substrates on which leafy liverworts can grow could limit the applicability of this approach, especially in lowland Britain, but it is a possible option where suitable substrates occur.

Similarly, a suite of bryophytes that is restricted to silty riparian trees such as *Cryphaea lamyana*, *Leskea polycarpa*, *Myrinia pulvinata* and *Syntrichia latifolia* could indicate variation in aspects of the influence of rivers on their floodplains. In particular, *L. polycarpa* and *S. latifolia* are sufficiently distinctive to be easily learned and are therefore suitable for identifying flood zone trees. The distance from the river to which these species extend and the height of the zone that they occupy could be used to monitor the amplitude of variation in water levels, at least on large lowland rivers.

Another possible approach to the use of bryophytes in monitoring involves the translocation of plants from unpolluted rivers or reaches of rivers into polluted reaches to assess survival (e.g. Martinez-Abaigar *et al.* 1993). Although this would require ex-situ experimental studies of the tolerances of the taxa involved to ensure that the conclusions actually relate to pollution. Another possibility (as noted by Glime and Saxena 1991, see Box 8.2) would be to use herbarium specimens of

Box 8.2 Aquatic bryophytes as indicators of water pollution

Glime and Saxena (1991) list 16 reasons (reproduced below) why bryophytes are often considered appropriate for pollution monitoring – all based on the idea of analysing bioaccumulated pollutants. They do not suggest that changes in presence or abundance of bryophytes occur as a consequence of pollution events, but emphasise that bryophytes will persist throughout a pollution event, enabling monitoring through bioaccumulation, something backed up by Lee *et al.* (1998). Herbarium specimens of bryophytes can also be used to provide a baseline for comparison from more than 100 years, particular in a country such as the UK, which has very good bryophyte herbaria.

1. Bryophytes attain high levels of accumulation.
2. The collection of suitable bryophyte material provides an easy method for people without access to water sampling facilities and sampling can be done by anyone who suspects that contamination has occurred.
3. Bryophytes can be harvested after a pulse of contaminated water has passed downstream, so it is possible to identify pollution hours or perhaps even days after it has occurred.
4. Bryophytes can provide an integrated record of pollution within a particular system.
5. Dried plant materials are much easier to keep than polluted water samples. They are compact and stable (most elements except mercury), so they are particularly suited for long-term storage.
6. They need minimum maintenance.
7. Water management bodies can retain samples indefinitely, permitting further elements to be analysed years later.
8. They are easy to send by post for inter-laboratory examinations.
9. One can acquire a data bank on pollution and pollutants by routine monitoring. Recovery of the plants can then be monitored and growth and morphology of the plants can be compared to similar data prior to treatment.
10. The content of a particular plant species is usually less heterogeneous at any one site than that of the sediments.
11. Plants are sessile and thus reflect conditions in a particular stretch of river, whereas sediments may be transported downstream, reflecting conditions some unknown distance upstream.
12. The methods of analysis are simple and many do not require sophisticated or expensive equipment.
13. In some countries, there are standard methods for the use of bryophytes to monitor heavy metals and other pollution.
14. Analysis of the pollutant and heavy metals composition of aquatic bryophytes provides a better approach than using water chemistry to analyse the water.
15. Bryophytes that have been used include *Leptodictyum riparium*, *Fontinalis antipyretica*, *Rhynchostegium riparioides* and *Scapania undulata*.
16. Death of the total biomass is slow; hence bryophytes have the ability to retain their toxic load after death. The slow release of accumulated substances means that they can be used as environmental specimen banks. Herbarium specimens can indicate historic pollutant loads.

bryophytes to measure changes in the chemistry of rivers over time. Few historic collections are sufficiently well localised to enable precise comparisons, but regular collection of specimens of a common species such as *Fontinalis antipyretica* from all British rivers from now on could provide a reliable and comprehensive basis for comparison with other chemical monitoring methods.

A switch to the use of a small set of well-researched indicator species would provide more meaningful data than the current "comprehensive" surveys, allowing more precise conclusions to be drawn. At the same time, removing the need to record every plant species in a long stretch of river would free-up surveyors' time to collect more meaningful information. For example noting precisely where in the river a patch of *Potamogeton nodosus* is, its extent and whether it is fruiting, or collecting *Schistidium* specimens for careful determination of *S. agassizii* could be more informative.

New systems are needed if we are to use aquatic plants as biological indicators. However, it is vital that in the future any testing of new methods are supported by autecological and ex situ experimental assessment of chemical tolerances to inform the selection of indicator species for monitoring.

Acknowledgements The authors would like to thank Benoit Demars for copies of his publications. We are also extremely grateful to Richard Chadd, Benoit Demars and Becky Lansdown for commenting on earlier drafts of this chapter.

References

Boeger MRT, Poulson ME (2003) Morphological adaptations and photosynthetic rates of amphibious *Veronica anagallis-aquatica* L. (Scrophulariaceae) under different flow regimes. Aquat Bot 75(2):123–135

Bosanquet SDS (2008) A survey and condition assessment of the bryophytes of Castlemartin Range (part of the Limestone Coast SAC). Pembrokeshire, CCW report, Pembroke Dock

Butcher RW (1927) A preliminary account of the vegetation of the River Itchen. J Ecol 15:55–65

Carbiener R, Trémolières M, Mercier JL, Ortscheit A (1990) Aquatic macrophyte communities as bioindicators of eutrophication in calcareous oligosaprobe stream waters (Upper Rhine Plain, Alsace). Vegetatio 86:71–88

Carbiener R, Trémolières M, Muller S (1995) Végétation des eaux courantes et qualité des eaux: une these, des débats, des perspectives. Acta Bot Gallica 142(6):489–531

Clarke SJ, Wharton G (2001) Using macrophytes for the environmental assessment of rivers: The role of sediment nutrients. Technical Report E1-S01/TR. Environment Agency, Bristol

Dawson FH, Newman JR, Gravelle MJ, Rouen KJ, Henville P (1999) Assessment of the trophic status of rivers using macrophytes: Evaluation of the mean trophic rank. R&D Technical Report E39. Environment Agency, Bristol

Demars BOL, Edwards AC (2009) Distribution of aquatic macrophytes in contrasting river systems: A critique of compositional-based assessment of water quality. Sci Total Environ 407:975–990

Demars BOL, Thiébault G (2008) Distribution of aquatic plants in the Northern Vosges rivers: Implications for biomonitoring and conservation. Aquat Conserv 18(5):619–632

Dormann CF (2007) Competition hierarchy, transitivity and additivity: Investigating the effects of fertilisation on plant–plant interactions using three common bryophytes. Plant Ecol 191(2):171–184

Edvarsen A, Økland RH (2006) Variation in plant species richness in and adjacent to 64 ponds in SE Norwegian agricultural landscapes. Aquat Bot 85(2):79–91
Giavarini VJ (2000) Priority lower plant surveys in Scotland: Collema dichotomum on the River Tweed. Unpublished Report to Scottish Natural Heritage
Gilbert OL (1996) The lichen vegetation of chalk and limestone streams in Britain. Lichenologist 28(2):145–160
Gilbert OL, Giavarini VJ (1997) The lichen vegetation of acid watercourses in England. Lichenologist 29(4):347–368
Glime JM, Saxena D (1991) Uses of bryophytes. Today and Tomorrow's Printers and Publishers, New Delhi
Hill MO, Preston CD, Roy DB (2004) PLANTATT – Attributes of British and Irish plants: Status, size, life history, geography and habitats. Centre for Ecology and Hydrology, Wallingford
Hill MO, Preston CD, Bosanquet SDS, Roy DB (2007) BRYOATT – Attributes of British and Irish Mosses, Liverworts and Hornworts: With information on native status, size, life form, life history, geography and habitat. Centre for Ecology and Hydrology, Wallingford
Holmes NTH (1983) Typing rivers according to their flora. Focus on Nature Conservation No. 4. Nature Conservancy Council, Peterborough
Holmes N, Boon P, Rowell T (1999a) Vegetation communities of British rivers: A revised classification. Joint Nature Conservation Committee, Peterborough
Holmes NT, Newman JR, Chadd S, Rouen KJ, Saint L, Dawson FH (1999b) Mean trophic rank: A user's manual. Environment Agency, Bristol
Holyoak DT (2001) Tiny fern-moss Fissidens exiguus: Report to plantlife on work carried out during 2000. Plantlife, Salisbury
Irfanullah HM, Moss B (2004) Factors influencing the return of submerged plants to a clear-water, shallow temperate lake. Aquat Bot 80(3):177–191
James CS, Eaton JW, Hardwick K (2006) Responses of three invasive aquatic macrophytes to nutrient enrichment do not explain their observed field displacements. Aquat Bot 84(4):347–353
Lansdown RV (2007) Assessment of the condition of the water courses of plain to montane levels with the Ranunculion fluitantis and Callitricho-Batrachion vegetation of the River Wye SAC. Unpublished Report to CCW, Bangor
Lee JA, Caporn SJM, Carroll J, Foot JP, Johnson D, Potter L, Taylor AFS (1998) Effects of ozone and atmospheric deposition on bryophytes. In: Bates JW, Ashton NW, Duckett JG (eds) Bryology for the twenty-first century. Maney and British Bryological Society, Leeds
Martinez-Abaigar J, Nuñez-Olivera E, Sanchez-Diaz M (1993) Effects of organic pollution on transplanted aquatic bryophytes. J Bryol 17:553–566
Mony C, Thiébaut G, Muller S (2007) Changes in morphological and physiological traits in the freshwater plant *Ranunculus peltatus* with phosphorus bioavailability. Plant Ecol 191(1):109–118
NRA (1992) River corridor surveys: Methods and procedures. Conservation Technical Handbook No. 1. Environment Agency, Bristol
O'Hare MT, Hutchinson KA, Clarke RT (2007) The drag and reconfiguration experienced by five macrophytes from a lowland river. Aquat Bot 86(3):253–259
Richard JH, Ivey CT (2004) Morphological plasticity of *Sagittaria lancifolia* in response to phosphorus. Aquat Bot 80(1):53–67
Sérgio C, Figueira R, Viegas Crespo AM (2000) Observations of heavy metal accumulation on the cell walls of *Fontinalis antipyretica*, in a Portuguese stream affected by mine effluent. J Bryol 22:251–255
Šraj-Krži N, Germ M, Urbanc-Ber i O, Kuhar U, Janauer GA, Gaberš ik A (2007) The quality of the aquatic environment and macrophytes of karstic watercourses. Plant Ecol 192(1):107–118
Thiébaut G, Muller S (1995) Nouvelles données relatives á la sequence de bioindication de l'eutrophisation dans les cours d'eau faiblement mineralises des Vosges du Nord. Acta Bot Gallica 142(6):627–638
Tylová E, Steinbachová L, Votrubová O, Lorenzen B, Brix H (2008) Different sensitivity of *Phragmites australis* and *Glyceria maxima* to high availability of ammonium-N. Aquat Bot 88(2):93–98

Vanderpoorten A, Klein J-P (1999) Variations of aquatic bryophyte assemblages in the Rhine Rift related to water quality: 2. The waterfalls of the Vosges and the Black Forest. J Bryol 21:109–115

Vanderpoorten A, Klein J-P, Stieperaere H, Trémolières M (1999) Variations of aquatic bryophyte assemblages in the Rhine Rift related to water quality: 1. The Alsatian Rhine floodplain. J Bryol 21:17–23

Watson W (1919) The bryophytes and lichens of fresh water. J Ecol 7:71–83

Willby NJ (2005) LEAFPACS methodology as provided by SEPA. Unpublished appendix to SEPA tendering documents

Zhu B, Mayer CM, Rudstam LG, Mills EL, Ritchies ME (2008) A comparison of irradiance and phosphorus effects on the growth of three submerged macrophytes. Aquat Bot 88(4):358–362

Chapter 9
Phytoplankton (Toxic Algae) as Biological Indicators

Héctor Rodríguez-Pérez, Victoria López-Rodas, and Eduardo Costas

Introduction

The cyanobacteria or cyanoprokaryota, formerly green–blue algae, (Class: Cyanophyta) is a cosmopolitan and diverse group of organisms, living in freshwater and seawater, running or quiet waters, in a benthic or planktonic manner. Furthermore, cyanobacteria can exist in habitats characterised by extremes of temperature, toxic mineral excess and high salinity. They can also survive long periods of drought thanks to resistant cells called akinetes.

Some cellular strains from several cyanobacteria genera (e.g. *Microcystis, Anabaena, Aphanizomenon, Oscillatoria, Pseudoanabaena, Nodularia, Nostoc, Cylindrospermopsis, Schizothrix, Trichodesmium* and others) have toxigenic properties. Of all toxic species of cyanobacteria, the hepatotoxin-producing *Microcystis aeruginosa* (Kützing) is the most important cause of toxic blooms affecting humans and animals in inland water systems (Skulberg *et al.* 1993).

Cyanobacteria produce potent toxins with a broad chemical nature and effects, like alkaloids and peptides with acute hepatotoxic, cytotoxic, neurotoxic and gastrointestinal disturbances, and respiratory and allergic reactions (neurotoxic alkaloids, saxitoxins, hepatotoxic peptides and alkaloids) (Carmichael 1992). Although other algal groups also produce toxins, cyanobacteria are responsible for most poisoning events in freshwater. However, toxic events do most harm when a toxic population blooms and forms a dense scum on the water surface. It is widely believed that eutrophication, exposure to intense light and the warming of waterbodies all contribute to the blooming of planktonic cyanobacteria populations.

Fatalities to humans due to intoxication by cyanotoxins have been reported several times (i.e. Jochimsen *et al.* 1998), though nowadays this type of incident is less common. By contrast, mass mortalities of livestock and wild animals are commonly

H. Rodríguez-Pérez (✉), V. López-Rodas, and E. Costas
Department of Animal Production, Veterinary Faculty, Complutense University of Madrid, Spain
e-mail: hector.hrp@gmail.com

C. Hurford et al. (eds.), *Conservation Monitoring in Freshwater Habitats: A Practical Guide and Case Studies*, DOI 10.1007/978-1-4020-9278-7_9,

reported worldwide (Stewart *et al.* 2008) and it has been also suggested that cyanobacteria caused mass mortalities in prehistoric times (Braun and Pfeiffer 2002). Most deaths happen when drinking water is contaminated with cyanotoxins. Cyanobacteria populations may also play a remarkable ecological role by disrupting food webs, through producing anoxia and shading the water column when a large bloom occurs (Scheffer 1998; Havens 2008), but studies of other ecological consequences than poisoning are uncommon.

Toxic cyanobacteria blooms are a major concern for reservoir managers, and corporations that manage supplies of drinking water usually implement complex monitoring programmes to prevent and predict these toxic events: investing large sums of money on minimising the effects of toxic blooms and keeping toxin levels below the legal thresholds (Davis and Mazumder 2003). This type of programme for monitoring toxic blooms is seldom implemented in Nature Reserves, as is the case at Doñana National Park. When a cyanobacteria bloom early warning system is triggered in a Natural Reserve, the main aim is to prevent mass wildlife mortalities.

The Experience of the Early Warning Network for Preventing Mass Wildlife Mortalities as a Consequence of Cyanobacteria Blooms in Doñana National Park

Doñana National Park is located in the Guadalquivir estuary (SW Spain), and this protected area holds a great wealth of waterbodies: temporary ponds, permanent ponds, artificial permanent ponds, a non-tidal temporary marshland (26,000 ha) and tidal channels (see García-Novo and Marín 2006 for further information). Most of these waterbodies dry out every year by late spring, before environmental conditions can facilitate cyanobacteria blooms, but in rainy years some keep water for longer. Furthermore, in some, the water level is maintained artificially, e.g. to ensure that waterbirds can breed or for fish farming.

Since Doñana was declared a National Park in 1969, 18 mass mortality events have been recorded. Before 2001, none of these mortalities was attributed to toxic algal poisoning (botulism, pesticide poisoning or unknown causes were blamed): the possibility of mass cyanotoxin poisoning was never explored. During the summer of 2001, however, more than 500 flamingos died in few days in a man-made pond called FAO pond, and researchers from the Veterinary Faculty of Complutense University of Madrid determined that the cyanobacteria *Microcystis aeruginosa* and *Anabaena flos-aquae* had caused the mortalities (Alonso-Andicoberry *et al.* 2002). Since then an early warning network has been implemented between the Doñana Natural Space managers (Junta de Andalucia, Regional Government of Andalusia), the Natural Processes Monitoring Team (Doñana Biological Station-CSIC), and the Genetic Laboratory of Veterinary Faculty (Complutense University of Madrid). The aim of this network is to prevent new mass mortalities and to revise annual procedures and protocols to improve efficiency. Although two other mass mortalities have happened since, others have probably been avoided because of the application of the early warning network recommendations.

In late July 2004, nearly 6,000 birds of 47 different species died within a 2-week period in Los Ánsares pond (López-Rodas *et al.* 2008), with a second mass mortality in June 2005, when 20 tons of fish died within 3 days in Caño del Guadimar channel (Costas and López-Rodas 2007). In 2004, large densities of *Microcystis aeruginosa* (up to 13,800 col/ml) and *Pseudoanabaena catenata* (up to 18,300 col/ml) were detected, and, in 2005, the neurotoxic saxitoxin produced by *Anabaena circinalis* was found to be the cause of the fish mortalities. Although toxic species of cyanobacteria are always present in the Doñana wetlands, they usually occur at low concentrations or with low toxicity. For example, in 2002, a year without noticeable mortalities, the range of colonies counted from May to September was 140–727 col/ml, with a peak of toxin production (microcystin-LR equivalent) of 0.06 μg/ml, whereas in 2004, during the bird mortality incident, up to 13.5 mg/ml of toxin (microcystin-LR equivalent) was detected in the water, 10^5 colonies/ml in the gut content of death birds, and 25.7–75.9 μg (microcystin-LR equivalent)/mg in the livers of dead birds.

Several genera of toxic cyanobacteria have been recorded in Doñana: *Microcystis*, *Anabaena*, *Pseudoanabaena*, *Gomphosphaeria* and *Oscillatoria*. But to date, only *Microcystis aeruginosa*, *Anabaena flos-aquae*, *A. circinalis* and *Pseudoanabaena catenata* have been identified as being responsible for the mass poisoning of wildlife.

The Design and Implementation of a Monitoring Programme to Detect Cyanobacteria Blooms

There is a huge difference between setting up a programme for monitoring cyanobacteria populations in reservoirs for human water supply and in natural waterbodies inside a nature reserve, particularly with regards to sampling time. To cover the distance between different waterbodies inside a nature reserve might be a very time consuming activity. For example, to sample all waterbodies in Doñana National Park during the summer (when the highest risk of toxic bloom occurs) takes more than a day of work for a single team of samplers. Therefore, as resources are usually limited, it is worth establishing a feasible protocol for sampling the sites, by identifying, through historical records or previous sampling, the waterbodies most at risk of cyanobacteria blooming on the basis of nutrient load, depth, and water column stability. Weekly sampling during the high-risk blooming season could be enough, increasing the sampling effort when the beginning of a bloom is detected.

Another consideration in the design of an early warning system is that the toxin production in cyanobacteria is controlled by its genetic basis; López-Rodas *et al.* (2006) found, under laboratory conditions, that the heritability value for the toxin production trait was over 0.7 for some strains of *Microcystis aeruginosa* collected in Doñana and elsewhere. This makes it almost impossible to predict if a cyanobacteria population will bloom, or whether a bloom will be toxic enough to cause poisoning.

As such, it is necessary to quantify toxin content and the toxicity of the samples in which cyanobacteria occur. Field measurements of environmental variables such as, pH, chlorophyll *a*, dissolved oxygen, turbidity, water temperature, nutrient load (NO_3 and NH_4), and colour and odour of the water can also help us to predict the potential for a cyanobacteria bloom.

Toxic blooms are usually very localised events and this can help management when a bloom occurs. If there is any kind of hydraulic control of the wetland it might be possible to isolate or drain the area: restricting mortalities to a limited area and allowing wildlife to use other parts of the wetland.

In the field, the water samples should be gathered using plastic gloves to avoid contact with skin. Samples must then be kept in the dark and refrigerated until they are analysed in the laboratory.

At the laboratory, samples must be homogenized by a gentle shake, and then put in a settling chamber to identify the algae under an inverted microscope. The taxonomic resolution must be at least to genus level. Later on, the concentration of each type of cyanobacteria should be estimated under a microscope by counting the number of colonies with the aid of a counting device, like an Uthermöll chamber, Neubauer chamber or the sort of chambers used for analysis in hospitals (e.g. Uriglass™).

Depending on the count of cyanobacteria and the risk level identified in the early warning system (see Table 9.1 for the risk levels in Doñana´s early warning network), the toxin should be quantified and assessed for its toxicity. The most common toxins in cyanobacteria blooms are microcystins and nodularines and commercial kits are available to quantify microcystin-LR equivalents based on enzyme linked immunosorbent assay (ELISA) or protein phosphatase inhibition assays. The toxicity should be assessed by a mouse bioassay (other species like *Artemia* sp. have been also used in toxin bioassays) following standard protocols and laboratory animal welfare laws. We can use the mouse bioassay to screen both for the neurotoxins and hepatotoxins of cyanobacteria, because the signs of poisoning for these two toxin groups are very different. A HPLC procedure is compulsory for precise toxin identification and quantification (Carmichael 1995).

When a toxic bloom occurs and its toxicity has been assessed, we can apply appropriate management measures, either alone or in combination, for example:

1. If there is hydraulic control of wetland flooding, we can drain the affected wetland and, if dykes, ditches and/or channels exist, isolate the affected area.
2. If the bloom happens in a coastal wetland, or in an estuary, we can flood the affected area with salt water during the high tide period and kill the population by osmotic shock, with the sea removing it on the falling tide.
3. We can stop the terrestrial fauna (mainly birds and mammals) using the wetland, either by fencing or by frightening the animals away while the bloom persists.
4. We can also use flocculant substances (if environmentally safety) to remove most of the bloom.

The use of algaecide chemicals it is not recommended.

Table 9.1 The warning levels and signs used by the early warning network for the prevention of mass mortalities of wildlife due to cyanobacteria blooms in Doñana National Park

0 Level **Weekly monitoring**	No scum No dead vertebrates or invertebrates pH < 8.5 Chl *a* < 150 μg/l No cyanobacteria	No risk of mass mortality	
1 Level **Weekly monitoring**	Little green scum to leeward No dead vertebrates or invertebrates pH ≈ 8.5 Chl *a* 150–500 μg/l <500 colonies/ml	No risk of mass mortality	It should be paid attention because it might be the beginning of a cyanobacteria bloom (toxic or not)
2 Level **Monitoring every 3 or 4 days**	Green scum to leeward Some dead fishes, amphibians or invertebrates pH > 8.5 Chl *a* > 500 μg/l 500–1,500 colonies/ml	Moderate risk of mass mortality	This situation is potentially dangerous because there is a bloom. It is compulsory to alert managers and to intensify the sampling If the toxin concentration (microcystin-LR equivalent) is above 1 μg/l, and the bioassay outcome is positive then go to level 3 of warning
3 Level **Intense monitoring**	Thick green scum to leeward Some dead fishes, amphibians, waterbirds or invertebrates pH > 8.5 Chl *a* > 750 μg/l >1,500 colonies/ml	High risk of mass mortality	There is a toxic bloom and a mass mortality may happen. It is compulsory to apply the management measurements included in the protocol to avoid the mortality.

References

Alonso-Andicoberry C, García-Villada L, López-Rodas V, Costas E (2002) Catastrophic mortality in flamingos in a Spanish National Park caused by cyanobacteria. Vet Rec 151:706–707

Braun A, Pfeiffer T (2002) Cyanobacterial blooms as the cause of a Pleistocene large mammal assemblage. Paleobiology 28(1):139–154

Carmichael WW (1992) Cyanobacteria secondary metabolites. The cyanotoxins. J Appl Bacteriol 72:445–459

Carmichael WW (1995) Cyanobacterial toxins. In: Hallegraef GM, Anderson DM, Cembella AD (eds) Manual on harmful marine microalgae. IOC manuals and guides Nº 33 UNESCO, pp 163–175

Costas E, López-Rodas V (2007) Cianobacterias tóxicas y mortandades masivas en el Parque Nacional de Doñana: Red de alerta temprana y estrategias de gestión. In: Ramírez L, Asensio B (eds) Proyectos de investigación en Parques Nacionales: 2003–2006. Organismo Autónomo de Parques Nacionales. Madrid: Ministerio de Medio Ambiente, pp 67–75 (in Spanish)

Davis JM, Mazumder A (2003) Health and environmental policy issues in Canada: The role of watershed management in sustaining clean drinking water quality at surface sources. J Environ Manage 68:273–286

García-Novo F, Marín C (eds) (2006) Doñana. Water and biosphere. Ministerio de Medio Ambiente, Madrid

Havens KE (2008) Cyanobacteria blooms: Effects on aquatic ecosystems. In: Hudnell HK (ed) Cyanobacterial harmful algal blooms: State of the science and research needs. Springer, New York, pp 733–747

Jochimsen EM, Carmichael WW, An JS, Cardo DM, Cookson ST, Holmes CEM, Antunes MB, de Melo Filho DA, Lyra TM, Barreto VST, Azuedo SMFO, Jarvis WR (1998) Liver failure and death after exposure to microcystins at a hemodialysis centre in Brazil. New Engl J Med 338:873–878

López-Rodas V, Maneiro E, Lanzarot MP, Perdigones N, Costas E (2008) Mass wildlife mortality due to cyanobacteria in the Doñana National Park, Spain. Vet Rec 162:317–318

López-Rodas V, Costas E, Bañares E, García-Villada L, Altamirano M, Rico M, Salgado C, Flores-Moya A (2006) Analysis of polygenetic traits of *Microcystis aeruginosa* (Cyanobacteria) strains by Restricted Maximum Likelihood (REML) procedures: 2 Microcystin net production, photosynthesis and respiration. Phycologia 45(3):243–248

Scheffer M (1998) Ecology of shallow lakes. Kluwer, Dordrecht

Skulberg OM, Carmichael WW, Codd GA, Skulberg R (1993) Toxigenic cyanophytes – Identification and taxonomy. In: Falconer IR (ed) Algal toxins in seafood and drinking water. Academic, London, pp 145–164

Stewart I, Seawright AA, Shaw GR (2008) Cyanobacterial poisoning in livestock, wild mammals and birds – An overview. In: Hudnell HK (ed) Cyanobacterial harmful algal blooms: State of the science and research needs. Springer, New York, pp 613–637

Chapter 10
Monitoring Biological Invasions in Freshwater Habitats

Montserrat Vilà and Emili García-Berthou

Introduction

Biological invasions of freshwater habitats are due to the accidental or deliberate introduction of organisms associated to aquaculture, sports fishing, passive transport by vessels, ornamental uses and man-made canals and corridors. In European freshwater habitats, about 296 species of alien invertebrate, mostly crustaceans, and 136 species of alien fish have been introduced (Gherardi *et al.* 2009). Inland surface waters have also been invaded by 444 species of alien plant (Lambdon *et al.* 2008). Fish have attracted most attention and the majority of these have originated from Asia, North America or within Europe (e.g. from southern to northern countries). Many aquatic alien species are widely distributed in Europe: the brook trout *Salvelinus fontinalis* in 26 countries; the crayfish *Pacifastacus leniusculus* in 25; the zebra mussel *Dreissena polymorpha* in 22. Among plants, the Canadian pondweed *Elodea canadensis* is the most widespread, being present in 36 countries (http://www.europe-aliens.org/).

Alien species invading freshwater systems are causing loss and degradation on all levels of biological organization from genes to populations and with cascading effects to entire ecosystems. One of the best-known European examples of genetic loss is the hybridisation between the native rare white-headed duck *Oxyura leucocephala* and the American ruddy-duck *O. jamaicensis*, the hybrids of which are very aggressive (Muñoz-Fuentes *et al.* 2007). Hybridization between native and alien species is occurring in other taxa such as plants (*Spartina*), crayfish (*Orconectes*), snails (*Melanoides*) and fish (*Salmo*). Invasions can also cause synergic

M. Vilà (✉)
Estación Biológica de Doñana (EBD-CSIC), Avda. Américo Vespucio s/n, Isla de la Cartuja, 41092, Sevilla, Spain
e-mail: montse.vila@ebd.csic.es

E. García-Berthou
Institut d'Ecologia Aquàtica, Universitat de Girona, E-17071, Girona, Catalonia, Spain

C. Hurford et al. (eds.), *Conservation Monitoring in Freshwater Habitats: A Practical Guide and Case Studies*, DOI 10.1007/978-1-4020-9278-7_10,

changes in nutrient cycling as well as habitat modification and alteration of hydrologic regimes. To cite a couple of examples, the red swamp crayfish *Procambarus clarkii* is usually the largest invertebrate in the invaded habitats and often exists in high densities; *P. clarkii* is an engineer species as its omnivory causes changes in carbon and nutrient pools and fluxes (Gherardi 2007). The coypu, *Myocastor coypu*, strongly disturbs the vegetation dynamics in freshwater habitats through grazing, and their burrowing activity can undermine riverbanks (Bertolino and Genovesi 2007). At the regional scale, invasions are causing the homogenisation of the flora and fauna. For example, in the last century, the fish fauna in the Iberian Peninsula river basins has increased in similarity by 17% compared to its native situation (Clavero and García-Berthou 2006).

Many alien species are causing socioeconomic impacts by interfering with water and fishery production, with commercial, industrial or even recreation activities (Vilà *et al.* in press). Aquatic plants forming thick floating carpets exclude the opportunity for angling, and interfere with navigation. Control of *Eichhornia crassipes* in the Guadiana river in Spain has cost 6.7 M€ in 1 year (Andreu *et al.* 2009). Invading aquatic molluscs also interfere with above-mentioned leisure activities, fouling fishing and pipe industry equipment. For example, agencies and individuals have spent about $1 billion over a decade mitigating the zebra mussel in the USA and the losses due to fish introductions (mostly freshwater) amount to $5.4 billion (Pimentel *et al.* 2005).

Strategies to overcome the impacts of invasions can be categorized as prevention, early detection, direct management and restoration. Prevention and early detection are recommended as the best strategy, as they promote a rapid response to invasion. Direct management by removing, controlling or containing alien species is very expensive and is mostly unsuccessful, especially when the introduced species is well established and widely distributed (Hobbs and Humphries 1995). Finally, active or passive restoration after invader removal needs to be assessed for the purposes of adaptive management. In all these strategies monitoring plays an essential role both from the applied point of view and for understanding the invasion process. In this chapter we discuss the reasons for monitoring invasive species beyond their use as ecological indicators, and offer some guidance on the design of appropriate long-term monitoring schemes.

Invasive Species as Ecological Indicators

One of the conceptual challenges of invasion biology is to understand whether invasive species are a driver of biodiversity loss or a consequence of other mechanisms of environmental degradation such as habitat loss or pollution (Didham *et al.* 2005). In practical terms, however, both cases are good reasons to monitor invasive species. An increase in the abundance and occupancy of invasive species can be indicative of environmental degradation (Kennard *et al.* 2005). Many freshwater fish species that are invasive in Europe, such as common carp *Cyprinus carpio*, mosquitofish *Gambusia holbrooki* or the pumpkinseed sunfish *Lepomis gibbosus*, are so called 'tolerant' species, i.e. species sensitive "to any common impact related to altered flow regime, nutrient regime, habitat structure and water chemistry" (Pont *et al.* 2006). Therefore, the presence, increase in abundance or the area of occupation of these species will be indicative of environmental degradation. At the same time, these invasive species can also have direct ecosystem effects and consequences for biodiversity, although these are less well known. For instance, the common carp decreases the abundance of macrophytes directly by uprooting them and indirectly by increasing water turbidity (Lougheed *et al.* 1998), and the mosquitofish has been directly linked to the decline of native fishes and amphibians (Alcaraz *et al.* 2008). Therefore, monitoring invasive species is a good tool both for detecting other drivers of environmental degradation and for understanding their direct impacts on biodiversity and ecosystem processes.

Another good reason to choose invasive species as target species for monitoring is that introduced populations are generally of no conservation value – but rather the opposite. To study or monitor dangerous chemical substances such as pesticides or heavy metals, which is mandatory according to several European laws, freshwater organisms (often fish) need to be sacrificed. To this end, it is of less ethical concern and more practical to use invasive rather than native species. This explains why many invasive species such as mosquitofish (Mulvey *et al.* 1995), common carp (Solé *et al.* 2003) or zebra mussel *Dreissena polymorpha* (Chevreuili *et al.* 1996) are among the most studied in aquatic toxicology and water pollution monitoring.

Other Reasons for Monitoring Invasive Species

Although invasive species monitoring is quite developed in North America (Table 10.1), and some European countries such as France or the UK have comprehensive monitoring schemes for water quality, fish populations and water pollutants, specific programs for invasive species in Europe are less established. For example, recently, in the UK the Department for Environment, Food and Rural Affairs (DEFRA 2008), has acknowledged that despite the comprehensive monitoring schemes in the UK, these are inadequate for invasive species and has identified the development of detection and monitoring mechanisms as a key objective of its invasive species strategy. The European Water Framework Directive (DIRECTIVE 2000/60/EC) also establishes that all countries will routinely have to monitor biological quality elements (algae, invertebrates and fish) with standard methods in most water bodies in order to achieve a good ecological status. As such, this Directive could provide an excellent tool for assessing the status of invasive species and understanding their impacts. Long-term monitoring of biological invasions, that integrates information before, during and after an invasion provides valuable information for both managers and scientists. Some of the basic and applied uses of monitoring programs on biological invasions could be summarized as follows.

Table 10.1 Worldwide examples of regional monitoring programs for freshwater invasive species

Name, Country and Internet link	Target invasive species	Objective
Invasive Plant Inventory, Monitoring and Mapping Protocol, U.S.D.A. Forest Service, USA, 1	Aquatic and terrestrial invasive plant species	Abundance and spread of invasive species; early detection and expansion
Pennsylvania Aquatic Invasive Species Monitoring Squad Pennsylania State, USA, 2	Zebra mussel (*Dreissena polymorpha*) Quagga mussel (*D. rostriformis bugensis*)	Abundance and spread of invasive mussels; early detection and expansion
Aquatic Invasive Species Network, Citizen Lake Monitoring Network Wisconsin state, USA, 3	Eurasian water-milfoil (*Myriophyllum spicatum*) Curly-leaf pondweed (*Potamogeton crispus*) Purple loosestrife (*Lythrum salicaria*) Jellyfish (*Craspedacusta sowerbyi*) Zebra mussel (*Dreissena polymorpha*) Chinese mystery snail (*Bellamya japonica*) Banded mystery snail (*Viviparus georgianus*) Rusty crayfish (*Orconectes rusticus*)	Abundance and spread of invasive species; early detection and expansion
Doñana Biological Station, Spain, 4	Mosquito fern (*Azolla filiculoides*) Red swamp crayfish (*Procambarus clarkii*) Red-eared slider (*Trachemys scripta elegans*)	Abundance and spread; early detection and expansion
Wiltshire River Monitoring Scheme Wiltshire, England, UK, 5	Giant hogweed (*Heracleum mantegazzium*) Himalayan balsam (*Impatiens glandulifera*) Japanese knotweed (*Fallopia japonica*) Signal crayfish (*Pacifastacus leniusculus*) Brown rat (*Rattus norvegicus*) American mink (*Mustela vison*)	Early detection and expansion

Zebra mussel monitoring, Catalan Water Agency and Hydrographic Confederation of the Ebro, Spain, 6 & 7	Zebra mussel (*Dreissena polymorpha*)	Early detection and expansion
Delivering Alien Species Inventories for Europe, Europe, 8	Freshwater species	Distribution and abundance of invasive species

1. http://www.wilderness.net/toolboxes/documents/invasive/FS_Inventory&Map_Guide.pdf
2 http://www.pserie.psu.edu/seagrant/zm/monitor/MonitoringManual2008.pdf
3. http://dnr.wi.gov/lakes/clmn/
4 http://icts.ebd.csic.es
5. http://wsbrc.wiltshirewildlife.org.uk/RecordingSchemes/RiverMonitoring/PageTemplate.aspx
6. http://oph.chebro.es/DOCUMENTACION/Calidad/mejillon/en_inicio.htm
7. http://aca-web.gencat.cat/aca/appmanager/aca/aca?_nfpb=true&_pageLabel=P1230354461208201714706
8. http://www.europe-aliens.org

Early Warning System

Taxonomic monitoring programs conducted by experts can detect the presence of alien species. For this purpose, an ad-hoc list of potential alien species to survey would increase the survey efficiency. Such a list is usually based on alien species present in nearby habitats or on alien species common in similar habitat types. Updating this list and training non-expert manager teams in species identification allows for early detection of alien species while conducting other routinely field surveys not necessarily related to invasions.

Early detection of new foci of invasion is extremely useful for management, as eradication at this stage is usually non-expensive and feasible. Invasion prevention programs should be linked to monitoring the putative vectors of transport and introduction such as vessel traffic, fishing activities and other anthropogenic impacts that can influence alien propagule pressure. For this reason it is extremely useful to know for each sampling locality its connection to nearby water bodies through piping, canals or natural corridors. The benefits of connecting invader and vector monitoring include that it might prevent the arrival of invaders by interfering with transport vectors.

Dynamics of Biological Invasions

Spread rates of biological invasions are poorly known in freshwater habitats compared to marine and terrestrial environments. To date, reconstructions of the spreading of invaders have mostly been inferred from chronosequence analyses of snapshots of the presence of the invader in different locations. Surveillance can provide a time series of invaded and non-invaded sites to demonstrate the spread of the invader in real time, determining the metapopulation dynamics, identifying species constraints, detecting boom and bust cycles and time-lags between appearance and establishment, etc.

Linking Invasions to Impacts and Habitat Resistance to Invasion

Probably one of the longest surveillance programs assessing the effects of an invader on community structure is the zebra mussel project that has been conducted for more than 60 yr in freshwater areas in the former Soviet Union (Karatayev *et al.* 1997). Quantifying the presence and abundance of invader and native species in conjunction with measuring water quality, disturbance level, habitat type, etc. allows for establishing correlations between the invasion degree, food-web alterations and environmental changes. Although such observational analysis does not demonstrate causality, as symptoms of invasion can be confounded by impacts, they are the baseline data from which to establish a hypothesis that can be tested by designing specific manipulative experiments.

Identification of Synergies with Environmental Global Change

Global change components (i.e. increased greenhouse gas emissions, climate change, water eutrophication, changes in land use, biotic homogenization) have typically been studied and managed in isolation. However, there are more and more examples indicating that water warming, increased hypoxia, altered flow regimes, higher salinity and changes in reservoir habitat temperatures can change invasion patterns (Rahel and Olden 2008). Monitoring the presence/absence and abundance of target invaders in combination with spatially-explicit information on CO_2 and CH_4 fluxes, climate parameters, nutrient loading and land use changes can be used to predict changes in invasion dynamics with environmental change (Lee *et al.* 2008) and make possible an appropriate management response.

Assessment of Habitat Restoration After Direct Management

Removal of an invader from an ecosystem represents a grand experiment testing the effect of a particular species on ecosystem processes. However, as removal usually requires some level of disturbance, we need long-term surveillance to avoid confusing recovery after invasion with habitat succession. Monitoring restoration should not only assess whether the population of the invader has been reduced but also if populations of native species have increased. Monitoring restoration is essential for adaptive management to set priorities among invaders, locations, timings and control treatments. Adaptive management may also be needed if the risk of invasion changes through time.

Reinforcement of Public Outreach and Environmental Education

Well-established long-term monitoring programs are based on standardised protocols. These protocols are sometimes inspired by common national or international initiatives, and allow partnership building and cooperation between agencies. This can provide an opportunity for replication and more powerful analysis. It can also offer the opportunity for rapidly training new professionals and for engaging non-experts in monitoring activities, therefore increasing the number of habitats sampled, resampling priority habitats or increasing the frequency of sampling (see several examples in Table 10.1). Monitoring programs can also be integrated into outreach conservation programs through education in summer schools, volunteer activities in natural areas, on-line education programs, etc.

Designing an Appropriate Monitoring Scheme

Obviously, the complexity of long-term monitoring programs will largely depend upon the taxa and the characteristics of the habitats. In many freshwater systems, such as large lakes and marshes, traditional field-based monitoring presents several challenges including inaccessibility to areas for sampling, temporal changes, and expensive equipment. Invasion by floating aquatic plants into huge areas best exemplifies this situation. The fern *Azolla filiculoides* is intermittently invading large areas of marsh in the Doñana National Park (Spain), an area very difficult to map due to its large size (25,000 ha). Recently, remote sensing by Landsat images has been used to identify the extent of *A. filiculoides* invasion and to detect seasonal and annual variability (Díaz-Delgado *et al.* 2008, Chapter 31 in this volume). However, to ascertain that the satellite multispectral imagery distinguishes between *A. filiculoides* and other vegetation such as sedges, ground-truthing by horse riding and aerial photograph analysis has been necessary.

Timing is also crucial both for monitoring and for management. Fish sampling, often by electrofishing, can be constrained by weather conditions or high flows. Fish populations are usually sampled during the same season each year, often in summer, to ensure the comparability of samples. Species-specific ecological features should be used to inform the sampling design: e.g. sampling for zebra mussel larvae is limited to the reproductive season (spring and summer); small organisms with shorter generation times (e.g. algae or invertebrates) should be sampled more often to detect early foci.

The methodology will also depend on whether the goal is to detect established invaders or to provide an early detection system. For example, early detection of the zebra mussel should focus on sampling the water column for the presence of larvae, rather than using fouling panels to count attached mussels. Taking reference samples of the first record is extremely important to provide a good estimation of the minimum residence time since introduction.

However, all of these monitoring and surveillance schemes are of little use if the data are not archived properly and cannot be easily retrieved in a "user-friendly" way. Whenever possible, the data should be integrated in larger biodiversity databases and information systems, preferably on the Internet.

Conclusions

Giving the unprecedented rate of introductions in freshwater ecosystems, monitoring should play a central role in the early detection of invaders and enabling a rapid management response. Furthermore, monitoring alien species in freshwater habitats allows us to explore the extent, dynamics, impacts and drivers of invasions. In order to relate invasion to impact and ecosystem vulnerability to invasion, the monitoring should be integrated with measurements of other physical, chemical and biological variables in the system. This information can also be related to spatially explicit GIS techniques to investigate the interactions with other drivers of global environmental change. As monitoring requires us to ascertain the extent of compliance with a pre-determined standard, it is as important to establish surveillance programs in locations where the invader is present as it is where it is absent, or across locations with a gradient of invader abundance. Scientifically, datasets accumulated by surveillance and monitoring are an important source of information valuable for answering many scientific questions, which can not be tackled by standard short-term research projects and limited scientific man-power.

Acknowledgements We thank J. Andreu, A. Montero and R. Díaz-Delgado for comments on an early version of the manuscript. This review was partially supported by the Integrated European Project ALARM (contract 506675, http://www.alarmproject.net) to MV and the Spanish Ministry of Education (Project CGL2006-11652-C02/BOS) to EG.

References

Alcaraz C, Bisazza A, García-Berthou E (2008) Salinity mediates the competitive interactions between invasive mosquitofish and an endangered fish. Oecologia 155:205–213

Andreu J, Vilà M, Hulme PE (2009) An assessment of stakeholder perceptions and management of noxious alien plants in Spain. Environ Manage 43:1244–1255

Bertolino S, Genovesi P (2007) Semiaquatic alien mammals introduced into Italy: Case studies in biological invasion. In: Gherardi F (ed) Biological invaders in inland waters: Profiles, distribution, and threats. Springer, Dordrecht, pp 175–192

Chevreuili M, Blanchard M, Teil MJ, Carru AM, Testard P, Chesterikoff A (1996) Evaluation of the pollution by organochlorinated compounds (polychlorobiphenyls and pesticides) and metals (Cd, Cr, Cu and Pb) in the water and in the zebra mussel (*Dreissena polymorpha* Pallas) of the river Seine. Water Air Soil Pollut 88:371–381

Clavero M, García-Berthou E (2006) Homogenization dynamics and introduction routes of invasive freshwater fish in the Iberian Peninsula. Ecol Appl 16:2313–2324

Department for Environment, Food and Rural Affairs (DEFRA) (2008) The invasive non-native species framework strategy for Great Britain. Department for Environment, Food and Rural Affairs, London, p 42

Díaz-Delgado R, Bustamante J, Aragonés D (2008) La teledetección como herramienta en la cartografía de especies invasoras: *Azolla filiculoides* en Doñana. In: Vilà M, Valladares F, Traveset A, Santamaría L, Castro P (eds) Invasiones Biológicas. Colección Divulgación CSIC, Madrid, pp 159–163 (in Spanish)

Didham RK, Tylianakis JM, Hutchison MA, Ewers RM, Gemmell NJ (2005) Are invasive species the drivers of ecological change? Trends Ecol Evol 20:470–474

Gherardi F (2007) Understanding the impact of invasive crayfish. In: Gherardi F (ed) Biological invaders in inland waters: Profiles, distribution and threats. Springer, Dordrecht, pp 507–542

Gherardi F, Gollasch S, Minchin D, Olenin S, Panov VE (2009) Alien invertebrates and fish in European inland waters. In: DAISIE (eds) Handbook of alien species in Europe. Springer, Dordrecht, pp 93–104

Hobbs RJ, Humphries SE (1995) An integrated approach to the ecology and management of plant invasions. Conserv Biol 9:761–770

Karatayev AY, Burlakova LE, Padilla DK (1997) The effects of *Dreissena polymorpha* (Pallas) invasion on aquatic communities in eastern Europe. J Shellfish Res 16:187–203

Kennard MJ, Arthington AH, Pusey BJ, Harch BD (2005) Are alien fish a reliable indicator of river health? Freshwater Biol 50:174–193

Lambdon PW, Pyšek P, Basnou C *et al* (2008) Alien flora of Europe: Species diversity, temporal trends, geographical patterns and research needs. Preslia 80:101–148

Lee H II, Reusser D, Olden JD *et al* (2008) Integrated monitoring and information systems for managing aquatic invasive species in a changing climate. Conserv Biol 22:575–584

Lougheed VL, Crosbie B, Chow-Fraser P (1998) Predictions on the effect of common carp (*Cyprinus carpio*) exclusion on water quality, zooplankton, and submergent macrophytes in a Great Lakes wetland. Can J Fish Aquat Sci 55:1189–1197

Mulvey M, Newman MC, Chazal A, Keklak MM, Heagler MG, Hales LJ (1995) Genetic and demographic responses of mosquitofish (*Gambusia holbrooki* Girard 1859) populations stressed by mercury. Environ Toxicol Chem 14:1411–1418

Muñoz-Fuentes V, Vilà C, Green AJ, Negro JJ, Sorensen MD (2007) Hybridization between white-headed ducks and introduced ruddy ducks in Spain. Mol Ecol 16:629–638

Pimentel D, Zuniga R, Morrison D (2005) Update on the environmental and economic costs associated with alien-invasive species in the United States. Ecol Econ 52:273–288

Pont D, Hugueny B, Beier U, Goffaux D, Melcher A, Noble R, Rogers C, Roset N, Schmutz S (2006) Assessing river biotic condition at a continental scale: A European approach using functional metrics and fish assemblages. J Appl Ecol 43:70–80

Rahel F, Olden JD (2008) Assessing the effects of climate change on aquatic invasive species. Conserv Biol 22:521–533

Solé M, Raldua D, Piferrer F, Barceló D, Porte C (2003) Feminization of wild carp, *Cyprinus carpio*, in a polluted environment: plasma steroid hormones, gonadal morphology and xenobiotic metabolizing system. Compar Biochem Physiol Part C Toxicol Pharmacol 136:145–156

Vilà M, Basnou C, Pyšek P, Josefsson M, Genovesi P, Gollasch S, Nentwig W, Olenin S, Roques A, Roy D, Hulme P, DAISIE partners. How well do we understand the impacts of alien species on ecosystem services? A pan-European cross-taxa assessment. Frontiers in Ecol Environ (in press) DOI: 10.1890/080083

Section 3
Rivers: Threats and Monitoring Issues

Chapter 11
Threats to River Habitats and Associated Plants and Animals

Nigel Holmes

Introduction

It might be stating the obvious, but rivers depend upon water, and it is the quality and quantity of water that has profound influences on the plants and animals a river can support. Add to this the physical dimension of habitat structure, and you have the third dimension of the key 'quality triangle' (Madsen 1995). Nature was, and thankfully still can be, the major influence in shaping river quality, but man's impacts in grossly changing rivers around the world are all too clearly evident. This is because man has wanted to change rivers directly to utilize their power and resources for themselves, and unwittingly damaged them indirectly because of exploitation within their catchments.

Three of the many reasons why rivers have been altered physically are (a) to 'control' the amount of land they flood, (b) to make them navigable for shipping, and (c) to generate power or to secure water supplies (via impounding for water storage). These directly destroy habitat, but indirect physical changes result also when land-use changes in the catchment and sediment is released to the channel. Quality of water changes when polluting discharges are piped directly into rivers or when mining or other catchment activities release chemicals from the land that have either been added deliberately to enhance production, or are a by-product of changes occurring to the soils. Quantity of water flowing through a river is highly important in shaping the physical structure of the channel, and is influenced by both natural forces and man. Increasingly climate change is being linked to both greater frequency and severity of droughts and floods, although careful research into such events over history suggests the link is less strong than the media portrays (Doe 2006). Whether true or not, the impacts of droughts and floods are far greater now on rivers because of the alterations made to their physical structure.

N. Holmes
The Almonds, Ramsey Road, Warboys, Cambs. PE28 2RW, Huntingdon, UK
e-mail: n.holmes3@btinternet.com

C. Hurford et al. (eds.), *Conservation Monitoring in Freshwater Habitats: A Practical Guide and Case Studies*, DOI 10.1007/978-1-4020-9278-7_11,

The character and size of rivers around the world vary vastly, as do the threats to them. All rivers may be altered due to direct influences of natural climate change, but this should not been seen as a 'threat', but a natural phase in their evolution. Rivers in the remotest, minimally populated, parts of the world fall into this category, but even their natural functions are under threat from global human impacts that then affect climate. Threats come from the ever increasing human population, and the amount of natural ecosystem they destroy and the stored carbon sources they use. In October 2008 the population was estimated at 6.7 billion, yet 200 years before it was less than a billion, and is forecast to reach 9.75 billion by 2050. However, direct threats from population increases to rivers will vary around the world. Europe's proportion of the world's population is estimated to drop four fold from c. 20% in 1750 to c. 5% in 2050, yet in Latin America the opposite is happening, with the population expected to more than quadruple as a percentage of world population. As rivers from one part of the world are so different from another, and the threats vary from country to country, a comprehensive global review of threats to rivers in such a short chapter is impossible.

Here, a few common threats to rivers world-wide will be illustrated primarily using examples in the UK and European Union. This is a necessary approach because whilst many of the threats are similar in nature, their potential magnitude and the way they are addressed vary greatly around the world. Everywhere the fate of rivers and the wildlife they support will be shaped by a combination of natural and anthropogenic factors, with the way in which water management is undertaken being increasingly important. This brings in the political and legislative dimension, where there has to be both the will, and laws, to enable water management to integrate the needs of naturally functioning rivers and their dependent ecosystems whilst delivering the requirements of human societies (Wescoat and White 2003). Unless the former is given greater focus in the future than it has in the past, the latter will fail.

Whilst thinking about how to gather different perspectives on 'threats', I thought it might be a good idea to pose the question to several groups of people heavily involved with practical measures to protect and enhance rivers in one small area of the UK: the Rivers Trusts/Foundations in Wales. In the blink of an eye one reply simply said "*In one chapter? – a major task – how long have you got?*". This shows that despite very positive improvements in many external factors that influence the health of UK rivers in recent decades, there are still great concerns out there and plenty to do to 'stop the rot'.

Hydromorphology

The term 'Hydromorphology' has been created for the Water Framework Directive (WFD) in the European Union to encompass the hydrological and geomorphological inter-related processes that shape waterbodies. In essence, therefore, it encompasses both the water quantity and the physical structure sides of the Madsen (1995) 'river health triangle'.

Physical degradation, such as straightening, deepening, widening, embanking and other constraints have grossly impacted the physical structure of rivers, and thus habitat diversity (Brookes 1988; Brookes and Shields 1996). Before looking at the huge changes that have occurred to many river channels themselves, it is important to note that even purely aquatic taxa, such as fish and some invertebrates, cannot thrive unless both the river channel and its immediate riparian zone is healthy. For centuries rivers have been modified to bring land-use change, allowing marshlands and other wetlands to be drained for either agriculture or urbanization (Purseglove 1989). In Europe this continued unabated to the mid-1980s, with multi-thread channels converged to one, oxbows and pools filled in, river sections deepened and widened, and bankside trees removed. As a result, many floodplain forests, wet grasslands, reedbeds and pools in floodplains were destroyed, yet now we are realizing that these lost wetlands are lost resources. In late 2008 this was clearly illustrated simply by the title of the tenth RAMSAR convention meeting in Asia – 'healthy wetlands, healthy people'.

Even totally undisturbed rivers that flow through greatly modified landscapes cannot support the natural range of birds, mammals, and invertebrates that utilize both the river channel and the river corridor at different stages in their life cycles, and/or at different times of the year. Drainage of floodplains has resulted in massive losses, of both plants and animals, but wading birds in the UK are perhaps the best group to illustrate the losses and that problems still exist. Typical species most affected are those of river floodplain marshes and wet meadows such as snipe (*Gallinago gallinago*), lapwing (*Vanellus vanellus*), redshank (*Tringa totanus*), and curlew (*Numenius arquata*). Lovegrove *et al.* (2009) report these were common along UK rivers up to the years immediately following the Second World War. Intensification of agriculture, especially that from the mid-1980s onwards, has resulted in massive losses, with lapwing declining 77% between 1987 and 1998 (Fig. 11.1).

Only a single section of one river floodplain in Wales now supports breeding snipe and redshank. Whilst in the UK the drive for more drainage has lapsed, adequate financial incentives to restore river and floodplain connectivity are limited. Worst still at a European Union scale, whilst finances are spent on reducing agricultural production with little or no environmental gain, funds are still available to destroy un-recreatable river floodplain wetlands through river engineering to improve agricultural production elsewhere.

When the diverse natural physical structure of rivers is degraded, the knock-on effect on plants and animals can be catastrophic. Most rivers around the world, big or small, in highly urbanized locations are 'straight-jacketed'. Where once there was a wide range of marginal and in-channel habitats, now there is simply a uniform flat bed and vertical armoured banks. This is particularly obvious where modifications have been made to allow rivers to become major highways for boat traffic (much of the Rhine for instance) or constrained to enable ancient and new cities to grow (e.g. London on the Thames or St Louis at the confluence of the great Missouri and Mississippi Rivers). We will never know what we lost, but research on relationships between invertebrates and habitats give us a clue. Harper *et al.* (1992, 1998) showed that in 100m lengths of rivers in the UK, the more diverse the

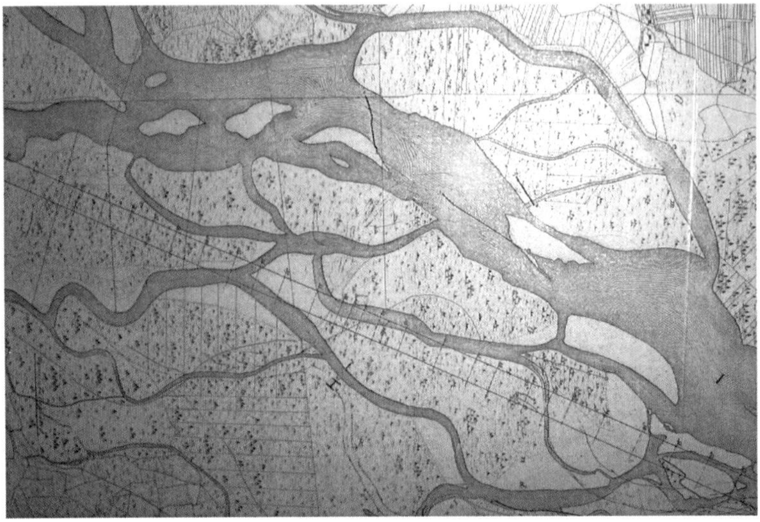

Fig. 11.1 Nineteenth century map of a plan view of a multi-threaded natural section of the Inn River, Bavaria, with a narrow single thread showing what it was converted to

physical structure (they called the building blocks of stream structure 'functional habitats'), the greater the aquatic invertebrate diversity. They elegantly showed that for sites with only one habitat, on average only a single caddis species was present; with five functional habitats present, the average number of species present was five; and so on until if 14 habitats were present, on average 14 caddis species were found.

Edge and riparian habitat diversity is as important for river ecology as submerged in-stream habitats. Periodically exposed river shingle and sand bar habitats on highly dynamic rivers are very important for many specialist invertebrates. In the past, most attention has been paid to aquatic invertebrates of rivers as water quality monitors, but in the UK attention in the past 15 years has also focused on those found in Exposed Riverine Sediments (ERS) – the shoals, bars and spits which are exposed for some time during periods of normal or lower than average flow. These habitats support nationally and internationally scarce invertebrates (Sadler and Bates 2008). True flies (Diptera), spiders (Araneae), ants (Formicidae), and bugs (Hemiptera) are typical groups, with beetles dominating in terms of numbers of specialist and rare species (Bates *et al.* 2005). Sadler and Bell (2002) have shown that past and present management has major implications, with the best ERS sites found on unregulated rivers, or on non- or minimally managed river systems.

Catchment impoundments not only affect river channels and their surroundings directly, but also have a major influence on discharge characteristics downstream, and hence habitat quality (i.e. hydromorphology). In much of the UK impounding reservoirs are in headwaters where valley sides are steep. Few have been built in recent decades, and many rivers remain relatively unaffected by them. In Spain, a very high proportion of river valleys suitable for containing water have been impounded, completely destroying the river and floodplain habitats they drown. On large rivers such as the Danube, long stretches of river have been dammed to generate huge amounts of electricity. Not only does this impact on the ecology, but also potential tensions exist between neighbouring countries. To promote cooperation in sustainable water management, conservation, pollution reduction, etc., the countries of the Danube River basin signed the Danube River Protection Convention in 1994. In Europe as a whole, the WFD requires member states of the European Union to not allow deterioration in river quality unless there are over-riding national interests. In theory this should provide for greater protection of natural valleys that remain, but how the 'over-riding national interest' is interpreted when secure water supplies, and greener energy, are required by an ever growing and ever more demanding public remains to be seen.

Worldwide, pressure for catchment impoundments for water supply and hydro-generation are immense. For example, many rivers flowing through tropical rainforest are still highly threatened. Increasing demands for 'green' energy are putting the world's rivers at risk with hydro-electric projects responsible for flooding vast areas of rainforest. The decay of felled forest wood adds more greenhouse gases to the atmosphere, contributing to global warming as well as nutrient inputs into the rivers. The flooding means floodplain and terrestrial species are lost, or at best migrate to new areas and fish and other river species must endure lake conditions or die. Dams also disrupt migration to headwaters that are important nursery grounds for many migratory fish species. It is irrefutable that large hydroelectric projects, funded by international aid and development organizations like the World Bank, have destroyed or damaged many rivers around the world. When this is associated with forest loss too, there are serious consequences not just for local people, but catchments downstream are impacted by siltation and eutrophication. It is good that 'mega-projects' in Latin America are facing more and more opposition today, as well as being re-examined by the sponsoring parties. Sadly this is not so in Asia, where on the Mekong alone, one of the biologically richest rivers in tropical Asia, 17 dams are presently planned.

Water Quality

During the Industrial revolution many UK rivers were virtually lifeless. Raw domestic and industrial waste poured into many of them as if they were simply viewed as open sewers. In Scotland, the River Tweed, where many mills were situated two centuries ago, was renowned for running the colours of the rainbow depending on which day of the week it was. The Mersey was another river that suffered severe pollution from a cocktail of pollutants from the textile industry, sewage and farm waste. It reached its 'dead' state in Victorian times, and only in the past 30 years have major improvements been made through the combination of legislation, major investment in effluent treatment, and the dedicated work of the Mersey Basin Trust. The Thames through London was little more than a giant open sewer during the industrial revolution: the last salmon was reportedly seen in 1833 before pollution made passage through the city impossible. No salmon, and few other fish species, were seen in the Thames for more than 100 years until the big clean-up took place. A salmon returned in 1974 and today 120 fish species are present.

In Wales, many small rivers were also made virtually lifeless through a variety of pollutants through most of the nineteenth and much of twentieth centuries. Sewage affected most rivers wherever there were settlements developed upon them; industrial discharges from metal mines affected rivers such as the Ystwyth and Rheidol; slate mines affected many rivers in the north whilst the coal and steel industries dramatically impacted many of the south Wales valley rivers and streams (e.g. Rivers Ebbw, Taff, Rhymney). In rural areas agricultural point sources also impact rivers, and many have been, and still are, adversely affected by 'diffuse' pollution. Some specific examples of the latter include acidification from acid rain and afforestation, agro-chemicals in the form of fertilizers or sheep-dips, and sediment released from the land to the watercourses through intensification of agricultural practices.

Where rivers are as obviously polluted as described above, severe impacts on plants and animals are clear-cut. However some species decline or are lost when the reasons for it are unclear, or take decades to unravel. The extreme decline in otter (*Lutra lutra*) throughout Europe in the mid-twentieth century is a case in point. The decline was shown to be due to PCBs and organochlorine pesticides used in sheep dips, paints, etc. (Mason and Macdonald 1993). Since the 1970s otters have made a remarkable recovery following the withdrawal of these pesticides.

The success story of the otter and the reason for it has been mirrored to some degree by the declines in some invertebrates due to the replacement of the organochlorines by synthetic pyrethroids such as cypermethrin, in sheep-dip. Links to impacts on the already severely threatened native crayfish (*Austropotamobius pallipes*) have also been made. Frequent and significant pollution incidents from such sources have occurred in the UK over the past 20 years, with Ormerod and Jüttner (2009) stating that numbers recently have increased. The use of cypermethrin in sheep dips in England and Wales has been banned since 2006, but vigilance will be required to ensure the alternatives do not have the unsuspected impacts that cypermethrin had when the killer ingredients of the otter where banned! Ormerod and Jüttner (in press) also report that in addition to the well known polluting substances that have affected river life for centuries, a wide range of pharmaceutical drugs and chemicals used in health care and beauty products now find their way to rivers via sewage treatment works. The potential effects of endocrine disrupting substances, such as those used in some female contraceptive pills, have been studied mostly in relation to fish and illustrate that new threats to river life exist even when major clean-ups occur on polluted rivers. All these threats are likely to increase in the future.

Alien Species

As man has moved around the world, alien species have been deliberately or accidentally carried too. Any web search with the words 'aliens' and 'impacts' with any named country in the world provides a mass of information on the spread of aliens across the world. Add the word 'river' and the story is the same. A headline in the UK's Daily Telegraph newspaper in February 2008 read: '*Alien species wreck worlds oceans and rivers*'. The fact that rivers have flowing water means they are very effective at spreading alien species should they once become established. Many alien plants threaten natural habitat characteristics and plant and animal assemblages of UK rivers. Rhododendron (*Rhododendron ponticum*) is invasive but only common in upland catchments. The same cannot be said of Himalayan balsam (*Impatiens glandulifera*) and Japanese knotweed (*Fallopia japonica* agg.); both are now very common throughout the UK and very invasive. The former is an annual and thrives where bare soil is exposed in spring whilst the latter is perennial and so tough it will grow through tarmac. Giant hogweed (*Heracleum mantegazzianum*), the third biggest menace of river banks in the UK, is mainly restricted to sandstone catchments. All threaten to change the natural assemblages of river banks, and so efforts are being made in many places to reduce this threat by eradicating the species from some catchments; for some it is too late, so keeping them out of catchments where they do not now occur is a priority. The problems are world wide, but involve many different species, so prevention of spread to pristine catchments is particularly important at a global scale.

So far aquatic alien species do not appear to have had major impacts on natural macrophyte assemblages of UK rivers, despite rooted species such as *Elodea canadensis* (Canadian waterweed) and *Elodea nuttallii* (Nuttall's water-thyme) being widespread. A free-floating species, floating pennywort (*Hydrocotyle ranunculoides*) is a more recent invader, and is spreading along many slow-flowing rivers. In terms of effects on river invertebrates, the clearest impacts from an alien species on a native animal has been the near eradication of the native white-clawed crayfish due to a plague carried into the UK by alien American signal crayfish (*Pacifastacus leniusculus*). In terms of impacts from other invertebrates on UK Rivers, zebra mussel (*Dreissena polymorpha*) has massively impacted some navigation channels such as the Erne and Shannon in Ireland, and an abundant population has developed in Cardiff Bay since it was built. Despite appearing to not cause extreme impacts on some mainland European rivers such as the Danube, further spreads cannot be ruled out in the UK, nor can the potential for habitats to be damaged by them. Chinese mitten crabs (*Eriocheir sinensis*) have also set up home and are rapidly colonizing (and damaging) slow-flowing rivers and coastal habitats. With a reported ability to travel up to 1,000 km in their lifetime, stopping their march seems impossible.

Water vole (*Arvicola terrestris*) is a mammal that has been massively impacted by an alien species. They were previously very common throughout rivers in much of Britain, even whilst the otter declined to its low in the 1960s. In most of the country the population has crashed in recent decades to make it Britain's most rapidly declining mammal. The threats to water voles are complex, but they are particularly vulnerable to predation by American mink (*Mustela vison*), an alien from the US (Macdonald and Strachan 1999). It is likely that water voles will not recover until mink are controlled.

Discussion

In the UK, but unfortunately not world-wide, chronic pollution is more or less a thing of the past and general improvements to river water quality have resulted from investment in effluent treatment. Around the world too, trusts have been formed by local interest groups to look after 'their' rivers. It is good news too that rivers and watercourses are managed much more sympathetically, and the era of drainage has long gone. It is thus not all doom and gloom. The tide has been turning, and there is much to be positive about, yet even where great advances have been made in restoration and protection throughout the world, threats to river biota remain from traditional sources and new ones alike, so vigilance is imperative.

Building barrages across estuaries is relatively new in the UK, with those in Middlesborough on the Tees, in Swansea on the River Tawe, and in Cardiff on the Rivers Taff and Ely being recent examples. Such barrages are common around the world and are built primarily to promote development. Not only do such modifications affect inter-tidal species, but freshwater migratory fish such as salmonids, lamprey and eels. Problems can be avoided, but impaired fish passage can occur, so future barrages could threaten freshwater fisheries in many catchments. At the opposite end of catchments, impoundments are built for flow regulation, water supply and hydro-power generation. Threats to rivers from these new types of development are likely to increase as politically there is a drive to produce more and more carbon-neutral energy. It may come at a price – the price being impacts to river structure and wildlife.

Due to climate change, fluvial flooding could become more frequent in the future, and be of greater magnitudes than seen before. There will be pressure from some quarters of the public for spontaneous reactions to dredge and canalise again; these real threats will need to be rebuffed for more holistic measures related to catchment land-use that lead to slower runoff. The production of Catchment Flood Management Plans and River Basin Plans under the WFD in Europe aim to reduce flood threats through more integrated management within catchments, and therefore should reduce threats to biota caused by unsympathetic river management. A key mantra is 'make space for water'.

Some people argue that in the future freshwater may be more valuable than oil. This is another example where climate change may increase threats to some plants and animals in rivers if summers are drier. With less secure water supplies, and an inexorable increase in demand, more rivers may be under threat of impoundment for reservoirs, leading to reduced naturalness of the hydrological regime downstream. Climate change is likely to produce 'winners' and 'losers', but without knowing what the change in weather will be in the long term, predicting the lucky species is difficult. In the UK, birds such as kingfisher (*Alcedo atthis*) and grey wagtail (*Motacilla cinerea*) could do well if winters are warmer, but will struggle to feed young if summer floods increase. Salmonids, on the other-hand, may be under even further pressure due to enhanced stream temperatures.

Alien species are known to have impacted mammals (e.g. water voles) and natural riparian vegetation in the UK. In Ireland, zebra mussels have spread to many closed water bodies with no boat links to habitats with the alien mussel present (Peter Hale; Northern Ireland Environment Agency pers. comm.). This shows that a species can be spread by several means. Natural dispersal by birds or mammals is a distinct possibility, but so is spread by anglers, or even ecologists surveying an uninfected water body soon after surveying an infected one. This shows that everyone must be vigilant to avoid avoidable threats by being more thoughtful and taking precautionary measures where possible.

Throughout Europe there are many reasons to be optimistic. The WFD requires that there be no deterioration in ecological status, with targets set for measures to ensure all rivers reach either 'good ecological status', or 'good ecological potential' (where previous impacts cannot be 'economically' reversed). Unfortunately, there is a threat from interpretation of the 'over-riding public interest' clause. In some parts of the world, rivers have either not been degraded or there have been successes in rehabilitating damaged ones. These are the success stories. Unfortunately the dumping of trash and human waste into rivers still continues in many overcrowded cities, resulting in loss of plant and animal life and rivers being no longer safe for human use. Catastrophic pollution is unfortunately not a thing of the past, even in Europe. In 2000 there was a serious cyanide and heavy metal contamination of the Tiza River in Hungary from a trans-boundary mining source.

World-wide, deforestation and chemical pollution from mining remain major threats to river habitats and species. Erosion is a well-known result of deforestation with serious consequences for river life. At the very least, increased sediment loads and reduced water flows usually seriously affect local fish populations. Species that rely primarily on sight decline the most, whilst the increased amount of suspended particles interferes with fish gills. If previous afforested areas are used for agriculture, fertilizers and pesticides are likely to be an additional threat. Unfortunately, impacts from mining operations on rivers continue unabated in many parts of the world. One such example that is all too familiar around the world is the copper pollution of streams in Indonesia.

Freeport-McMoRan, based in New Orleans, has operated the Mount Ertsberg gold, silver, and copper mine in Indonesia for over 20 years and has converted the mountain into a 600-m hole (Perlez and Bonner 2005). Perlez and Bonner report that the mining company has dumped appalling amounts of waste into local streams, rendering downstream waterways and wetlands "unsuitable for aquatic life." Government surveys have found that discharges from the mines have produced levels of copper and sediment so high that almost all fish have disappeared from nearly 90 sq. mi. of wetlands downstream from the operation.

Clearly threats to rivers, their catchments and their wildlife remain.

Yin and Yang

In ancient Chinese philosophy, the concept of yin and yang describes seemingly opposing forces, often bound together with interdependence in the natural world. In this short paper it is clear that there have been 'winners' and 'losers' as a result of anthropogenic interference with the natural functioning of rivers. Perhaps more interestingly it is clear that some species have recovered as a result of a change in policy or product use, but the targeted gains achieved may be off-set by equally serious impacts on other interests. The classic example is the changes to the active ingredients in sheep dip that played an important role in the recovery of the otter, but led to a decline in sensitive invertebrates in many rivers, including native crayfish.

Management of watercourses in the UK to assist in salmonid recovery after serious declines throughout Europe, including fencing (Wye and Usk Foundation 2008) may well have opposing drawbacks to some highly interesting and sensitive riparian ERS invertebrate species (Sadler and Bates 2008). Similarly coppicing of bankside trees to reduce shade to bring short-term benefit for trout could be detrimental to 'oceanic bryophytes' that demand shade and high humidity.

Future threats to the rich ecological, aesthetic, recreational and economic assets of rivers exist throughout the world and in some places are both serious and imminent – avoidance of the impacts these threats might have is better than having to reverse them. For others the clear message is that management, if needed, and regulations, should be based on sound knowledge underpinned by evidence, where the 'yin and yang' are complementary.

References

Bates AJ, Sadler JP, Fowles AP, Butcher CR (2005) Spatial dynamics of beetles living on exposed riverine sediments in the upper River Severn: Method development and preliminary results. Aquat Conserv Marine Freshwater Ecosyst 15:159–174

Brookes A (1988) Channelized rivers: Perspectives for environmental management. Wiley, Chichester pp. 326

Brookes A, Shields DF (eds) (1996) River channel restoration. Wiley, Chichester, pp. 433

Doe R (2006) Extreme floods: A history in a changing climate. Sutton Publishing, Stroud, Gloucestershire, pp. 228

Harper DM, Smith CD, Crosa GA (1992) Habitats as the building blocks for river conservation assessment. In: Boon PJ, Calow P, Petts GE (eds) River conservation and management. Wiley, Chichester, pp. 311–319

Harper DM, Smith CD, Kemp JL (1998) The use of 'functional habitats' in the conservation, management and rehabilitation of rivers. In: Bretschko G, Helesic J (eds) Advances in river bottom ecology. Backhuys Publishers, Leiden, pp. 315–326

Lovegrove R, Strachan R, Crawford A, Tyler S, Halliwell L, Slater F (2009) Life in Welsh rivers: Other vertebrates. In: Williams DD, Duigan CA (eds) The rivers of Wales. A natural resource of international and historic significance. Backhuys, Leiden

Macdonald SM, Strachan R (1999) The mink and the water vole: Analyses for conservation. Wildlife Conservation Research Unit, University of Oxford, Oxford

Madsen BL (1995) Danish watercourses – Ten years with the New Watercourse Act. Danish EPA, Copenhagen

Mason CF, Macdonald SM (1993) PCBs and organochlorine pesticide residues in otter (*Lutra lutra*) spraints from Welsh catchments and their significance to otter conservation strategies. Aquat Conserv Marine Freshwater Ecosyst 3:43–51

Ormerod S, Jüttner I (2009) Water quality effects on Welsh rivers: A polluted past, an uncertain future? In: Williams DD, Duigan CA (eds) The rivers of Wales. A natural resource of international and historic significance. Backhuys, Leiden

Perlez J, Bonner R (2005) Below a mountain of wealth, a river of waste. New York Times December 27th 2005

Purseglove J (1989) Taming the flood. Oxford University Press, Oxford pp. 307

Sadler JP, Bates AJ (2008) The ecohydrology of invertebrates associated with exposed riverine sediments. In: Wood PJ, Hannah DM, Sadler JP (eds) Hydroecology and ecohydrology: Past, present and future. Wiley, Chichester, pp. 37–56

Sadler J, Bell D (2002) Invertebrates of exposed riverine sediments phase 3 – Baseline faunas. Technical Report W1-034/TR. Environment Agency, Bristol

Wescoat JL, White GE (2003) Water for life: Water management and environmental policy. Cambridge University Press, Cambridge, p. 321

Wye and Usk Foundation (2008) Problems on our rivers ... and some solutions. Available at: http://www.wyeuskfoundation.org/problems/index.php (accessed on 29 May 2009)

Chapter 12
The Development and Application of Mean Trophic Rank (MTR)

Nigel Holmes

Introduction

In October 2000, the European Union adopted the Water Framework Directive (WFD: EU 2000). The purpose of the directive, amongst others, is to establish a framework for the protection of inland surface waters (rivers and lakes). Unless heavily modified beyond economic rehabilitation, water bodies are to be restored to 'good status', and the best protected from declining. The Directive requires Member States to establish river basin districts and for each of these river basin management plans are produced. The Directive envisages a cyclical process where river basins management plans are prepared, implemented and reviewed every 6 years. Within a planning cycle there is a need for: (i) characterisation and assessment of impacts; (ii) environmental monitoring to quantify the impacts; (iii) the setting of environmental objectives; (iv) designing and implementing the programme of measures needed to achieve the environmental objects. At its core the WFD is driven by ecological quality and status, with fish, invertebrates, phytoplankton, phytobenthos and macrophytes used in the assessment process. Having confidence in survey and assessment of these biological elements is thus at the heart of implementing the WFD; the MTR is a survey method that fulfils the requirements for the macrophyte element of the Directive.

Biological monitoring of water quality has a long history, but it is only relatively recently that the use of macrophytes has been added to the suite of existing systems based on animals. In Europe, as in many other parts of the world, surveys of invertebrates in rivers dominated the early work on biological water quality assessments. One of the earliest was developed for the River Trent in the UK, with pioneering biological river water quality indices being developed, the Trent Biotic Index (TBI; Woodiwiss 1964). Many different systems then were developed in different countries,

---

N. Holmes
The Almonds, Ramsey Road, Warboys, Cambs. PE28 2RW, Huntingdon, UK
e-mail: n.holmes3@btinternet.com

C. Hurford et al. (eds.), *Conservation Monitoring in Freshwater Habitats: A Practical Guide and Case Studies*, DOI 10.1007/978-1-4020-9278-7_12,

so in the late 1970s an attempt was made to develop a system that could be used throughout Europe. Several countries got together to assess commonalities within their systems, and the result was 'The Biological Monitoring Working Party' (BMWP) procedure for measuring water quality using macro-invertebrates to create quality indices. It was adopted in the UK in the early 1980s (Armitage *et al.* 1983) and the basic form has been used in many other countries since.

The BMWP method is based on the principle that different aquatic invertebrates have varying tolerances to nutrients or pollutants. Many mayfly (Ephemeroptera) and stonefly (Plecoptera) species demand high quality water, and so in the BMWP system they are accorded a score of 10. In contrast, worms (Oligochaeta) can withstand gross pollution and they are given a score of one. Site BMWP scores are determined by adding the tolerance scores of all macroinvertebrate families in the sample. Thus, a high BMWP score reflects good water quality. However, naturally species-rich sites have higher scores than naturally poor sites even if the water quality is the same. To over-come this problem, Average Score Per Taxon (ASPT) provides a uniform scale from 1 to 10 for any sample; it is the average of the tolerance scores of all macroinvertebrate families found in a sample.

Papers describing the value of macro-invertebrates as water quality indicators often quote their relatively large size, long life cycles and low mobility as key attributes; this ensures that the presence of a given taxon at the time of sampling reflects past conditions (Hellawell 1986). Additional reasons quoted include the low cost and ease of sampling and relatively easy identification at family level. Brabec and Szoszkiewicz (2006) give equally good reasons why aquatic macrophytes and algae can reflect changing nutrient regimes and flow conditions. Phytobenthos are often good early warning indicators of change as they grow rapidly, but appear transiently. Most macrophytes on the other hand, due to their longer life cycles, indicate more persistent impairment. General advantages of using macrophytes for quality assessment are that they are stationary (cannot flee from a pollution incident), visible to the naked eye, quantity can be assessed, and relatively low manpower is needed to identify core species.

In 1987, Haslam reviewed some of the literature on the use of macrophytes as indicators of enrichment in Europe. Some conclusions she came to included the following.

- "Species are influenced by nutrient regime, and can be ranked according to their distribution in relation to trophic status."
- "Ranking order differs slightly between countries."
- "Tolerance range [of species] may be wide, or narrow."
- "Correlations with nutrient status may be due to the species' nutrient requirements, but may also be due to its requirement for another factor which occurs in only part of the nutrient range. Therefore the correlation may or may not be causal."
- "There is general agreement that species can be ranked and grouped in relation to habitat nutrient".

Early correlation of macrophytes with water chemistry, for both running and standing waters, paid more attention to relationships between species and base status than nutrient status. For example whilst Seddon (1972) recognised relationships between species and nutrient levels, his paper on macrophytes as limnological indicators concentrated heavily on parameters such as hardness, conductivity, etc. Both Seddon and Hutchinson (1970) noted that the distribution of *Myriophyllum alterniflorum* & *M. spicatum* could be determined in relation to pH, calcium and bicarbonate. Whilst this is certainly true, the latter is far more tolerant of nutrient enrichment than the former (Haslam 1987). Weber-Oldecop (1976) and Wiegleb (1978) in Germany have associated nutrient-poor species and communities with calcium-poor chemistry (and vice-versa for nutrient-rich ones), as did Meriaux (1982) in France. This highlights that the presence and abundance of many species, not just macrophytes, is usually a reflection of many (often inter-related) factors, not one.

The work of Kohler (1975) developed clear associations between individual species and communities with nutrients as opposed to cation associations – e.g. Ammonium-N and Phosphate. Kelly (1990), Ellenberg (1973) and many more have all expounded on the value of macrophytes as indicators of their chemical environments but their use in water quality assessments found little favour until relatively recently. In Ireland a system of combining macrophyte data with standard invertebrate methods was also developed in the 1980s (Caffrey 1987). The system is a five point biotic Index – Macrophyte Index Scheme (MIS) based upon assessment of data from 140 sites on 27 Irish rivers – which enables grouping of pollution-sensitive and pollution-favoured species.

What is MTR and How Does it Work?

The MTR system was developed in the early 1990s to enable macrophytes to be used alongside invertebrates and other methods to assess river water quality in the UK. The prime driver for its development was to assist the UK in implementing the European Union's Urban Wastewater Directive (UWWD: EU 1991).

The term 'macrophyte' in MTR embraces higher plants that grow submerged or partly submerged, vascular cryptograms and bryophytes, together with macroscopically distinct algae that are typically composed predominantly of a single species. Through several years of testing from 1994 to 1996 the system was refined, and a checklist of taxa for inclusion in the system produced. As this was based on UK rivers, the checklist included only species that are widely distributed in rivers of all types within the UK. It, however, excluded species that are very commonly found along the waterlogged margins or on the banks, and others that were equally considered unlikely to respond to even major changes in nutrient status. A total of 129 taxa were included on the list, with the possibility of recording indeterminate specimens such as *Ranunculus*, *Veronica* and *Callitriche* that cannot always be identified to species. The list included: seven algae; seven liverworts; 23 mosses; three vascular cryptogams and 89 flowering plants. The use of non-flowering plants within the system was considered essential if all river types could be assessed using the system; this was confirmed by Baattrup-Pedersen *et al.* (2006) who reported on

European-wide river surveys that small or even large mountains streams are usually dominated totally by mosses and liverworts, yet the communities of lowland streams tend to be dominated by vascular plants.

Each species on the checklist was assigned a Species Trophic Rank (STR) of 1–10, depending on its perceived tolerance to eutrophication (1 = tolerant; 10 = intolerant). This is similar to the invertebrate BMWP system where invertebrates requiring the highest quality of water are assigned a rank of 10, and those with the highest tolerance to pollution are given a score of 1.

The sampling unit used in MTR is 100 m in length. For small rivers this would include the whole channel width, but for wide or very deep rivers the survey unit may be reduced to a 5 m strip down one side or other for 100 m. Macrophyte abundance is expressed in terms of the percentage of the survey length covered by each taxon present. The scoring system is on the following scale: 1: <0.1%, 2: 0.1–1%, 3: 1–2.5%, 4: 2.5–5%, 5: 5–10%, 6: 10–25%, 7: 25–50%, 8: 50–75%, 9: >75% cover. The percentage cover for each taxon in each site is referred to as the Species Cover Value (SCV). When macrophyte species overlap each other, or there are floating species over submerged species, total cover values can exceed 100%. In shallow rivers, survey may be done first from the bank and then by wading the channel. If channels are too deep for wading, boats may be used as well as grapnels to collect material from the riverbed; in such cases the percentage cover values are less reliable and this should be noted.

To determine the nutrient status of a site, the SCVs of species present in a site are used in combination with the STR to enable a site MTR score to be derived. To illustrate how the system works, two hypothetical sites are compared in Tables 12.1 and 12.2. Both have exactly the same species present within the sites, yet their abundance values are very different (expressed as SCVs). In the hypothetical upstream site, SCVs are low for species with a low STR but high for species with a higher STR. The converse is the case for the hypothetical site downstream of the nutrient-rich discharge. MTR site scores are determined by first multiplying the SCV scores for each species present with their STRs to give a Cover Value Score (CVS). Next the individual SCV and CVS scores are added together and the latter

Table 12.1 Hypothetical site 1

Taxon	STR	SCV	Scoring	CVS
Enteromorpha	1	1	1 × 1	1
Cladophora	1	1	1 × 1	1
Nuphar lutea	3	1	1 × 3	3
Lemna minor	4	7	7 × 4	28
Potamogeton pectinatus	1	1	1 × 1	1
Zannichellia palustris	2	1	1 × 2	2
Leptodictyum riparium	1	1	1 × 1	1
Ranunculus fluitans	7	7	7 × 7	49
Column totals for MTR scoring		20		86
MTR Score	86 (CVS total) 4/20 (SCV total) = 4.3 × 10 = 43			

Table 12.2 Hypothetical site 2

Taxon	STR	SCV	Scoring	CVS
Enteromorpha	1	7	1 × 7	7
Cladophora	1	7	1 × 7	7
Nuphar lutea	3	1	1 × 3	3
Lemna minor	4	1	1 × 4	28
Potamogeton pectinatus	1	7	7 × 1	7
Zannichellia palustris	2	7	7 × 2	14
Leptodictyum riparium	1	7	1 × 7	7
Ranunculus fluitans	7	1	1 × 7	7
Column totals for MTR scoring		38		56
MTR Score	56 (CVS total) 4/38 (SCV total) = 1.5 × 10 = 15			

total is divided by the former total score. To obtain an index from 10 to 100, this score is multiplied by 10. From this example it can be seen how important the accurate estimation of cover within a site is. In the examples given, the MTR scores of 43 and 15 are analogous to invertebrate ASPT scores of 4.3 and 1.5.

A draft manual was produced in 1996 to trial the method, followed by a thorough independent review of its performance. The review of the method involved matching sets of water quality data with the distribution of individual species on the UK checklist, and MTR scores derived from surveys carried out where there were matching water quality stations. The review concluded the system generally reflected well differences in nutrient status, and made recommendations for its wide use, unaltered (Dawson *et al.* 1999). A detailed manual (Holmes *et al.* 1999) was then produced to describe the method in detail, and outline applications for its use.

The MTR method is similar to several other macrophyte survey and assessment methods developed at broadly similar times in Europe. Two of the most applied and tested are the Trophic Index of Macrophytes (TIM) and The Biological Indice Macrophytique (IBMR). The former was developed in Bavaria, and the latter in France. The principles of both are similar to the MTR method, but in the IBMR system there are 207 checklist taxa, and trophic ranks range from 0 to 20 (with the highest numbers representing oligotrophy, and therefore is on an opposite scale to MTR and BMWP).

Applications of MTR

The MTR system was developed specifically to assist in delivering the requirements of the UWWD in the UK. In England and Wales, wherever there were concerns that qualifying discharges (serving >10,000 people) may be impacting a river, MTR surveys were carried out upstream and downstream of the discharge. Care was always taken to sample sites of similar physical characteristics as flow velocity and

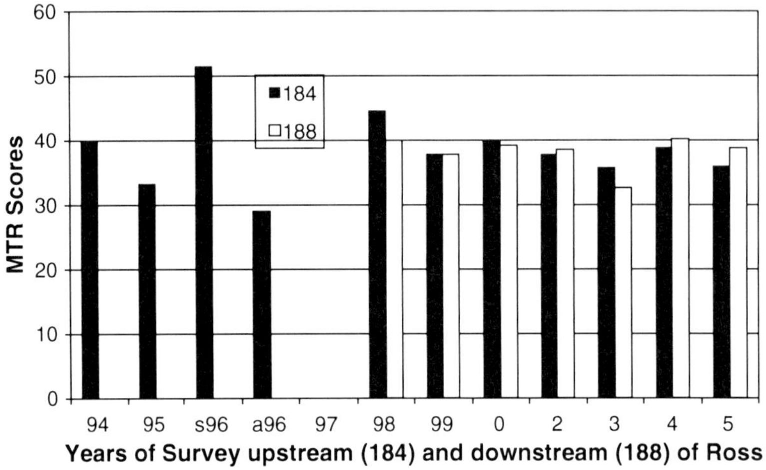

Fig. 12.1 MTR Scores for surveys upstream and downstream of Ross STW (data courtesy of the Environment Agency): two surveys were undertaken in 1996 – one in summer (s) and the other in autumn (a)

sediment can have considerable influence on the assemblage of macrophytes present. If surveys consistently indicated much lower MTR scores downstream of the discharge, and the reason could be shown to be due to the discharge, phosphorus stripping was introduced to reduce the nutrient levels. Surveys usually continued following this to monitor the response of the macrophytes through MTR scores.

An example of the application of MTR in relation to sewage treatment work (STW) discharges is illustrated in Fig. 12.1. This figure shows the MTR scores for sites surveyed on the River Wye, a large river on the Welsh/English border. Surveys were carried out upstream of Ross-on-Wye STW discharge for several years from 1994 to 1996. From 1998 to 2005, surveys were undertaken both upstream and downstream of the discharge.

In 2000 phosphate stripping was introduced, and upstream and downstream macrophyte surveys continued for 5 years; MTR results clearly suggest that there is now no measurable impact on the macrophyte community from the STW discharge.

The example of using MTR surveys to determine whether nutrient stripping is required, and then monitoring the response in the river of the management action taken, has been repeated on many rivers in England and Wales. These are examples of MTR being used specifically to determine the impact of single point sources of potential enrichment in rivers. MTR can also be used to identify the cumulative effect of diffuse sources of eutrophication within catchments, or characterise the nutrient status of whole rivers, or sections within larger ones. Figure 12.2 illustrates this application, again from the Wye. This shows 5 years of MTR surveys at 18 sites down the Wye from 2000 to 2005. The figure shows that each survey year produced very similar results for each site; given the dynamic nature of rivers, these annual variations in scores are very small. The MTR scores in the two headwater sites were

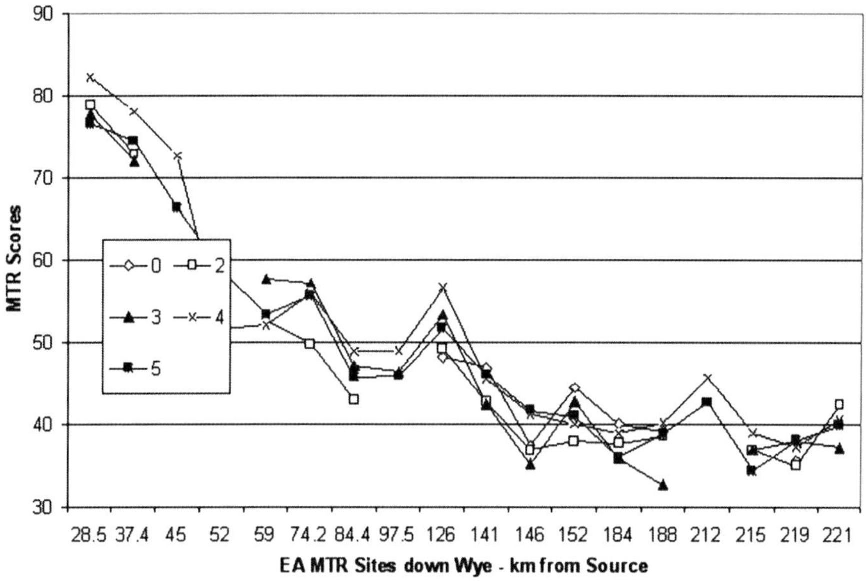

Fig. 12.2 Changes in MTR scores down the Wye 2000–2005 (data courtesy of the Environment Agency); lines represent surveys in years 2000, 2002, 2003, 2004 and 2005

always above 70, reflecting the very clean water due to the nutrient poor moorland landscape and base-poor rocks in the upper catchment. In the lower reaches, from km 146 downstream, MTR scores are around 40, reflecting a much more sluggish lowland river flowing over more basic and richer rocks, as well as receiving the cumulative nutrients from all upstream diffuse and point sources. The figure shows clearly that MTR scores rapidly drop around 50 km from the river's source. Reference to geological maps shows this is where the base-poor rocks change to more basic sandstones and limestone. Not only does the natural chemistry of the water become more enriched, but also land-use becomes more intensive – MTR surveys can show evidence of both.

The European Star Project

The STAR (STAndardisation of River Classifications) research project was an initiative funded by the European Commission with links to the implementation of the Key Action "Sustainable Management and Quality of Water" within the Energy, Environment and Sustainable Development Programme. The project had formal links to CEN (European Committee for Standardisation – http://www.cen.eu/cenorm/homepage.htm) and a key aim was to provide relevant environmental CEN working groups with draft methods for surveys and assessments. Results from the STAR project were published in a special issue of the journal *Hydrobiologia* in 2006 (Furse *et al.* 2006).

As part of the project, over 100 sites in four different WFD regions were surveyed using the MTR method alongside any existing national macrophyte survey methods (Szoszkiewicz *et al.* 2006). From the analysis of data, they reported that MTR (and IBMR) data could be used to produce metrics for quality assessment. Relationships between several macrophyte metrics and nutrient enrichment were found to be highly significant, although not sufficiently well correlated to be used as a uniform system for all the river types across the whole of Europe. Analysis based on the STAR MTR macrophyte database enabled the project to propose a re-designed MTR for lowland rivers and a slightly different version for the mountain streams. Regional checklists were recommended to improve its application and ease of use in different parts of Europe, and it was proposed that the checklist of taxa be enlarged so the same basic system could be used as a pan-European macrophyte-based bio-assessment method. Thus they recommended MTR could be a useful tool for determining ecological status of rivers for the WFD, but such an application of the method will be most reliable when linked to regional lists developed through locally developed programmes using an international protocol.

The Polish MTR System

Following research into the application and performance of MTR in the UK (Szoszkiewicz 2004), macrophyte surveys were undertaken in Poland using the MTR method, but other macrophytes additional to the UK checklist taxa were included during the survey. Using the data from these surveys, a Polish MTR system was developed that includes some additional taxa that are common in rivers there, and altering slightly STRs for some taxa. Based on correlations with nutrient status, some edge taxa have also been added that do not appear in the UK system. This modification is consistent with the recommendations of the STAR project.

In 2007 the author undertook surveys on 12 Polish river sites to compare the MTR scores derived by using: (i) the unmodified UK MTR method; (ii) the UK check-list but using any modified STRs of the Polish system; (iii) the Polish checklist, with the modified STRS, and the weighting modifications in the Polish system. Results from the different scoring systems are depicted in Fig. 12.3. The first is using the UK check-list of taxa, and the original UK STR assigned to each taxon; the second used the same UK check-list, but with STRs assigned in the Polish (PL) system; the third shows results using the Polish check-list and their SDTR system, whilst the last system is a variation of the standard Polish system allowing for weighting of selected species. The overall scores derived from the systems are very similar. The Polish MTR system (i.e. including the additional taxa and using their scoring system) resulted in scores either staying the same as that without using the additional taxa, or slightly increasing. These comparisons show that differences in STR result in greater changes to MTR scores than the addition of taxa.

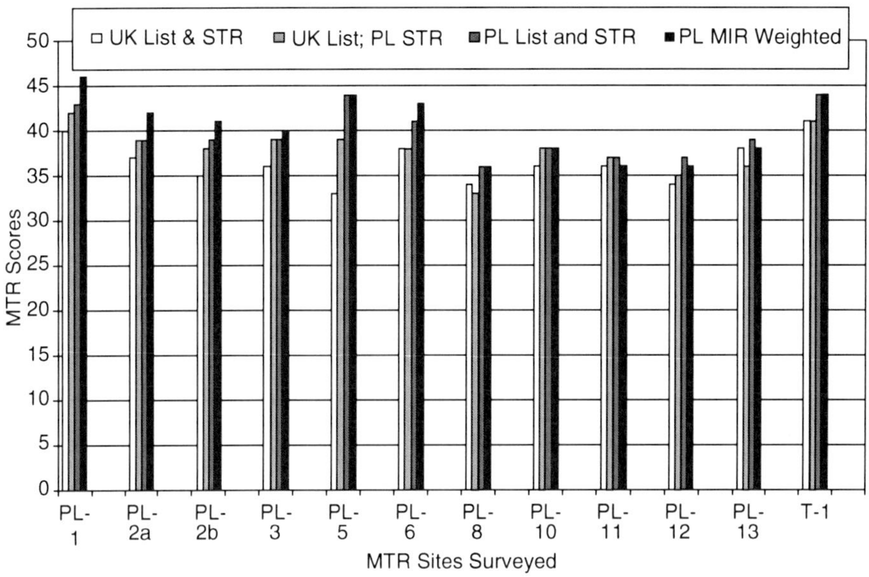

Fig. 12.3 Comparison of MTR Scores derived from 12 sites surveyed in Poland

Conclusions

The use of macrophytes in river quality assessment has a very recent history. The MTR method has been tested at a pan-European scale and shown to have many useful applications, not least helping member EU countries to meet the requirements of the WFD. By wide use and testing in many countries in Europe it has been shown to be a robust method with a core system that can be refined for use in different geographical areas to improve its effectiveness as an assessment tool.

References

Armitage PD, Moss D, Wright JF, Furse MT (1983) The performance of a new biological water quality score system based on macroinvertebrates over a wide range of unpolluted running-water sites. Water Res 17:333–347

Baattrup-Pedersen A, Szoszkiewicz K, Nijboer R, O'Hare M, Ferreira T (2006) Macrophyte communities in unimpacted European streams: Variability in assemblage patterns, abundance and diversity. Hydrobiologia 566:179–196

Brabec K, Szoszkiewicz K (2006) Macrophytes and diatoms – Major results and conclusions from the STAR project. Hydrobiologia 566:175–178

Caffrey J (1987) Macrophytes as indicators of organic pollution in Irish rivers. In: Richardson D (ed) Biological indicators of pollution. Royal Irish Academy, Dublin, pp 77–87

Dawson FH, Newman JR, Gravelle MJ, Rouen KJ, Henville P (1999) Assessment of the trophic status of rivers using macrophytes. Evaluation of the mean trophic rank. R&D Technical Report E39. Environment Agency, Bristol

Ellenberg H (1973) Chemical data and aquatic vascular plants as indicators for pollution in the Moosach river system near Munich. Archiv für Hydrobiologie 72:533–549
EU (1991) Directive 91/271/EEC 21 May 1991. Urban Wastewater Directive
EU (2000) Directive 2000/60/EC 23 October 2000. Water Framework Directive
Furse MT, Hering D, Brabec K, Buffagni A, Sandin L, Verdonschot PFM (eds) (2006) The ecological status of European rivers: Evaluation and intercalibration of assessment methods. Hydrobiologia 566:1–555
Haslam SM (1987) River plants of western Europe. Cambridge University Press, Cambridge
Hellawell JM (1986) Biological indicators of freshwater pollution and environmental management. Elsevier, Amsterdam
Holmes NTH, Newman JR, Chadd S, Rouen KJ, Saint L, Dawson FH (1999) Mean trophic rank: A users manual. R&D Technical Report E38. Environment Agency. Bristol, p 134
Hutchinson GE (1970) The chemical ecology of three species of *Myriophyllum*. Limnol Oceanogr 15:1–5
Kelly M (1990) Monitoring water pollution: The role of plants. Plants Today 2:96–100
Kohler A (1975) Makrophytische Wasserpflanzen als Bioindikatoren für Belastungen von Flieβgewässer-Ökosytemen. Verhandlungen der Gesellschaft für Ökologie, Wien 1975:255–276 (in German)
Meriaux JL (1982) L'utilisation des macrophytes et des phytocoenoses aquatiques comme indicateurs de la qualite des eaux. Les Naturalistes Belges 63:12–28 (in French)
Seddon B (1972) Aquatic macrophytes as limnological indicators. Freshwater Biol 2:107–130
Szoszkiewicz K (2004) Vegetation as an indicator of the trophic status of running waters, based on rivers of Great Britain and Northern Ireland. University of Poznan, Thesis
Szoszkiewicz K, Ferreira T, Korte T, Baattrup-Pedersen A, Davy-Bowker J, O'Hare M (2006) European river plant communities: The importance of organic pollution and the usefulness of existing macrophyte metrics. Hydrobiologia 566:211–234
Weber-Oldecop DW (1976) Flieβgewässertypologie in Niedersachsen auf floristisch-soziologischer Grundlage. Göttinger Floristische Rundbriefe 10:73–79 (in German)
Wiegleb G (1978) Untersuchungen über den Zusammenhang zwischen hydrochemischen Umweltfaktoren und Makrophytenvegetation in stehenden Gewässern. Archiv für Hydrobologie 83:443–484 (in German)
Woodiwiss FS (1964) A biological system to stream classification used by Trent River Board. Chem Ind 11:443–447

Chapter 13
Monitoring the *Ranunculion* Habitat on the River Itchen: Practical Application and Constraints

Richard Lansdown and Tim Pankhurst

Introduction

The River Itchen is approximately 28 km long and runs entirely within the UK county of Hampshire, from New Cheriton, through Winchester to join the Solent in Southampton. It is one of the "classic" chalk rivers (Butcher 1927), rising on the chalk of the South Downs and flowing either over chalk or alluvial deposits throughout its length; consequently wherever there is faster flow, the bed is dominated by flints. Partly due to this uniform geology and partly because of the buffering effect of slow release of water from the aquifers from which they derive, chalk rivers are seen as being among the most stable river systems in the UK, lacking the massive level of flow variation or scour typical of upland rivers on hard geology. Chalk rivers are also among the most managed riverine systems in the UK, most having been modified and maintained for fishing, watercress farming and, to a lesser extent, as a source of power to run mills for many centuries.

The River Itchen was designated as a Special Area of Conservation (SAC) under the Habitats Directive. EUNIS habitat 3260 "Water courses of plain to montane levels with the *Ranunculion fluitantis* and *Callitricho-Batrachion* vegetation" is cited as the only habitat serving as a primary reason for designation. In the Joint Nature Conservation Committee (JNCC) description, the Itchen is described as "a classic example of a sub-type 1 chalk river [..] dominated throughout by aquatic *Ranunculus* spp. The headwaters contain pond water-crowfoot *Ranunculus peltatus*, while two *Ranunculus* species occur further downstream: stream water-crowfoot *R. penicillatus* ssp. *pseudofluitans* (a species especially characteristic of calcium-rich

R. Lansdown (✉)
45 The Bridle, Stroud, Glos, GL5 4SQ, UK
e-mail: rlansdown@ardeola.demon.co.uk

T. Pankhurst
Plantlife, Cambridge University Botanic Gardens, Cory Lodge,
Bateman Street, Cambridge, CB2 1JF, UK

C. Hurford et al. (eds.), *Conservation Monitoring in Freshwater Habitats: A Practical Guide and Case Studies*, DOI 10.1007/978-1-4020-9278-7_13,

rivers) and river water-crowfoot *R. fluitans*" (www.jncc.gov.uk). The description of Habitat 3260 reads: "Watercourses of plain to montane levels, with submerged or floating vegetation of the *Ranunculion fluitantis* and *Callitricho-Batrachion* (low water level during summer) or aquatic mosses", with the following taxa considered to be characteristic: *Callitriche* subspp., *Fontinalis antipyretica, Myriophyllum* spp., *Potamogeton* spp., *Ranunculus aquatilis, R. fluitans, R. peltatus, R. penicillatus* subspp. *penicillatus* and *pseudofluitans, R. saniculifolius, R. trichophyllus, Sium erectum* (*Berula erecta*) and *Zannichellia palustris* (eunis.eea.europa.eu). It is interpreted into UK practice by Hatton-Ellis and Grieve (2003).

It is a requirement of the Habitats Directive to monitor the condition of habitat 3260. Following designation, extensive macrophyte surveys were carried out in 2002–2003 to assess the condition of the *Ranunculion fluitantis* and *Callitricho-Batrachion* habitat, the conclusions of which were that insufficient information was available on the macrophytes to fully assess the condition of the Annex I habitat in the river. This chapter reports on the use of survey data to assess the condition of the *Ranunculion fluitantis* and *Callitricho-Batrachion* habitat.

The Study

A project was initiated to collect data on the aquatic plants in the river employing the Mean Trophic Rank (MTR) protocol (Holmes *et al.* 1999b) (see Chapter 12). Whilst this method was not designed for this purpose, the Environment Agency (EA) was keen to assess whether it had value for monitoring the condition of the *Ranunculion fluitantis* and *Callitricho-Batrachion* habitat. For this assessment, we made the following deviations from the standard method:

- A single 100 m section was surveyed for each of the 28 km of the river, whether or not there was a known inflow or discharge. Figure 13.1 provides a topological representation of the location of sampling sites on the river.
- Surveys were repeated in March, June/July and September each year to measure seasonal effects.
- Rather than record taxa against the standard list specified by the MTR method, all taxa were recorded as accurately as possible and, in this, the method employed accords better with LEAFPACS (Willby 2005).

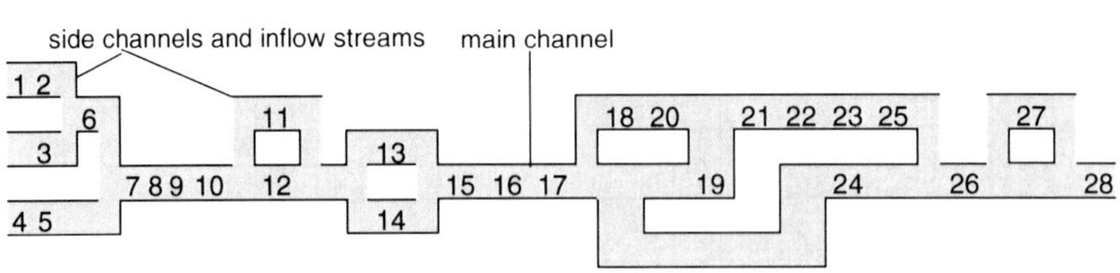

Fig. 13.1 Topological representation of the location of sampling sites on the River Itchen

- Rather than record taxa to a nine point cover scale, the cover of each taxon was estimated to within one percentage point (except estimates below one per cent which were estimated as being <0.1% or 0.1–1%), although in practice there is a tendency for the surveyor to assign cover values to integers below 20% and to the nearest 5 or 10% above this.

Use of the data collected was based on a number of assumptions (Box 13.1), some of which apply to all interpretations and some only to calculation of MTR scores. Equally, it was not possible to simply employ the data collected without modification, and decisions on interpretation of species data are explained in Box 13.2.

Data were analysed to establish whether any of the following could serve to indicate the condition of the *Ranunculion fluitantis* and *Callitricho-Batrachion* vegetation:

- MTR scores
- Traditional concepts of conservation value (such as rarity, diversity, etc.)
- Recorded characteristics of the *Ranunculus* population

Box 13.1 For the purposes of analysis, the following notes must be taken into account

- The first four datasets (September 2004 and March, June and September 2006) were collected by R.V. Lansdown or T.J. Pankhurst, subsequently all data were collected by R.V. Lansdown, to reduce the effects of observer differences.
- Data collected from surveys in 2002–2003 were included in the analysis of MTR scores where they were known to be directly comparable.
- During the period for which data are analysed here, the only obvious changes in the river were (i) a period of increased turbidity in the summer of 2005, probably due to excavation of an online pond, and (ii) an increase in flow throughout the survey period.
- On no occasion were data from previous surveys consulted before a survey, the idea being that each recording event should be as independent as possible and therefore to resemble the standard one-off recording visit.
- The stability of the system (considered a characteristic of chalk rivers) should ensure that vegetation would remain more or less the same in the absence of perturbation.
- The absolute percentage cover estimates were converted to a 9-point scale for MTR score calculation, as recommended by Holmes *et al.* (1999a), as follows:
  - In all cases, if a recorded cover estimate fell between two ranks, it was assigned to the lower of the two.
  - When combining scores (for example various names used for the Batrachian *Ranunculus*), the highest value has been selected or integers added.

Box 13.2 Decisions relating to the analysis of data

In the process of preparing data for analysis, a number of decisions were taken and some artefacts of recording recognised; these are described below. The authors are unaware of any record made by other workers of the reasoning behind such decisions and yet such decisions are made, consciously or otherwise, every time that MTR survey is carried out on a river. Without a record of the decisions taken and the implications of these decisions for data interpretation, it is not possible to make a reliable comparison of datasets collected by different surveyors. In addition, differences in identification and nomenclature will strongly affect comparability of data.

- Records of "*Cladophora* agg." were combined as *Cladophora glomerata*, because according to John *et al.* (2002) only three species of *Cladophora* occur in UK waters, of which *C. fracta* occurs in nutrient-rich ponds and ditches, while *C. aegagropila* is easily distinguished from the other two and does not occur on the River Itchen.
- In the early part of the study, submerged *Pellia* populations were recorded as "Thallose liverwort". Following confirmation that they were actually *Pellia* species but because they lacked reproductive structures could not be identified to species, they were recorded as "*Pellia* sp.". These and records of *Pellia epiphylla* which does not occur on the River Itchen, have been treated as *P. endiviifolia*.
- All Batrachian *Ranunculus* records are treated as *R. penicillatus* subsp. *pseudofluitans* because it has been demonstrated that all populations in the river can reasonably be assigned to this taxon (Lansdown 2007). Where two Batrachian *Ranunculus* taxa were recorded, the lowest possible score was used (e.g. 4 + 5 = 7.5 = 5).
- All records of *Lemna minor* and *L. minor*/*gibba* are treated as *L. gibba* because rigorous searching showed that *L. minor* does not appear to occur on the Itchen.
- All records of *Callitriche* populations that were not flowering and recorded as *Callitriche platycarpa*/*obtusangula* have been omitted from analysis. However this means that the records of *C. obtusangula* used actually represent the proportion of all *Callitriche* plants present which were flowering and could be confirmed as that species, rather than any measure of the abundance of *Callitriche* species in general.
- Records of *Rorippa nasturtium aquaticum* in the 2003–2003 data were assigned a score and have been interpreted as *R. nasturtium-aquaticum sensu stricto*, although it is likely that this does include records of the other two taxa. Conversely, in 2004–2008 all populations were identified to species wherever possible and non-flowering populations recorded as *Rorippa nasturtium-aquaticum* agg. which have been omitted from analysis.

(continued)

Box 13.2 (continued)

- As is the case with the *Rorippa nasturtium-aquaticum* agg. and *Callitriche* species, non-flowering *Veronica* populations cannot be identified more precisely than to genus. *V.* × *lackschewitzii* was not recorded in 2002–2003, although it seems very likely that this is the dominant taxon present on the river. Records of *V. anagallis-aquatica* or *V. catenata* from 2002 to 2003 are interpreted as *sensu stricto*, although these records certainly include *V.* × *lackschewitzii*.

As noted by Hatton-Ellis and Grieve (2003) "The definition of watercourses characterised by *Ranunculion fluitantis* and *Callitricho-Batrachion* (CB) communities is very wide, in practice covering the majority of rivers and streams with aquatic plant communities of note". It cannot be applied with any rigour to different sections on a river and any variation sufficient to exclude a section from this class would need to be extreme. Hatton-Ellis and Grieve (2003) provide a means of distinguishing between six different types of the *Ranunculion fluitantis* and *Callitricho-Batrachion* communities but these seem to apply only at the whole river level. There seems to be no means by which to recognise a degraded community or to define where the community begins or ends.

Consequently, although the data collected for this project can be used to assign the Itchen to a type CB2 "Base-rich *Ranunculus penicillatus* subsp. *pseudofluitans* - *Callitriche obtusangula* rivers, including chalk streams", it is not possible to assess whether individual sections may also be so assigned.

MTR

The MTR scores from sections upstream and downstream of the selected licensed discharges or groups of discharges were considered. However, there are hundreds of licensed discharges into the River Itchen and it was not practical or meaningful to address all of them in the analysis. Therefore, only 17 large discharges identified by the Environment Agency's SIMCAT model were compared. These are represented by two Sewage Treatment Works (STW), eight intensive and two unintensive watercress beds and five fish farms or hatcheries.

Any consideration of these data must recognise that there are licensed discharges not considered here, as well as large numbers of unlicensed discharges and hard-surface run-off into the river and its tributaries which are generally subject to minimal treatment. In addition to various inorganic substances, these may carry organic substances which will have an effect on nutrient levels within the river. This is particularly the case with the River Itchen, which receives input of run-off from the

Table 13.1 Difference between MTR scores from summer surveys on the main channel between sections up- and downstream of licensed discharges. A positive number indicates a rise in MTR score as one progresses downstream and a negative score a decline. Bold cells with a shaded background indicate a fall in MTR score of more than 4 points and bold cells with a white background indicate a rise in MTR score of more than 4 points

Section numbers	2:6	4:5	9:10	14:15	16:17	25:26
2002	−1.7					−1.0
2003	−0.7				0.0	−2.5
2005	2.4	2.5	−0.4	−2.7	**9.3**	−3.0
2006	3.3	3.8	**5.5**	1.3	-0.9	**−4.4**
2007	−1.9	**9.9**	3.6		−3.6	0.0
2008	3.2	−0.6	**6.1**	0.7	**−7.4**	−3.8

Table 13.2 The difference between MTR scores from autumn surveys on the main channel between sections upstream and downstream of licensed discharges (presentation follows Table 13.1)

Section numbers	2:6	4:5	9:10	14:15	16:17	25:26
2002	2.0					−1.9
2003	2.0				0.4	−3.2
2004	−0.9	**6.9**		**6.0**	1.7	2.3
2005	0.5	−0.3	2.5	2.7	**−9.1**	−0.6
2006	1.3	3.2	**7.0**	**−5.0**	**−4.9**	**−4.3**
2007	**7.3**	3.1	**14.0**	**5.0**	−1.5	−1.8
2008	**7.0**	1.6	**13.6**	−0.9	−0.7	

M3 motorway in a number of places. There are also important inputs of diffuse pollution which cannot be quantified for the purposes of this analysis.

Overall the MTR scores (Tables 13.1 and 13.2) are difficult to explain. Downstream of some discharges there are significant decreases in MTR score implying nutrient enrichment. However, downstream of other discharges MTR actually increases by more than 4 units, which may imply a significant improvement in trophic status. Equally MTR scores may increase or decrease upstream or downstream of the same discharge on different surveys. Significant increases in MTR scores downstream of discharges are unlikely and the large number of these results is of some concern.

It may be that the MTR method is not able to adequately reflect trophic changes in the state of the River Itchen and this requires further investigation. Also, there is no evidence to suggest that observed variation in MTR scores reflects changes in trophic state in a way that relates to *Ranunculion fluitantis* and *Callitricho-Batrachion* communities.

If such problems have been observed in other areas of the country then it would seem appropriate to carry out a full review of the MTR method. Until a critical review is completed, however, it seems sensible to suggest that a difference of 14 MTR points (which equates to the greatest increase observed) between two sections must be considered to represent normal variation and only a difference well in excess of this should be considered "significant".

Conservation Value

An established method of assessing the conservation value of sites is to use the criteria developed by Ratcliffe (1977) (Box 13.3). However, data collected for MTR analysis do not include any means of assessing or comparing the naturalness, fragility, typicalness, position in an ecological unit, potential value or aesthetic appeal.

Similarly, although the Itchen has a very important recorded history, dating at least from work by Butcher (1927), the data collected for this study provide no information relating to this. All the sections surveyed are the same length and therefore vary in size only by virtue of channel width, which is not known to have any particular association with conservation value. Therefore, only two of the criteria can meaningfully be assessed using the data collected: diversity and rarity.

Diversity

A total of 246 plant taxa (including marginal species) was recorded from the River Itchen during the study. Of these, 142 (58.9%) were recorded from fewer than 5 (18%) sections and only 33 (13.7%) were recorded from more than 25 sections (89%).

Box 13.3 The "Ratcliffe" criteria

D.A. Ratcliffe (1977) published a review of nature conservation in Britain which included a set of criteria for comparison of the relative merits of different sites. These can be summarised as follows:

Size – self-explanatory.

Diversity – variety in number of both plant associations and species which are usually closely related and in turn depend largely on diversity of habitat.

Naturalness – the degree of modification by human influence.

Rarity – the presence of rare communities, habitats or species.

Fragility – the degree of sensitivity of habitats, communities and species to environmental change.

Typicalness – the degree to which a site can be seen to represent the best example of its habitat.

Recorded history – the extent to which a site has been used for scientific research.

Position in an ecological unit – the degree of connectivity to other sites, communities or habitats of conservation value.

Potential value – covers sites which may not have high value at the time of assessment, but which could achieve high value through appropriate management.

Intrinsic appeal – public perception of value.

Table 13.3 The number of species per section and number of species per survey by section

	2004	2005			2006			2007			2008			
Section	Sept.	Mar.	Jun.	Sept.	Mar.	Jun.	Sept.	Mar.	Jun.	Sept.	Mar.	Jun.	Sept.	Total
1	15	16	18	29	20	22	20	20	18	19	23	15	16	72
2	28	25	21	26	25	20	28	20	26	22	18	29	19	78
3	23	31	29	28	27	32	26	32	27	26	31	32	32	78
4	22	22	25	23	20	25	21	21	22	22	24	34	21	74
5	40	31	34	31	24	28	27	23	29	33	22	32	27	81
6	24	23	30	33	22	28	29	22	31	28	24	29	33	77
7	33	22	35	35	33	35	35	26	28	34	26	35	39	90
8	45	35	33	32	33	35	34	21	34	34	33	39	31	106
9	24	31	26	22	16	30	29	34	29	30	31	27	26	84
10	31	27	30	35	22	30	25	22	28	36	25	30	30	77
11	38	12	23	36	20	27	32	27	29	34	25	32	23	84
12	35	23	30	29	21	31	25	17	34	30	30	30	25	77
13	29	15	24	32	15	30	29	18	31	30	33	34	32	79
14	33	28	29	33	30	36	30	24	37	34	25	35	27	78
15	18	20	22	22	16	28	21	13		18		22	23	65
16	52	50	41	38	40	42	34	26	40	45		40	34	130
17	44	28	27	30	34	37	33	24	30	40		28	26	98
18	30	26	32	30	22	32	32	19		31		34		85
19	29	27	31	31	22	29	31	25	24	28	29	32	26	75
20	37	36	33	34	25	37	41	24	30	24	30	25	25	86
21	33	29	35	33	18	34	34	28	38	36		35	34	89
22	41	41	40	33	20	41	38	25		33		33	27	98
23	23	25	32	33	22	36	34	23		40		29	33	81
24	44	35	34	31	26	34	28	13	27	38		32	33	100
25	34	23	30	24	16	30	35	26	28	40		32	30	78
26	39	25	37	27	26	30	32	18	26	35		32		88
27	23	24	36	33	15	36	43	20	22	34		31		79
28	40	21	35	35	15	27	25			33		23		74
Min	15	12	18	22	15	20	20	13	18	18	18	15	16	65
Max	52	50	41	38	40	42	43	34	40	45	33	40	39	130
Average	32.4	26.8	30.4	30.6	23.0	31.5	30.4	22.6	29.0	31.7	26.8	30.8	28.0	84.3

Thus it can be seen that if the whole river represents a single community (as suggested by Holmes *et al.* 1999a, etc.) then either it is a community with little homogeneity or many taxa are recorded that do not actually fall within the community and there is no replicable means of distinguishing these additional taxa.

Table 13.3 shows the number of species recorded on each survey on each section, together with the total number of species recorded on the section throughout the project.

This clearly shows wide variation, not only in the number of species recorded on each visit, but in the total number of species recorded on different sections. This wide variation is not limited to taxa in the channel but is most strongly influenced by the number of marginal and terrestrial taxa occurring within the recording zone. It is particularly striking that all sections on which 100 or more species have been recorded, have exposed clays on the margin, with the consequence that species more typical of disturbed habitats were recorded. However, there are other sections which also have exposed clays in the margins but which have a lower total list. There appears to be no simple explanation of the variation in the total number of taxa recorded.

MTR is normally recorded against a standard checklist; using such a system, a measure of diversity might be indicated by the proportion of listed taxa recorded on each section. However, analysis shows that variation in the number and range of MTR taxa mirrors that of all taxa recorded and shows no logical relationship with the condition of the river or vegetation.

Rarity

Not many "rare" taxa were recorded on the river; records of *Amblystegium humile*, *Orthotrichum tenellum* and *O. striatum* represented the only known populations of these species in North Hampshire at the time of recording. However, although almost certainly associated with the humidity of the river corridor, *A. humile* is the only one of these that occurred within the MTR recording zone.

Catabrosa aquatica was recorded on two sections, apparently associated at both with a reduction in competition caused by cattle-grazing. *Oenanthe fluviatilis* was consistently recorded in river sections from Winchester downstream and on a single occasion, a short distance upstream of Winchester; although scarce and declining, this species is often associated with filamentous algae and silt deposits and may not be an indicator of good condition. These represent only three species directly dependent upon the recording zone to serve as indicators of national or regional rarity, which is inadequate to enable scoring of sites.

As described above (and considered typical of vegetation sampling), more than half of the taxa recorded occurred on less than 20% of the sections and observed variation was more closely linked to marginal habitats than to the channel. It therefore seems unlikely that assessments of rarity on the river are of use in indicating the condition of the *Ranunculion fluitantis* and *Callitricho-Batrachion* vegetation using the MTR method.

Ranunculus Subgenus *Batrachium*

Lansdown (2007) demonstrated that it is highly likely that the populations of Batrachian *Ranunculus* in the River Itchen are represented by a suite of polyploid clones of *R. penicillatus* subsp. *pseudofluitans*. These clones may respond differently to different environmental influences and therefore it is impossible to interpret changes in populations at the species level using the data collected. Collection of data on standard lengths, rather than in relation to stands or individuals plants of *Ranunculus,* means that it is not even possible to measure changes affecting different clones.

Ranunculus subgenus *Batrachium* populations have been the focus of much attention over the last 30 years, probably more so than any other aquatic plants. However, it is of note that C.D. Preston (in Preston *et al.* 2002) does not support the idea that members of the genus are in decline and instead makes statements such as "The aquatic *Ranunculus* species are difficult to identify…". "Changes in distribution are difficult to assess", "subsp. *pseudofluitans* [..] is probably still under-recorded", etc., while Cheffings and Farrell (2005) treat all taxa as of Least Concern. Equally, Cranston and Darby (2004) carried out an extremely thorough and extensive review of information available on the subgenus and refer only to "apparent decline", also noting "major recoveries". On the Itchen, grazing by swans appears to have a significant effect on the biomass of *Ranunculus*, but even here it is by no means clear whether the number or distribution of plants is affected. It must be a priority to establish, in fact, whether or not there has been a real decline in Batrachian *Ranunculus* and what form that decline might take.

Conclusions

In summary, plant surveys on 100 m standard-lengths of river do not appear to provide data that can be used to monitor the condition of the *Ranunculion fluitantis* and *Callitricho-Batrachion* community in the River Itchen, for five main reasons.

- Significant (i.e. greater than 4 MTR points) positive and negative changes in MTR score occur even when there is no other evidence to suggest a change in trophic state.
- There is no evidence to suggest that observed variation in MTR scores reflects changes in trophic state in a way that relates to *Ranunculion fluitantis* and *Callitricho-Batrachion* communities.
- Of the "Ratcliffe criteria", the data collected can only be assessed against rarity and diversity. Not enough taxa of national or local conservation concern were recorded to be used as a measure of the quality of the communities, while actual diversity is obscured by adherence to a checklist and, when all observed taxa are recorded, by massive variation in the number of taxa observed between visits.
- The *Ranunculion fluitantis* and *Callitricho-Batrachion* communities are too poorly defined, both by EUNIS and Hatton-Ellis and Grieve (2003), to act as a standard against which the quality of the river may be measured.
- The taxonomy and identification of Batrachian *Ranunculus* species has not been adequately clarified. Consequently, it is not yet possible to demonstrate unambiguously that species or populations are in decline. It is highly likely that neither *Ranunculus fluitans* nor *R. peltatus* occur in the River Itchen and therefore that the description of the River Itchen SAC is inaccurate. It follows that monitoring against this description must inevitably, and probably erroneously, indicate a decline.

There is no immediately obvious means of assessing the condition of the *Ranunculion fluitantis* and *Callitricho-Batrachion* vegetation of the River Itchen. Given that it seems likely that the Batrachian *Ranunculus* is one of the more tolerant plants in the river, focus on the subgenus may actually distract from more urgent conservation needs. Clearly, there is a need to measure and describe the condition of the vegetation of the river both for the purpose of reporting to Europe and simply to be able to respond to the needs of conservation and exploitation of the river. One of the problems is poor definition of exactly what is intended by the term "condition" of the habitat as far as it relates to the vegetation.

The following recommendations raise issues that would appear to merit further consideration and which might enable a more meaningful indication of conservation value:

- It is clear that certain vegetation associations are repeated along the river (probably more so on a chalk river than other river types). It may be possible to characterise these on a phytosociological basis and then compare (a) the range of species in different examples of each association and (b) representation of these associations in different parts of the river, to derive indices of comparative conservation value and even change in response to anthropogenic influences.
- All available methods of assessment of the conservation value of the river appear to treat the channel and marginal vegetation in isolation. However, the relationship between the river and its floodplain is critical for the survival of a very large number of plant species. A thorough assessment of the condition of the river must take into account habitats away from the channel, particularly as it is likely that the instream vegetation is very strongly affected by changes to the management and condition of floodplain habitats.

Acknowledgements The authors would like to thank the Environment Agency for permission to use data from the River Itchen study and Kevin Exley for help throughout the project, as well as commenting on an earlier draft of this chapter. We are also grateful to Adam Fulton for comments, discussions and feedback and to his colleagues Kerry Evans and Grant Miller for help with surveys in 2008.

References

Butcher RW (1927) A preliminary account of the vegetation of the River Itchen. J Ecol 15:55–65

Cheffings CM, Farrell L (eds) (2005) The vascular plant Red Data List for Great Britain. Species Status 7:1–116

Cranston E, Darby E (2004) *Ranunculus* in chalk rivers: Phase 2. R & D Technical Report W1-042/TR. Environment Agency, Bristol

Hatton-Ellis TW, Grieve N (2003) Ecology of watercourses characterised by *Ranunculion fluitantis* and *Callitricho-Batrachion* vegetation. Conserving Natura 2000 rivers ecology series no. 11. English Nature, Peterborough

Holmes N, Boon P, Rowell T (1999a) Vegetation communities of British rivers: A revised classification. Joint Nature Conservation Committee, Peterborough

Holmes NT, Newman JR, Chadd S, Rouen KJ, Saint L, Dawson FH (1999b) Mean trophic rank: A user's manual. Environment Agency, Bristol

John DM, Whitton BA, Brook AJ (2002) The freshwater algal flora of the British Isles: An identification guide to freshwater and terrestrial algae. Cambridge University Press, Cambridge

Lansdown RV (2007) The identity of *Ranunculus* subgenus *Batrachium* in the River Itchen. Environment Agency, Southern Region

Preston CD, Pearman DA, Dines TD (2002) New atlas of the British and Irish flora. An atlas of the vascular plants of Britain, Ireland, the Isle of Man and the Channel Islands. Oxford University Press, Oxford

Ratcliffe DA (1977) A nature conservation review, vol 1. Cambridge University Press, Cambridge

Willby NJ (2005) LEAFPACS methodology as provided by SEPA. Unpublished appendix to SEPA tendering documents

Chapter 14
Observer Variation in River Macrophyte Surveys

The Results of Multiple-Observer Sampling Trials on the Western Cleddau

Clive Hurford

Introduction

In 2008 we carried out multiple-observer sampling trials on two stretches of the Western Cleddau, a lowland river in south-west Wales. These trials were set up to determine the rates of observer variation between surveyors, specifically in relation to the detection of macrophyte species and estimates of vegetation cover. These are key components of all forms of river macrophyte survey and monitoring projects in UK rivers.

The Sampling Trial Locations

The sampling trials took place in two sections of the Western Cleddau, the first extending for 500 m downstream of St Catherine's Bridge (Site 5 in Fig. 17.1), and the second extending for 100 m upstream from the small road bridge in Wolf's Castle (Site 3 in Fig. 17.1). These sites are separated by a distance of more than 8 km.

The 500 m Section at St Catherine's Bridge

This section comprises shallow riffles (<1 m deep) and deeper pools (up to 1.5 m deep). The river is about 13 m wide in this section (Fig. 14.1), and it was possible, with careful navigation, to complete the recording without leaving the river. Despite this, most surveyors got out of the river on at least one occasion to avoid having to negotiate the deeper sections of channel.

C. Hurford (✉)
Countryside Council for Wales, Woodside, Orielton, Pembroke, SA71 5EZ, UK
e-mail: clive.hurford@btinternet.com

C. Hurford et al. (eds.), *Conservation Monitoring in Freshwater Habitats: A Practical Guide and Case Studies*, DOI 10.1007/978-1-4020-9278-7_14,

Fig. 14.1 Part of the 500 m sampling trial section at St Catherine's Bridge. *Photo by Clive Hurford*

Fig. 14.2 Part of the 100 m sampling trial section at Wolf's Castle. *Photo by Clive Hurford*

The 100 m Section at Wolf's Castle

The river is about 8 m wide in this section and is shallow riffle habitat <1 m deep throughout its length (Fig. 14.2). All surveyors recorded the full 100 m section without leaving the river.

Methods

The data were collected on seven dates from 23rd June to 30th September 2008, within the recommended period for macrophyte recording in the UK. On four dates, two surveyors were in the river at the same time, though working independently. On the other three dates, a lone surveyor was accompanied by a non-participating colleague for health and safety purposes.

Eleven experienced professional botanists participated in the sampling trial. These included three accredited freshwater specialist surveyors, six 'non-specialist' surveyors who had previously received Mean Trophic Rank (MTR) training, and two surveyors who had never received training in freshwater macrophyte sampling methods. All participants were familiar with the aquatic flora of the region.

For the data collection exercise, we asked the surveyors to record cover estimates for all higher and lower plants (mosses and liverworts), lichens and algae present in the river channel. To minimise the scope for observer interpretation of the term 'river channel', we asked the surveyors to record 'all macrophytes submerged or partly submerged in the river within the survey length'. We modified an existing macrophyte recording form to meet the purposes of the survey. The surveyors were also asked to record how long was spent surveying each stretch of river.

As the exercise focused primarily on species detection, as opposed to species identification, the surveyors could take samples away with them and submit their recording forms after difficult specimens had been verified (by an appropriate referee if necessary). Similarly, there was no time limit on the exercise; we asked the surveyors to stay on site until they were satisfied that the sample was complete.

We excluded the bank flora from the sampling trials, as the levels of observer variation associated with recording terrestrial vegetation are already well documented (Leach and Doarks 1991; Hurford 2006).

Finally, we asked the surveyors to collect data only under optimal conditions during the recommended sampling period for river macrophyte surveys i.e. during periods of low flow and good water clarity in the period from June to September.

Issues Associated with Collating the Data

As the sampling trials focused on recording the diversity and cover of aquatic and emergent species, we removed all other species from the dataset following advice provided by Nigel Holmes and Richard Lansdown. We then collated all of the surveyors' data into a master dataset.

If there was any doubt over the presence of a species in the river sections, we removed that species from the master dataset that we used to determine species detection rates. We did not, however, remove any aquatic or emergent species from the individual surveyors' datasets, as these are the data that they would have presented under normal circumstances.

Finally, as we believe that all of the Batrachian *Ranunculus* vegetation in the river was hybridised, we lumped all *Ranunculus* records into a single indeterminate group. The *Ranunculus* was variably recorded by the surveyors as *R. fluitans*, two subtypes of *R. penicillatus*, *Ranunculus* hybrid and as indeterminate *Ranunculus*.

Results

The sections below outline the results of the sampling trials, focusing on the species detection rates and cover estimates in the 500 m and 100 m sections of river.

Detection Rates for Aquatic and Emergent Plants in the 500 m Section

At least 59 aquatic and emergent species were present in the 500 m section, comprising 12 species of algae and lichen, 14 species of bryophyte and horsetail, and 33 species of vascular plant. Table 14.1 highlights those species (excluding algae) with high detection rates (>75%), and those with low detection rates (<20%).

Detection Rates for Aquatic and Emergent Plants in the 100 m Section

At least 48 species of aquatic and emergent plant were present in the 100 m stretch of river. These comprised 13 species of algae and lichen, 14 species of bryophyte and 21 species of vascular plant. Table 14.1 highlights those species with high and low detection rates. Figure 14.3 shows the distribution of detection rates for species in this section.

Cover Estimates

The range of observer variation associated with the estimates of vegetation cover in each section is outlined below.

Cover Estimates for Aquatic Species in the 500 m Section

Only two species were recorded as achieving more than 1% cover in this section, these were the various forms of *Ranunculus* recorded (grouped here as *Ranunculus* sp.) and *Verrucaria* sp. The range of observer variation associated with the cover

Table 14.1 Vascular plant and bryophyte species with detection rates >80% (*left*) and <20% in the 500 m and 100 m river sections. Note that *Juncus effusus* and *Iris pseudacorus* appear on both sides of the table, reflecting their relative abundance in the two sections

Species with a high detection rate of >80%	Detection rate (%) 500 m	100 m	Species with a low detection rate of <20%	Detection rate (%) 500 m	100 m
Agrostis stolonifera	82		*Alnus glutinosa*	9	
Apium nodiflorum	91	91	*Conocephalum conicum*	9	9
Callitriche brutia		91	*Epilobium parviflorum*	9	
Chiloscyphus polyanthos	91	100	*Equisetum palustre*	18	
Fontinalis antipyretica	100	100	*Fissidens curnovii*		18
Fontinalis squamosa	100	100	*Fontinalis antipyretica* var. *gracilis*		9
Glyceria fluitans	100	91	*Glyceria notata*	18	9
Iris pseudacorus	91		*Glyceria x pedicellata*	9	
Juncus effusus	82		*Amblystegium* sp.		18
Myriophyllum alterniflorum	91	91	*Hygrohypnum* sp.	18	9
Oenanthe crocata		91	*Iris pseudacorus*		18
Persicaria hydropiper	100		*Juncus acutiflorus*	9	
Phalaris arundinacea	100	100	*Juncus effusus*		9
Ranunculus sp.	100	100	*Lejeunea lamacerina*	9	
Sparganium emersum		82	*Lemna minor*		9
Sparganium erectum	100		*Lunularia cruciata*		9
			Mentha aquatica	18	
			Myosotis sp.	18	18
			Pellia epiphylla	9	9
			Porella pinnata	9	18
			Salix cinerea	9	9
			Salix viminalis		9
			Scirpus sylvaticus	9	
			Stachys palustris		18
			Veronica beccabunga		9

estimates for these species is shown in Table 14.2. No seasonal pattern was evident in the dataset: if there was a significant reduction of *Ranunculus* cover throughout the recording period, it was hidden within the range of observer variation present in the dataset.

Cover Estimates for Aquatic Species in the 100 m Section

Six species were recorded as achieving more than 1% cover in the 100 m section at Wolf's Castle: these were *Vaucheria* sp., *Chiloscyphus polyanthos*, *Fontinalis antipyretica*, *Fontinalis squamosa*, *Myriophyllum alterniflorum* and *Ranunculus* sp.

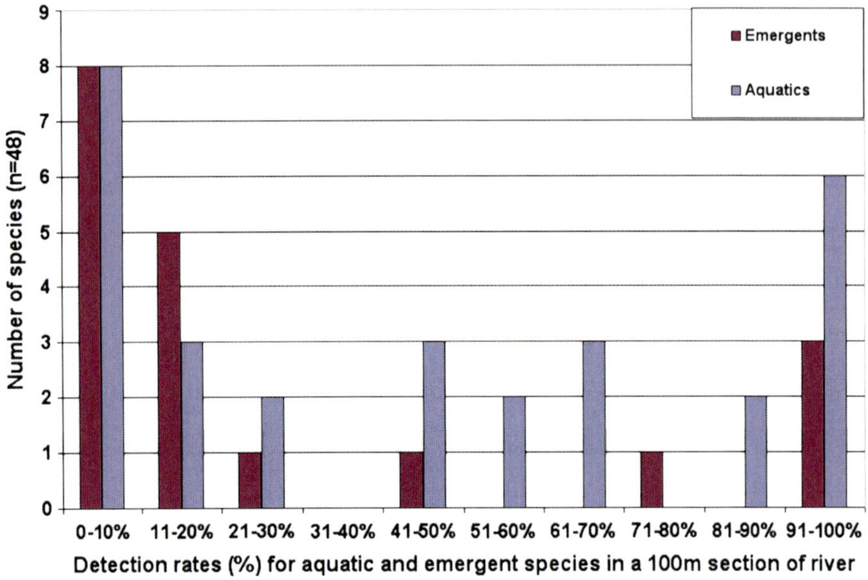

Fig. 14.3 The detection rates for aquatic and emergent species in the 100 m section at Wolf's Castle. Half of the species have a less than 20% chance of being detected

Table 14.2 The range of observer variation associated with the cover estimates recorded for aquatic species in the 500 m section of river at St Catherine's Bridge

	Minimum cover		Maximum cover		Range of variation	
Species	%	Area (m^2)	%	Area (m^2)	%	Area (m^2)
Ranunculus sp.	4	260	40	2,600	36	2,340
Verrucaria sp.	<1	<65	60	3,900	59	3,835

Of these, the major cover-formers were the two species of *Fontinalis*, *M. alterniflorum* and *Ranunculus* sp. The range of observer variation associated with the cover estimates for these species is shown in Table 14.3.

Estimating the cover of the two species of *Fontinalis* presented a difficult challenge for the surveyors, who not only had to attempt to separate the species from the other bryophytes in the channel, but also had to try and separate them from each other.

The cover estimates for *Ranunculus* sp. ranged from 0.1–1% (<8 m^2) to 30% (240 m^2), with specialist surveyors recording both of these estimates on the same day (2nd July). All of the subsequent estimates fell within this range. There was no obvious seasonal pattern to the cover estimates, with 30% cover being recorded on 2nd July and 25% cover recorded on 30th September (see Fig. 14.4).

Table 14.3 The range of observer variation associated with the cover estimates recorded by the surveyors for aquatic species in the 100 m section at Wolf's Castle

	Minimum cover		Maximum cover		Range of variation	
Species	%	Area (m^2)	%	Area (m^2)	%	Area (m^2)
Ranunculus sp.	0.1–1	8	30	240	29	232
Myriophyllum alterniflorum	0.1–1	<8	25	200	24	192
Fontinalis antipyretica	<1	<8	12	96	11	88
Fontinalis squamosa	<1	<8	10	80	9	72
Fontinalis sp.	<1	<8	15	120	14	112
Chiloscyphus	<1	<8	3	24	2	16
Vaucheria sp.	<1	<8	5	40	4	32

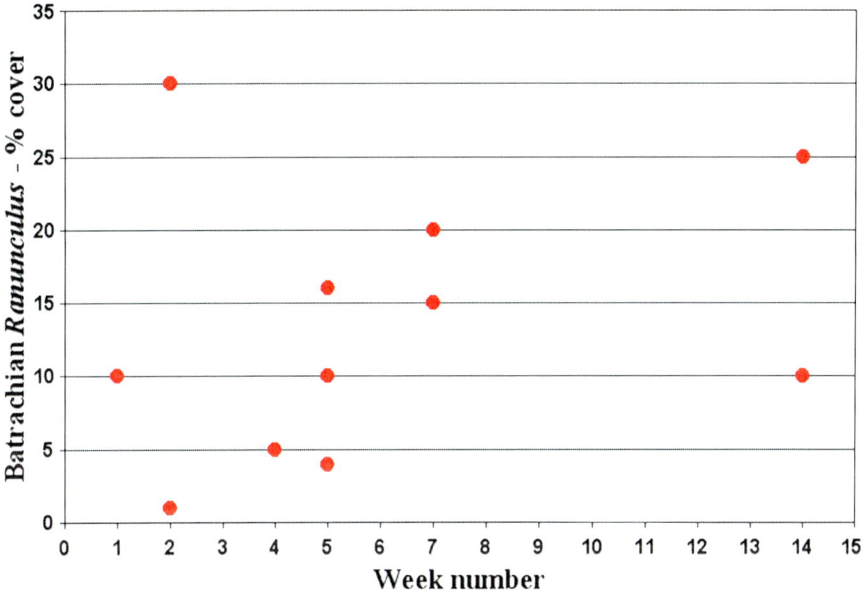

Fig. 14.4 The cover estimates recorded by the surveyors for Batrachian *Ranunculus* species in the 100 m section of river at Wolf's Castle. Note that the range of cover estimates recorded by accredited surveyors on the same day in Week 2 encompassed all of the cover values recorded by the other surveyors in the 14-week period from 23 June to 30 September

Variation in the Time Spent Collecting Data

The time that the surveyors spent recording the 500 m section ranged from 1 h 20 min to 3 h, while the time spent recording the 100 m section ranged from 35 min to 1 h 15 min. This equates to a minimum of 36 m^2 per minute for the 500 m section and a minimum of 11 m^2 per minute for the 100 m section. For comparison, an experienced terrestrial surveyor might spend 90 min recording a 2 × 2 m quadrat in grassland.

Discussion

The Western Cleddau sampling trials were carried out to assess the levels of observer variation associated with recording river macrophytes. All of the data used for assessing observer variation were collected under optimum conditions in the period recommended for collecting macrophyte data in the UK.

Although the data were collected on various dates in the period from 24th June to 30th September 2008, on four occasions data were collected by two different surveyors on the same day: these datasets allowed us to isolate examples of true observer variation from seasonal change.

Surveyor Performance in Recording Species Diversity

The results from the Western Cleddau sampling trials show a similar pattern to the results from similar exercises carried out in terrestrial habitats. For example, Leach and Doarks (1991) found that no surveyor recorded more than 73% of the species in a fixed 1 × 1 m quadrat in grassland vegetation, and that no surveyor recorded more than 63% of the species in a 10 × 10 m quadrat. By comparison, no surveyor recorded more than 64% of the aquatic and emergent species in the 500 m section on the Western Cleddau, and no surveyor recorded more than 54% of the species in the 100 m section.

With the exception of the freshwater algae and lichens, where the specialist surveyors consistently recorded more species than the non-specialists, there was no obvious difference between the specialist and non-specialist surveyors (Table 14.4).

However, despite their increased awareness of freshwater algae, there was little agreement between the specialist surveyors as to which species of algae were present: only four of the 11 species were recorded by more than one specialist surveyor.

Surveyor Performance in Recording Cover Estimates

With regards to observer variation in cover estimates, the results from the Western Cleddau trials again showed a similar pattern to the results from sampling trials in terrestrial habitats. Sampling trials in blanket bog vegetation (Hurford 2006) found that estimates of ericoid cover varied by a mean of 36% between observers. By comparison, estimates of vegetation cover for the main cover-forming species in the Western Cleddau varied by a mean of 24% between observers, with no difference between the non-specialist and specialist surveyors (Table 14.5).

Table 14.4 The mean numbers of species recorded in each section by the non-specialist surveyors and the accredited specialist surveyors

Species group	Mean number of species recorded in the 500 m section	
	Non-specialist surveyors	Accredited specialists
Aquatic algae and lichens	3	5
Aquatic bryophytes	5	5
Aquatic vascular plants	8	7
Emergents	12	15
Total	28/59 (47%)	32/59 (54%)
Species group	Mean number of species recorded in the 100 m section	
	Non-specialist surveyors	Accredited specialists
Aquatic algae and lichens	4	7
Aquatic bryophytes	5	6
Aquatic vascular plants	6	5
Emergents	7	4
Total	22/48 (46%)	22/48 (46%)

Table 14.5 The range of cover estimates recorded by the non-specialist and specialist surveyors for the major cover-forming species in the 100 m and 500 m sections of river

Species	Range of cover estimates recorded in the 500 m section	
	Non-specialist surveyors	Accredited specialists
Batrachian *Ranunculus*	4–40%	7–25%
Species	Range of cover estimates recorded in the 100 m section	
	Non-specialist surveyors	Accredited specialists
Batrachian *Ranunculus*	4–25%	1–30%
Myriophyllum alterniflorum	4–20%	1–25%

Conclusions

The sections on surveyor performance illustrate that there was no appreciable difference between the accredited specialist and non-specialist surveyors, either in terms of their ability to record species diversity or estimate vegetation cover. The fact that no surveyor recorded more than 54% of the aquatic and emergent species in a shallow 100 m section of river suggests that macrophyte data collected from both 100 m and 500 m sections of river are likely to be seriously compromised by observer variation.

Half of the species in the 100 m section had a low detection rate (<20%), including *Iris pseudacorus*, *Juncus effusus*, *Lemna minor*, *Stachys palustris* and *Veronica beccabunga*: these species do not present identification difficulties. Neither does *Potamogeton perfoliatus* (Fig. 14.5), which was overlooked by almost half of the surveyors (including two of the three specialists) in the 500 m section of river.

Fig. 14.5 Only 55% of the surveyors recorded this 7 × 2 m patch of *Potamogeton perfoliatus* in the 500 m section at St Catherine's Bridge, including only one of the three specialist surveyors. *Photo by Clive Hurford*

In practice, if a species was locally distributed in the survey sections it had a <20% chance of being detected. These detection rates reflect the inability of the surveyors, no matter how experienced, to critically survey the required areas of search: c. 6,500 m^2 for the 500 m section of river and c.800 m^2 for the 100 m section of river. In effect, the expertise of specialist surveyors is being nullified by the methods that they are being asked to use.

The implications of these sampling trial results for established macrophyte recording methods are discussed in Chapter 15.

Acknowledgements Thanks are due to all of the surveyors who risked their reputations to participate in the sampling trials, and in particular, to the accredited freshwater surveyors. Thanks also to Terry Rowell and Dan Guest for commenting on the early drafts of this chapter.

References

Hurford C (2006) Minimising observer error. In: Hurford C, Schneider M (eds) Monitoring nature conservation in cultural habitats: A practical guide and case studies. Springer, Dordrecht, the Netherlands, pp 79–92

Leach SJ, Doarks C (1991) Site quality monitoring methods and approaches (with particular reference to grasslands). Project No. 135. English Field Unit, Nature Conservancy Council, Peterborough

Chapter 15
The Implications of Observer Variation for Existing Macrophyte Recording Methods

Clive Hurford and Richard Lansdown

Introduction

The scale of observer variation recorded during the Western Cleddau sampling trials has implications for all of the established methods of river macrophyte recording, and in this chapter we look at the implications for two of the methods associated with Water Framework Directive (WFD) monitoring in the UK, namely Common Standards Monitoring (JNCC 2005) and the LEAFPACS macrophyte methodology (Willby 2005). We also look at the implications for Mean Trophic Rank (MTR) recording (Holmes *et al.* 1999a; Holmes *et al.* 1999b).

The Implications of Observer Variation for Common Standards Monitoring (CSM) River Macrophyte Data

Although we didn't record riverbank species in the Western Cleddau sampling trials, our datasets still allowed us to assess the potential impact of observer variation on the Common Standards Monitoring (CSM) methods for macrophyte survey.

The CSM approach differs to the MTR and LEAFPACS methods in four key ways:

1. The condition assessment relies heavily on reference to river community type, and on the data gathered during 'river-type defining' surveys in the early 1980s.
2. The surveyors must record all species, as they cannot be certain which species they will be assessing the river section against – again this will depend on the river type defined in the 1980s.

C. Hurford (✉)
Countryside Council for Wales, Woodside, Orielton, Pembroke, SA71 5EZ, UK
e-mail: clive.hurford@btinternet.com

R. Lansdown
45 The Bridle, Stroud, Glos, GL5 4SQ, UK

C. Hurford et al. (eds.), *Conservation Monitoring in Freshwater Habitats: A Practical Guide and Case Studies*, DOI 10.1007/978-1-4020-9278-7_15,

3. The area of search is a 500 m length of river, as opposed to a 100 m length.
4. The cover estimates are recorded on a three-point scale, as opposed to a nine-point scale.

The sample trial data from the St Catherine's Bridge river section were collected over the recommended 500 m survey length, so we have used these data to illustrate the potential impact of observer variation on CSM results.

Assessment of River Type

Based on data collected in 1980 (Holmes 1983), the river at St Catherine's Bridge was classified as community type BVd (Western, stable rivers on sandstone and shales). Perhaps a little surprisingly, during the 1980s surveys the absence of a species within a 500 m section of river was treated with the same significance as the presence of a species.

The master dataset from the 2008 sampling trials (derived from pooling the data from all 11 surveyors) also suggested that the river community type at St Catherine's Bridge was BVd. However, when we processed the individual datasets, we found that, depending on which surveyor's data we used, the river could have been classified as three different communities (and five different sub-communities), as follows:

- One surveyor would have classified the section as river type BVd.
- Seven surveyors would have classified the section as river type AIVa (Base-rich/neutral impoverished rivers, normally close to source).
- One surveyor would have classified the section as river type AIVb (Base-poor impoverished ditch communities).
- One surveyor would have classified the section as river type BVe (Lowland large rivers in south-west England and Wales).
- One surveyor would have classified the section as river type CVIIIa (Steep-gradient, low-altitude sand/shale rivers).

During the surveys of the 1980s, four community types were recorded along the length of the Western Cleddau. By contrast, during the 2008 sampling trials, depending on which surveyor's data you used to classify the section, three of these four river types were identified within the same 500 m section of river at St Catherine's Bridge. This suggests that observer bias can have a major influence on the identification of the river type (biased both by species familiarity and which parts of the river channel the surveyor looks at). It also suggests that all of the community and sub-community types identified by the sampling trial surveyors are different species groupings within the same broad river community.

Furthermore, as most rivers were classified on the evidence of a single dataset collected by a single surveyor, observer variation could have had a significant bearing both on the selection of sites for protection (Type AIVb does not qualify for protection) and, more recently, the results of the CSM condition assessments – which use the original 1980s river type to determine which species should be present in any given section of river. It is not difficult to see how a section of river conforming to one river type could be assessed against the criteria for a different river type.

A final thought on this matter. In 1980, a total of 39 aquatic and emergent species was recorded in two contiguous 500 m sections at Wolf's Castle: these two reference datasets were then used to define the river type. By contrast, in 2008 a total of 48 aquatic and emergent species was recorded in a 100 m section at Wolf's Castle, including all but one of the aquatic species recorded in the two 500 m sections in 1980. This suggests that the 1980s datasets used to classify the river were incomplete, and the dataset from St Catherine's Bridge illustrates all too clearly the impact that using incomplete datasets can have on the identification of the river type.

Cover Estimates in CSM

The cover estimates for CSM condition assessments are recorded on a three-point scale, where:

- 1 = <0.1% cover of the channel (river) or at its wetted margins (bank)
- 2 = 0.1 – 5.0% cover
- 3 = >5.0% cover

The cover estimates recorded by the surveyors during the 2008 sampling trials for *Ranunculus penicillatus* ssp. *penicillatus* in the 500 m section at St Catherine's Bridge are shown in Fig. 15.1. If we convert these cover estimates to scores on the

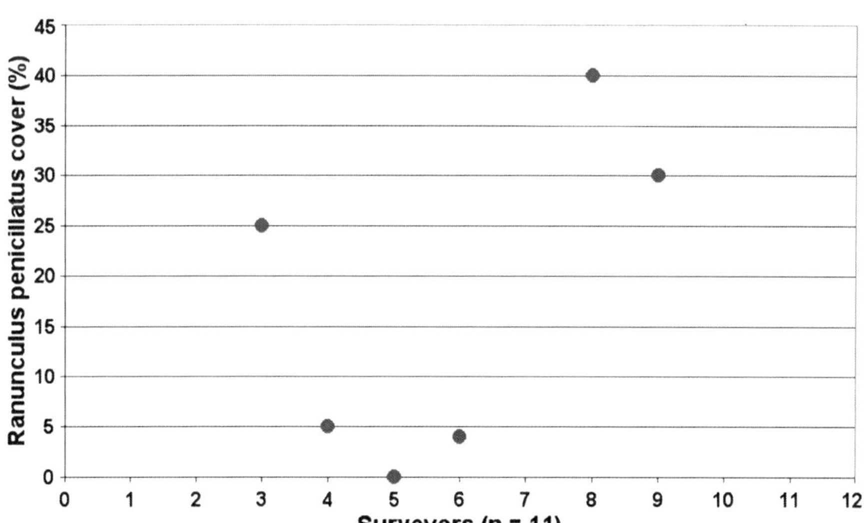

Fig. 15.1 Cover estimates recorded by surveyors for *Ranunculus penicillatus* in the 500 m section at St Catherine's Bridge. Surveyors 1, 2, 7, 10 and 11 recorded the *Ranunculus* under different names

three-point scale outlined above, we find that three surveyors would have posted a score of 2, and that three surveyors would have recorded a score of 3: the remaining surveyors recorded the *Ranunculus* under different names.

This suggests that, for practical purposes at least, the three point scale is actually a two point scale, where 1 = <0.1% and 2 = >0.1%. The obvious limitation of this is that major cover-forming species can undergo significant declines, e.g. from 40% channel cover to 0.1% channel cover without any means of detecting the loss.

CSM Assessment Results

Using the master dataset, even without recording the bank species, the 500 m section of river at St Catherine's Bridge passed the CSM criteria for a BVd river type. By contrast, no individual surveyor would have passed the river section against the CSM criteria. Major discrepancies on this scale could lead to unnecessary expenditure on management – as a direct consequence of observer variation.

Had we recorded the bank vegetation during the sampling trial, it is feasible that some of the surveyors' datasets would have passed the CSM criteria for macrophytes. However, this raises the separate issue of whether the presence or perceived absence of species such as *Agrostis stolonifera*, *Solanum dulcamara* and *Epilobium hirsutum* should determine favourable conservation status for the Ranunculion habitat, and by association, good ecological status for a Natura 2000 river.

The Impact of Observer Variation on LEAFPACS Data

The LEAFPACS methodology was developed to meet the requirements of the Water Framework Directive in the UK, and in some respects, it is similar to the MTR method, for example:

- The macrophyte data are collected from 100 m sections of river.
- Cover estimates are recorded on the same nine-point scale.

LEAFPACS differs from MTR in requiring surveyors to be familiar with and record 270 species, including many taxa that present significant identification problems.

During the Western Cleddau sampling trial, the effect of increasing the number of species contributing to the assessment was to increase the scale of observer variation associated with the dataset. For example:

Table 15.1 Headline results from the 100 m section at Wolf's Castle, after converting data from the Western Cleddau sampling trials into LEAFPACS and MTR data

	LEAFPACS	MTR
Minimum no. of species present	41	22
Maximum no. of species detected	21 (51%)	15 (68%)
Mean no. of species detected by specialists	18 (44%)	14 (64%)
Mean no. of species detected by non-specialists	17 (41%)	13 (59%)
No. of species with a detection rate >75%	11 (27%)	9 (41%)
No. of species with a detection rate of <20%	19 (46%)	6 (27%)
Maximum MTR score recorded		65
Minimum MTR score recorded		45

- No surveyor recorded more than 51% of the LEAFPACS species in the 100 m section.
- The specialist surveyors recorded a mean of 44% of these species.
- The non-specialist surveyors recorded a mean of 41% of the species.

The small disparity between the numbers of species recorded by the specialist and non-specialist surveyors (Table 15.1) can be attributed solely to the familiarity of the specialist surveyors with the increased number of algae species in the LEAFPACS list.

This increased awareness resulted in 11 species of algae being recorded in the section. However, only four species of algae were recorded by more than one of the specialist surveyors; these were *Cladophora glomerata, Lemanea fluviatilis, Nostoc commune* and *Vaucheria* sp.

Of the 41 LEAFPACS species recorded in the 100 m section, only 11 (27%) had a detection rate of >70%; these were *Vaucheria* sp., *Chiloscyphus polyanthos, Fontinalis antipyretica, Fontinalis squamosa, Apium nodiflorum, Callitriche brutia, Glyceria fluitans, Myriophyllum alterniflorum, Oenanthe crocata, Phalaris arundinacea* and *Sparganium emersum*. At the opposite end of the scale, 20 species (49%) had a detection rate of <20% (Fig. 15.2).

The range of values recorded for the cover-forming species in the 100 m section of river is shown in Table 15.2. These results suggest that we should allow a shift of five points on the nine-point scale to neutralise the effects of observer variation.

As the method of conversion from raw data to ecological quality ratio (EQR LEAFPACS) is not transparent, it is not possible to generate the EQR scores based on the sample trial data, and therefore not possible to assess the impact that these levels of observer variation would have on the EQR.

However, any statistical analysis is only as good as the data underpinning it, and it would be difficult to justify any statistical analysis that could accommodate apparent losses of >50% in species diversity and losses of >30% vegetation cover – simply as a consequence of observer variation. Shifts of this magnitude could have a significant impact on the conservation value of the habitat.

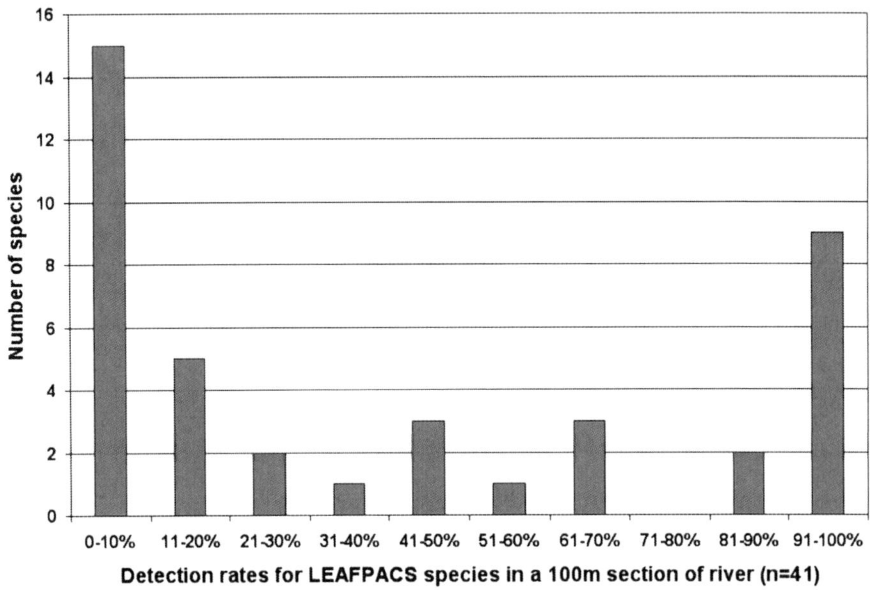

Fig. 15.2 Detection rates for LEAFPACS macrophyte species in the 100 m section at Wolf's Castle. Almost half of the species have a less than 20% chance of being detected by surveyors

Table 15.2 The range of cover value scores (CVS) recorded during the Western Cleddau sampling trials for the cover-forming species in the 100 m section at Wolf's Castle. This system (a nine-point scale) is used by both the LEAFPACS and MTR methods

Species	Minimum CVS	Maximum CVS
Vaucheria sp.	1	4
Fontinalis antipyretica	1	6
Fontinalis squamosa	1	5
Myriophyllum alterniflorum	2	6
Ranunculus sp.	2	7

The Implications of Observer Variation for MTR Data

Because MTR methods use a restricted suite of selected species to assess changes in trophic status, we would expect the effects of observer variation to be less pronounced than in more wide-ranging survey methods that require the detection of a larger number of species, e.g. LEAFPACS and Common Standards Monitoring (CSM).

The methods used to collect data during the Western Cleddau sampling trial allowed us to extract the relevant information and convert it to MTR data. The analysis of these MTR data is given below. As all 11 surveyors recorded eight or more of the key MTR indicator species in the 100 m stretch of river, all of the datasets would meet the criteria for 'high confidence' assessments.

MTR Data for the 100 m Section at Wolf's Castle

The critical points from the analysis of the MTR data for the 100 m section are shown in Table 15.1, in summary:

- At least 22 MTR species were present in this section of the river.
- No surveyor recorded >15 (68%) of the MTR species known to be present.
- The mean number of species recorded by the specialist surveyors was 14 (64%), compared to a mean of 13 (59%) recorded by the non-specialists.
- The MTR cover value scores (CVS) for cover-forming species varied by up to five points between observers (Table 15.2).
- The MTR scores recorded by the surveyors ranged from MTR 45 to MTR 65 (Fig. 15.3).

Only nine (41%) of the 22 MTR species had a detection rate of >75%. These comprised cover-forming aquatic species, obvious emergent species and species that were abundantly distributed, i.e. *Chiloscyphus polyanthos*, *Fontinalis antipyretica*, *Fontinalis squamosa*, *Rhynchostegium riparioides*, *Apium nodiflorum*, *Callitriche brutia*, *Myriophyllum alterniflorum*, *Oenanthe crocata* and *Sparganium emersum*.

At the opposite end of the scale, six (27%) of the MTR species had a detection rate of <20%: *Iris pseudacorus*, for example, was detected by less than 20% of the surveyors, and was overlooked by all of the specialist surveyors.

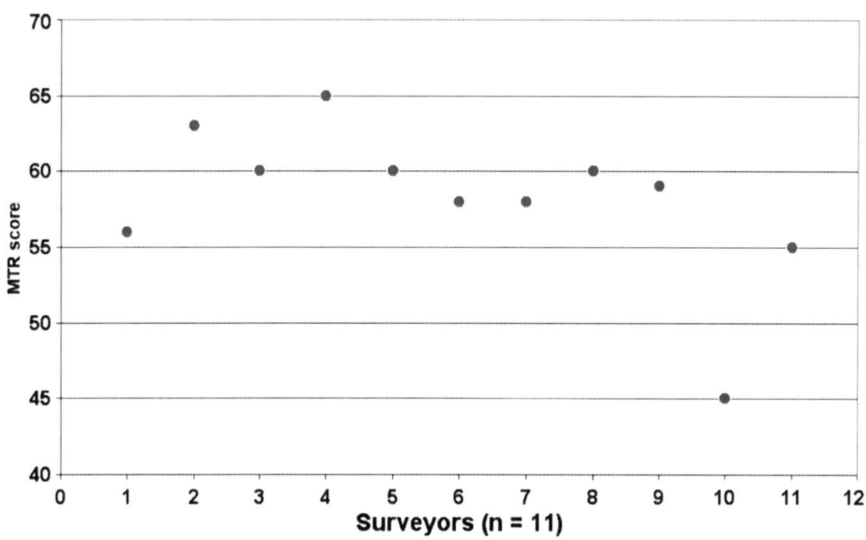

Fig. 15.3 The range of MTR scores recorded by surveyors for the 100 m section at Wolf's Castle: surveyors 9, 10 and 11 were accredited freshwater specialists

Discussion on the Impact of Observer Variation on MTR Results

The MTR scores generated by the surveyors for the 100 m section of the Western Cleddau varied by up to 21 points on the MTR scale (Fig. 15.3), from MTR 45 to MTR 65, with the specialist surveyor scores varying by up to 19 points (from MTR 45 to MTR 63). In the latter case, the surveyors who posted these scores carried out the exercise on the same day, so we can be certain that this was purely a consequence of observer variation, and not influenced by seasonal change.

Even if we exclude the outlier dataset (although recorded by a specialist surveyor), these results suggest that, to accommodate the range of observer variation associated with collecting MTR data, we need to allow 10 points either side of the recorded MTR score to be certain that any change at all has occurred. Consequently, if we are relying on a single dataset to determine the MTR we have no way of knowing whether the surveyor was:

- Scoring at the high end of the range (e.g. MTR 65), in which case, we would need to allow for variation of about 10 points lower than the recorded score
- Scoring at the low end of the range (e.g. MTR 55), in which case, we would need to allow for variation of about 10 points higher than the recorded score
- Scoring somewhere between these two values

The implications of this are not trivial. For example, if a surveyor recorded a score of MTR 60 in the first monitoring event, then we could only suspect that the true score was somewhere in the range of MTR 50 to MTR 70. If the surveyor in the second monitoring event then recorded a score of MTR 50, we could not know with any certainty whether the trophic status had:

- Shifted from MTR 70 to MTR 40, and hence become more eutrophic
- Shifted from MTR 50 to MTR 60, and hence become less eutrophic
- Remained more or less stable at MTR 60

To be certain of a meaningful change in the trophic status, we would need to see a change in the region of more than 15 points on the previous MTR score.

The Strengths and Weaknesses of the MTR Method

The main strength of the MTR method is that it is relatively robust. Despite considerable variation between observers in both the number of species recorded and in their respective cover values, most of the MTR scores fell within a range of 10 MTR points. In effect, this means that all of the sample trial results would have suggested that the section of river surveyed was 'mesotrophic, but at risk of eutrophication'. This consistency is primarily a consequence of the weightings allocated to the individual species.

Many of the species recorded in the section are deemed to be typical of mesotrophic conditions, and therefore have a similar Species Trophic Rank (STR) score. Consequently, adding up the STR scores and then dividing the total by the number of species recorded will always generate a similar figure – no matter how many species are recorded.

At the risk of stating the obvious, if you record ten species all with an STR of 7, the sum of the STR scores will be 70. If you then divide this by the number of species recorded you will be left with a score of 7. By contrast, if you record five species with an STR of 7, the sum of the STR scores will be 35. If you then divide this by the number of species recorded, again you are left with a score of 7.

Therefore, it makes no difference whether you record five species or ten species, the result will still be the same. In effect, the only thing that will have a significant impact on the MTR score is if several species with mesotrophic weightings are replaced by species with eutrophic or oligotrophic weightings.

The sampling trial results showed that even if you lost 55% of the MTR species in the river, the MTR score might move only two points – from MTR 56 to MTR 58. Similarly, the cover of a key species such as *Myriophyllum alterniflorum* could decline from 25% to 0.1%, with the MTR score moving only three points from MTR 58 to MTR 55. Therefore, while MTR may provide an indication of trophic status, it is not sensitive to changes in either species diversity or cover (real or apparent).

Summary

The Western Cleddau sampling trials suggest that high levels of observer variation are associated with all of the methods currently used for monitoring river macrophytes in the UK. The key findings were that:

- Running a river classification on the surveyor datasets from the 500 m section resulted in the section being classified as five different river types, with only one of the eleven datasets corresponding to the river type defined during the 1980s.
- The master dataset passed the CSM macrophyte criteria for Favourable Conservation Status, whereas the datasets of all eleven surveyors failed.
- The observer variation underpinning the LEAFPACS method was greater than that recorded for MTR.
- MTR scores for the 100 m section of river covered a range of 21 points.
- Only the more obvious species in the two survey sections were detected by more than 75% of the surveyors and, in both cases, almost half of the species in were detected by less than 20% of the surveyors.

These results should be a concern for anybody associated with recording river macrophytes, particularly as the 100 m data were collected from a stretch of riffle habitat that was <1 m deep, and in a straight section of river no more than 8 m wide. Furthermore, as the sampling trials took place under optimal conditions, the results represent the minimum range of variation that we might expect to find in macrophyte data collected by professional surveyors. Many river surveys are not carried out in optimal conditions.

Recommendations for Minimising Observer Variation in River Projects

There are several options available to us that will reduce the levels of observer variation associated with recording macrophytes in rivers.

1. Significantly reduce the area of search, by replacing 500 m and 100 m sections with much shorter sections, ideally located in shallow riffle habitats.
2. Clearly define both (a) the optimum recording period (Fig. 15.4) and (b) the optimum river conditions for repeat surveys – accepting that optimal conditions for macrophyte surveys are unlikely to occur annually.
3. Use quantitative methods to assess the area of channel covered by key macrophytes.
4. Use the co-occurrence of readily identifiable species (Fig. 15.5), combined with minimum area targets, as a means of habitat condition assessment.
5. Focus on a small number of species with high detection rates (fauna and flora) that are sensitive to the factors known to be impacting on the river.
6. Move away from the 'one size fits all' approach to macrophyte surveys and target different parts of the river for habitat-specific and species-specific monitoring projects.

The results of the Western Cleddau sampling trials raise serious questions over the suitability of all macrophyte methods in common usage to deliver either reliable or replicable monitoring results. If we persist in using these methods of data collection, all of the indications are that we are simply gathering data on the personal biases of the surveyors, which will mask all but the most dramatic changes in macrophyte extent and diversity in our rivers.

In the event of a much-needed review, the following should be considered good practice when developing methods for river macrophyte monitoring:

1. Carry out independent multiple-observer sampling trials to identify sources of observer variation before publishing guidance for collecting macrophyte data.
2. Use the sampling trial results to modify the methods and minimise the impact of observer variation, then carry out further multiple-observer trials to test the revised method.
3. State the margin of error needed to accommodate the levels of observer variation associated with the method when interpreting the survey/monitoring results.

Fig. 15.4 The top image shows the Western Cleddau at Welsh Hook in June 2006, while the image below shows the same section of river in September 2006. Both of these photographs were taken under optimal surveying conditions, at times of low flow, and within the recommended recording period for river macrophytes. If surveyors used quantitative measures to assess the vegetation cover, there would be a significant difference due to seasonal growth: the Western Cleddau sampling trial results suggest that the level of observer variation associated with using subjective cover estimates would mask these differences. *Photos by Clive Hurford*

Fig. 15.5 Although identification was not the main focus the Western Cleddau sampling trials, it was clear from the results that the identification of difficult species groups such as Batrachian *Ranunculus* (seen here) and *Callitriche* remains a problem. *Photo by Clive Hurford*

Acknowledgements Thanks are due to Terry Rowell and Dan Guest for commenting on the numerous draft versions of this chapter, and to Rhian Thomas for providing the 1980s datasets for the Western Cleddau.

References

Holmes NTHH (1983) Typing rivers according to their flora. Focus on nature conservation no. 4. Nature Conservancy Council, Peterborough

Holmes N, Boon P, Rowell T (1999a) Vegetation communities of British rivers: A revised classification. Joint Nature Conservation Committee, Peterborough

Holmes NT, Newman JR, Chadd S, Rouen KJ, Saint L, Dawson FH (1999b) Mean trophic rank: A user's manual. Environment Agency, Bristol

JNCC (2005) Common standards monitoring guidance for rivers. Joint Nature Conservation Committee, Peterborough

Willby NJ (2005) LEAFPACS methodology as provided by SEPA. Unpublished appendix to SEPA tendering documents

Chapter 16
Unitisation of Protected Rivers

Including Special Areas of Conservation and Sites of Special Scientific Interest

Chris Dyson

Introduction

Unitisation is the term used by the UK nature conservation agencies for the process of dividing a protected area into smaller compartments to allow management and by association, monitoring issues to be addressed at an appropriate geographical scale. For many terrestrial sites (e.g. mires, grasslands), ownership, field boundaries and broad habitat types are important factors in the unitisation process. However, in the case of rivers these factors are less applicable due to the typically large number of owners, each having comparatively little influence on site condition, and the general continuity of habitat, lacking clear internal boundaries. A suggested approach to the unitisation of river sites is given here using the example of the River Usk Special Area of Conservation (SAC) in south east Wales. The River Usk SAC is designated principally for its important populations of twaite shad *Alosa fallax*, sea lamprey *Petromyzon marinus*, river lamprey *Lampetra fluviatilis*, brook lamprey *L. planeri*, Atlantic salmon *Salmo salar*, bullhead *Cottus gobio* and European otter *Lutra lutra*. Secondary reasons for designation are the presence of allis shad *A. alosa* and river habitats dominated by *Ranunculus* vegetation.

An important general principle is that unitisation enables more effective management of harmful impacts on the protected area and is therefore not a purely academic exercise. The process also enables the site manager to focus the monitoring effort for the key species and habitats in the most appropriate sections of the river. The site manager's knowledge of the distribution of species and the variation of ecosystems across the site should be combined with information on the spatial occurrence of adverse impacts to define appropriate management units. As unitisation is intended to provide a practical framework for management and monitoring, understanding when to stop once an optimal number of units has been identified is important. In a large river site, the aggregation of several tributaries within a single unit provides

C. Dyson (✉)
Countryside Council for Wales, Eden House, Ithon Road,
Llandrindod Wells, Powys, LD1 6AS, UK
e-mail: c.dyson@ccw.gov.uk

C. Hurford et al. (eds.), *Conservation Monitoring in Freshwater Habitats: A Practical Guide and Case Studies*, DOI 10.1007/978-1-4020-9278-7_16,

useful information on broad-scale similarity, which may be obscured by finer subdivision. On the other hand, there should be no prescribed maximum number of units; the decisions should be made on a site-specific basis. A coding system that uses a hierarchy of 'units' and 'sub-units' (e.g. unit 10a, unit 10b) can help to retain useful information about larger scale patterns. As each 'unit' and 'sub-unit' has a unique identifier within the site management system, there can be no spatial overlap between these nominal types.

The Basis for Setting Unit Boundaries

Site unit boundaries may be based on a variety of factors affecting management including the location of adverse impacts, the known and predicted distribution of designated species and habitats, and administrative considerations such as the boundaries of other protected areas included within the site. The rationale for applying each of these factors to unitisation is considered in more detail below.

Artificial Impacts on the Site

Adverse artificial impacts on site condition are generally the most important factor in defining units. In the River Usk SAC, these impacts comprise major weirs forming barriers to the movement of fish, and major water supply abstraction points impacting on flows in the river downstream. Further units or sub-units may be defined where additional, more localised management issues come to light, for example reaches that are heavily impacted by diffuse pollution.

Figure 16.1 shows the River Usk SAC units and the man-made impacts and other factors used to define unit boundaries in the main river. Large weirs and major water abstractions significantly affect the ecology of the river both upstream and downstream. Weirs and bridge sills form a physical barrier to the upstream migration of anadromous fish such as shad and Atlantic salmon, while the abstraction points are capable of significantly depleting flows and water levels downstream with resulting effects on fish and in-stream vegetation. The bridge sill at Crickhowell for example, is identified as a cause of the unfavourable status of shad in the SAC, specifically in Unit 5; while the public water supply abstraction at Usk is at risk of affecting the upstream migration of shad and reducing the area of suitable habitat for lamprey ammocoetes in the river margins downstream. The nature of the adverse effects, or the positive results from their mitigation, differ upstream and downstream, so it is helpful to consider the ecological status of these reaches separately. If mitigation measures are successful, the units may be combined where they have broadly the same ecosystem characteristics (see the following section).

The tributary rivers are grouped into units according to their position on the main stem, i.e. with reference to the boundaries described above for the main stem, and

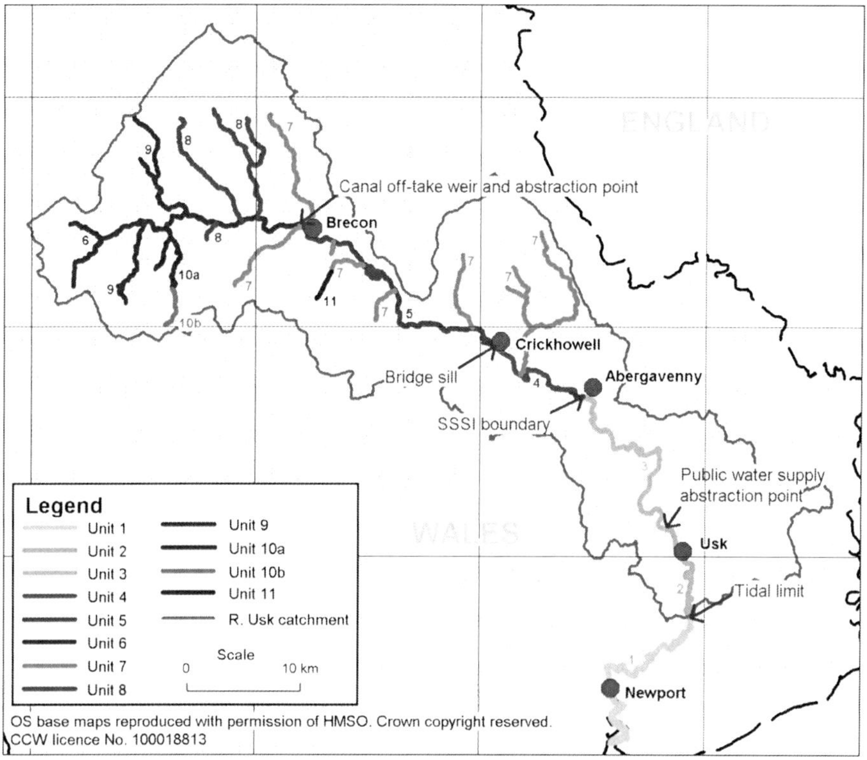

Fig. 16.1 River Usk SAC units and unit boundaries defined by man-made impacts, SSSI boundaries, confluences and the tidal limit. © *Crown Copyright and/or database right. All rights reserved. Licence number 100043571*

also on the basis of a provisional assessment of broad-scale differences in ecosystem characteristics between tributaries. Unit 7 includes all tributaries downstream of the weir at Brecon, on the basis that the impact of the weir on mobile species such as Atlantic salmon will be the same in adjacent tributaries as in the main stem.

Variation in Natural Ecosystems Across the Site

Within the River Usk SAC, units 8–10 are differentiated tentatively on the basis of ecosystem characteristics. Units 9 and 10 are separated from Unit 8 on the basis of the results from habitat suitability analysis shown in Fig. 16.2, supported by monitoring results (Hull International Fisheries Institute 2006), which suggest that the former tributaries (and limited areas within Unit 7) may be especially important for brook and river lampreys. The habitat suitability analysis is based on a

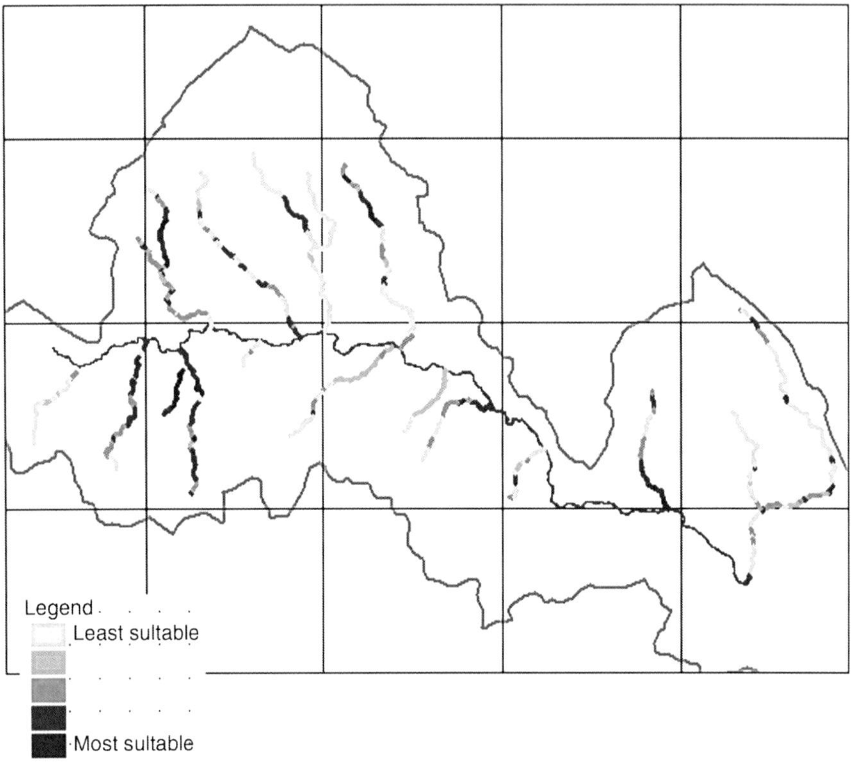

Fig. 16.2 Reach suitability for brook/river lamprey derived from hydromorphological data (GeoData, 2005).

combination of river habitat elements of importance to the species (Maitland 2003), including substrate and flow types, to give an overall suitability score at the reach scale. Monitoring for lampreys consists principally of recording the abundance of ammocoetes (i.e. juvenile lampreys) in a sample of silt deposits on the river bed (Harvey and Cowx 2003). Tributary-scale habitat differences need to be validated by further investigation, but it is considered beneficial to highlight them at this stage within the SAC Management Plan (http://www.ccw.gov.uk/landscape-wildlife/protecting-our-landscape/special-sites-project/river-to-usk-sac-list/river-usk-sac.aspx) so that the potential for differing management and monitoring requirements can be taken into account.

At a more detailed 'within tributary' or reach level, Unit 10b (the Upper Senni) represents the only reach in the Usk tributaries known to contain extensive stands of *Ranunculus*-dominated vegetation, suggesting it has relatively suitable hydromorphological conditions for the 'Ranunculion' Natura 2000 habitat (H3620). Large stands of *Ranunculus* occur in a meandering alluvial channel characterised by riffle and pool

Fig. 16.3 *Ranunculus* vegetation in upper Afon Senni, River Usk SAC. *Photo by Chris Dyson*

morphology and a shallow gradient (Fig. 16.3) compared to the steeper, higher energy streams that predominate in the Usk catchment. The relative stability of gravel bed material and water depth may be important factors favouring the vegetation here: there is no known difference in water quality between this and other tributaries.

River habitat types can be distinguished at the reach scale based on erosional and depositional features (e.g. gorge sections, floodplains) and these can be used to define units or sub-units where key species show a close association with a particular habitat type or where management actions are targeted at the reach scale to address a particular localised impact. The distribution of each designated species and habitat in the River Usk SAC has been linked with site units at a broad scale with varying levels of confidence.

Further work is needed to better understand the actual and potential distribution of the species and habitats at the reach scale, principally to help design more efficient monitoring programmes. In general, relatively little *direct* management of habitats and species is carried out in river sites compared to terrestrial sites. Management proposals are more likely to involve the management of areas of land adjacent to the river as part of catchment management initiatives. An analysis of land use and habitats on the Usk floodplain by the author has identified several reaches including the Upper Senni (Unit 10b), Upper and Mid Cilieni (part of Unit 9) and Pencelli floodplain (part of Unit 5) where there is a high cover and/or diversity of semi-natural habitats, which may be managed to help restore catchment hydrological processes. In these units, further sub-units may be defined to help focus the management on key catchment and riparian areas (CCW 2009).

Administrative Factors

Administrative considerations in ascribing SAC unit boundaries include the requirement for all component SSSI to be treated as separate units for reporting purposes: this predetermines the choice of some unit boundaries even though the ecological justification may be relatively weak (e.g. Units 4 and 11). In many cases the unit boundaries also coincide with water bodies designated under the Water Framework Directive (WFD): at the time of writing there is a greater number of WFD water bodies than SAC units defined for the River Usk. WFD water bodies are defined mainly by confluences in the river network. While these represent significant subdivisions in terms of flow and water quality, the ecological basis for this approach in terms of the distribution of suitable habitat for the designated species is less clear. Future revisions of the SAC unit boundaries may rely less on tributary confluences once the distribution of suitable habitat for the key species is better understood. Some species may prove to be relatively evenly distributed through the site, in which case the division into units would be more arbitrary in nature, or based only on impacts. Other species may show close associations with particular reaches within tributaries or the main river, such as the brook and river lampreys in Units 9 and 10.

The examples above illustrate how the assessment of site condition and ecological status depend on the selection of appropriate reference conditions with regard to hydromorphology in a spatially heterogenous river system. As techniques develop and are more widely applied, improved understanding of the hydromorphological basis for species distribution and potential distribution within a river system should lead to more confident assessments of condition and better targeting of management action and monitoring. This demands a multi-disciplinary approach involving

Fig. 16.4 Less affected by barriers in the main channel than Twaite Shad, the Atlantic Salmon *Salmo salar* is the best known of the fish species designated in the River Usk SAC. *Photo by Clive Hurford*

relatively new hydromorphological survey techniques (e.g. fluvial audits) combined with studies of species biology and ecology, alongside well-established techniques of environmental quality monitoring such as benthic invertebrate community indices.

Unitisation for Monitoring and Reporting

The integration of site unitisation into monitoring and reporting protocols raises issues concerning sampling design, and the analysis and presentation of results at the sample, unit and whole site levels. Site condition will be reported to stakeholders at the unit level in order to show the positive or adverse impacts of the prevailing management at that level, and to prompt positive management action where required. However, reports of habitat and species status at the whole site level must also be made periodically to the UK Joint Nature Conservation Committee. Therefore, the way in which localised reports of unfavourable condition at the unit level are handled in the context of the whole site must be considered. Where species population statistics are averaged over the whole site, an adverse result in one or more units may not affect the overall result for the SAC. On the other hand, certain localised impacts may need to be mitigated before favourable conservation status can be reported for some SAC habitats or species, in particular where they are considered to have an adverse impact on site integrity. Taking again the example of barriers to the migration of shad in the Usk, even though the population in downstream units may exceed threshold values individually or in combination, the status of the overall SAC population is likely to be considered unfavourable due to a reduction in species distribution and in the accessibility of suitable habitat. Restoration of shad to favourable condition in Unit 5 is therefore one of the objectives of management in the Usk SAC.

Conclusions

The approach taken to unitisation of rivers in south east Wales is not based on a rigid hierarchy of criteria (with the exception of compliance with component SSSI boundaries), but instead represents a flexible, evolving system designed to facilitate the achievement of favourable conservation status for the range of designated habitats and species and the accurate reporting of their status. On-going work is required to validate, further subdivide or aggregate the management units as knowledge and changing circumstances dictate. In doing so it is essential to consider the overall aims of habitat connectivity and coherence of ecosystem structure and function across the whole site. To this end, a decreasing number of site units may be considered an indicator of favourable management. Conversely, as knowledge of ecosystem heterogeneity across the site improves, the number of units may increase. The resulting optimal number of site units will aid the delivery of efficient programmes of conservation management and monitoring.

References

CCW (2009) Floodplain habitats in Wales and their potential for delivering multiple benefits: a pilot study in the Usk catchment using GIS. Published on CCW website http://www.ccw.gov.uk/idoc.ashx?docid=7ff147f6-6c61-4da5-aa66-54e2d65f6bdc&version=-1

Geodata Institute (2005) Fluvial audit of the Upper Usk tributaries. Unpublished report to CCW

Harvey JP, Cowx IG (2003) Monitoring the river, brook and sea lamprey, *Lampetra fluviatilis*, L. planeri and *Petromyzon marinus*. Conserving Natura 2000 Rivers Monitoring Series No. 5. English Nature, Peterborough

Hull International Fisheries Institute (2006) Monitoring of lamprey in the rivers Wye and Usk SACs 2005–2006. Unpublished report to CCW

Maitland PS (2003) Ecology of the river, brook and sea lamprey. Conserving Natura 2000 rivers ecology series no. 5. English Nature, Peterborough

Section 4
River Case Studies

Chapter 17
Monitoring the Ranunculion Habitat of the Western Cleddau: A Case Study

Clive Hurford and Dan Guest

Introduction

The Western Cleddau is a short river in the county of Pembrokeshire in south-west Wales. The main stem of the river extends for only 37.8 km, and has a catchment area of 294.5 km^2. The river is mesotrophic in nature and is underlain by siliceous geology. It rises at about 115 m above sea level and shows the characteristics of a lowland river throughout its length. The upper stretch of the river flows through a wide marshy valley, with numerous fens and small alluvial woodlands, while the lower stretches pass through land that is predominantly agricultural. In the past, the water authorities straightened and deepened the upper reaches, and further bank modifications have occurred along its length through to the town of Haverfordwest. A weir is situated immediately downstream of the town, below which the river becomes tidal.

The Western Cleddau is of international importance and forms part of the Afonydd Cleddau Special Area of Conservation (SAC) under the Habitats Directive regulations. There are three Annex I habitats within the SAC boundary:

- Watercourses of plain to montane levels with Ranunculion fluitantis and Callitricho batrachion vegetation, which is designated (H3620)
- Alluvial forests with *Alnus glutinosa* and *Fraxinus excelsior*, which is present but not designated (91EO)
- Active raised bog (7710)

The river also supports six Annex II species within the SAC boundary, these are:

- Otter *Lutra lutra*, sea lamprey *Petromyzon marinus*, brook lamprey *Lampetra planeri*, river lamprey *Lampetra fluviatilis* and bullhead *Cottus gobio*, all of which are designated
- Salmon *Salmo salar*, which uses the river but is not designated

C. Hurford (✉)
Countryside Council for Wales, Woodside, Orielton, Pembroke, SA71 5EZ, UK
e-mail: clive.hurford@btinternet.com

D. Guest
Countryside Council for Wales, Llys Tawe, King's Road, Swansea, SA1 8PG, UK

C. Hurford et al. (eds.), *Conservation Monitoring in Freshwater Habitats: A Practical Guide and Case Studies*, DOI 10.1007/978-1-4020-9278-7_17,

Defining the Ranunculion Habitat of the Western Cleddau

The initial phase of the project involved compiling background information on the flora and fauna of the river. This information, coupled with knowledge gathered from site visits, contributed to the following site-specific definition of the Ranunculion habitat of the Western Cleddau:

Fig. 17.1 The distribution of sampling sites along the main stem of the Western Cleddau: Site 1 in the north is near Letterston, Site 2 is at Welsh Hook, Site 3 is at Wolf's Castle, Site 4 is at Treffgarne and Site 5 is at St Catherine's Bridge.

Fig. 17.2 The macrophyte flora at Site 1, near Letterston. *Photo by Clive Hurford*

- Sections of river where water crowfoots *Ranunculus* spp. of the Batrachium sub-genus are dominant; or where
- Batrachian *Ranunculus* spp. are co-dominant with other aquatic macrophytes such as alternate-flowered water milfoil *Myriophyllum alternifolium*, water starworts *Callitriche* spp., floating bur-reed *Sparganium emersum*, and the aquatic bryophytes *Fontinalis antipyretica* and *Fontinalis squamosa* (Fig.17.2)

Typical Species Associated with the Ranunculion of the Western Cleddau

We decided that any condition assessment of the Ranunculion habitat should monitor both the aquatic flora and the fauna that we would expect to be associated with it. We considered several options during the process of selecting our list of 'typical species'. For example, kingfishers *Alcedo atthis* are unusually abundant on the river and dippers *Cinclus cinclus* also breed at several locations. However, while being typical of the river, neither of these species is strongly associated with the Ranunculion habitat, so neither would be appropriate for assessing its condition.

Similarly, despite regularly finding lamprey *Lampetra* sp. ammocoetes in the vicinity of the *Ranunculus* beds, this was simply because there were patches of silt nearby.

Fig. 17.3 Some of the attributes that we used to monitor the typical species associated with the Ranunculion habitat of the Western Cleddau. Clockwise from top left, these are brown trout *Salmo trutta*, mayfly (Heptageniidae) larvae, adult banded demoiselle *Calopteryx splendens* damselfly and an otter *Lutra lutra* spraint. *Photos by Clive Hurford*

It would be misleading, therefore, to suggest that these fish are directly associated with the Ranunculion habitat.

We would, however, expect to find bullheads *Cottus gobio* and salmonids, e.g. brown trout *Salmo trutta*, using the *Ranunculus* beds as a source of cover (Davies *et al.* 2004), and to find signs of *L. lutra* activity nearby (Fig. 17.3).

We also felt that the freshwater invertebrates associated with the Ranunculion are an integral component of the habitat when it is in good condition.

The most visually obvious of these are the adult banded demoiselle *Calopteryx splendens* and beautiful demoiselle *Calopteryx virgo* damselflies, which use the beds of aquatic vegetation for display and egg laying, as well as for cover and feeding during the two years that they spend as larvae. Furthermore, as both of these species are sensitive to pollution (Brooks and Lewington 1997), they will provide an additional layer of confidence when assessing the condition of the habitat.

The list of typical species was completed with the inclusion of representatives from the benthic invertebrate fauna: Environment Agency staff advised us on which species are most indicative of clean water environments in the region.

The species we selected were all top-scoring, pure water, indicators from the RIVPACS method (Wright *et al.* 1984; Wright 2000). Logically, if enough of these highly sensitive species are present in the river, we can reasonably expect the less demanding species to be there.

The Baseline Monitoring

After defining the Ranunculion habitat of the Western Cleddau and selecting its typical species, we decided to sample the attributes listed in Table 17.1.

Sampling the Vegetation

After considering the options, we decided that Mean Trophic Rank (MTR) (Holmes *et al.* 1999) was the most appropriate of the existing macrophyte sampling methods for monitoring the Ranunculion vegetation. The main reasons for this were that:

Table 17.1 The attributes selected for monitoring the condition of the Ranunculion habitat of the Western Cleddau

Attribute	Definition	Assessment method
Vegetation	Aquatic vegetation dominated by *Ranunculus* spp., *Myriophyllum* spp., *Callitriche* spp., *Sparganium emersum* or *Fontinalis* spp.	MTR survey
Mammals	*Lutra lutra*	Check for signs of activity
Fish	*Salmo* spp.	Check for presence
	Cottus gobio	Check for presence
Damselflies	*Calopteryx splendens*	Check for presence
	Calopteryx virgo	Check for presence
Benthic invertebrate families	*Leuctridae* *Perlodidae* *Chloroperlidae* *Ephemerellidae* *Heptageniidae* *Odontoceridae* *Goeridae* *Brachycentridae* *Sericostomatidae*	Kick sample for presence

- Using the 100 m (as opposed to 500 m) recording unit would reduce the scope for observer variation and increase sampling efficiency.
- Using a restricted suite of 129 ecologically meaningful species would also reduce the scope for observer variation.
- The method recommends a semi-quantitative approach to assessing vegetation cover.

Despite reservations about the area of search, and the potential for overlooking locally distributed species in the channel, we believed that the MTR method offered the best chance of obtaining a replicable monitoring result.

As the Western Cleddau is generally a slow-flowing mesotrophic river throughout its length, we treated the main stem of the river as a single unit (see Chapter 16). This decision was supported by a series of site visits – where we found that Ranunculion vegetation was well distributed along the length of the river, wherever suitable conditions prevailed.

Against this background, we decided to sample five 100 m sections, distributed along the length of the river: downstream from near Letterston in the north to St Catherine's Bridge near Haverfordwest in the south (see Fig. 17.1). We used the following criteria to select the sections for monitoring:

- Ranunculion vegetation is abundant.
- The section is mostly unshaded and predominantly shallow riffle habitat.
- The site is easy to relocate.

During the course of sampling we used a high accuracy differential global positioning system (GPS) to record the OS grid reference for the start and end of each section. We then measured the width of the channel, and worked in pairs to collect the vegetation data, recording the presence of species as we walked upstream and making a semi-quantitative assessment of the area of channel covered by each species on our return downstream: this area estimate was subsequently assigned to the appropriate value on the nine-point cover/abundance scale used to generate an MTR score.

Sampling for the Typical Species

Some of the typical species were recorded during the course of the vegetation survey, e.g. we always recorded the presence of salmonids as we were sampling the aquatic vegetation. The same was often true of the *Calopteryx* damselflies, unless sub-optimal flying conditions prevailed. In this instance, we also checked the emergent vegetation for resting individuals.

After completing the vegetation survey, we sampled for the benthic invertebrates, using the standard method applied by the Environment Agency.

This involved:

- Carrying out three 1-min kick samples in riffle habitat, with the pond net held immediately downstream of the person doing the kick sample
- Carrying out a 1-min sweep of the other habitats in the section

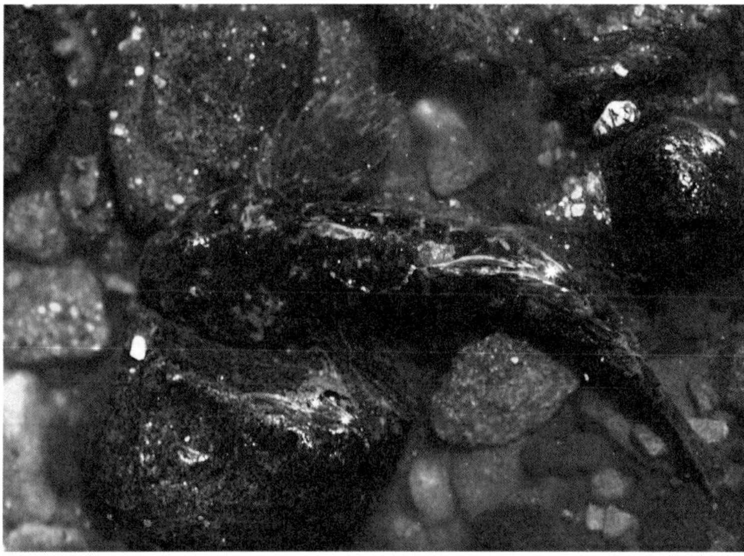

Fig. 17.4 Bullheads *Cottus gobio* are abundant in the Western Cleddau and are commonly found sheltering in the *Ranunculus* beds. *Photo by Clive Hurford*

- Emptying the pond net into a shallow white tray containing river water after each 1-min sample
- Undertaking the species identification after completing the sampling

At this point, one of the two surveyors would carry out the identification of the benthic invertebrates while the other checked the section for signs of *L. lutra* activity (spraints or tracks) and for *C. gobio* (if necessary). At most sites, *C. gobio* (Fig.17.4) was found when sampling for the benthic invertebrates, often while sweeping the *Ranunculus* beds.

On average, each section took two surveyors 2 h and 30 min to sample.

Results

The baseline sampling was carried out in July 2006, during an extended period of hot and dry weather that resulted in low flows and optimal sampling conditions. Table 17.2 shows the sampling results for each of the five sections.

The vegetation survey generated a mean MTR score of 58 across the five 100 m sections, suggesting that the river is generally of a mesotrophic nature, but at risk of eutrophication. On average, more than five of the pure-water benthic invertebrate families were recorded at the sampling sites, and representatives of the typical species groups were found at every site.

Table 17.2 The results of the Ranunculion sampling for the Western Cleddau (+ = Present)

Attribute	Site 1	Site 2	Site 3	Site 4	Site 5
Aquatic flora	MTR 57	MTR 55	MTR 53	MTR 65	MTR 58
Benthic invertebrates					
Leuctridae	+	+	+	+	+
Perlodidae		+	+		+
Chloroperlidae			+	+	+
Ephemerellidae	+	+	+	+	+
Heptageniidae			+	+	+
Odontoceridae			+		
Goeridae	+			+	+
Brachycentridae		+	+		+
Sericostomatidae			+		
Damselflies					
Calopteryx virgo	+	+	+	+	+
Calopteryx splendens					+
Mammals					
L. lutra signs	+	+	+	+	+
Fish					
Salmonids	+	+	+	+	+
Cottus gobio	+	+	+	+	+

Discussion

The main purpose of the baseline sampling exercise was to collect enough information to develop a set of condition indicators for the Ranunculion habitat of the Western Cleddau: these are shown in Table 17.3. In our opinion, the sampling results suggested that the Ranunculion habitat of the Western Cleddau was in a favourable state, and that the attributes we sampled could be used to provide evidence-based condition assessments for the habitat in the future.

Rationale Underpinning the Condition Indicators

The condition indicators in Table 17.3 take account not only of the data collected in July 2006, but also of the results from the observer variation sampling trials carried out in the summer of 2008 (Chapters 14 and 15).

The targets were set on the basis that the sampling would take place under optimum conditions in the month of July. In effect, the condition indicators state that the Ranunculion habitat of the Western Cleddau will be in favourable condition if:

1. There is sufficient channel cover of Ranunculion macrophytes.
2. The macrophyte species present suggest that the trophic status of the river is stable.

Table 17.3 The condition indicators for the Ranunculion habitat of the Western Cleddau

Condition indicator table	The Ranunculion habitat of the Western Cleddau will be in favourable condition when,	
Habitat extent	Lower limit	In each of Sections 1–5, during periods of low flow and good water clarity in the month of July: The major cover-forming aquatic plants cover >150 m^2 of river channel
Habitat quality	Lower limit	(1) Four or more aquatic mesotrophic indicator species are present in each of Sections 1–5 (2) On average, five or more clean-water benthic invertebrate families should be present in Sections 1–5, with no less than three families present in any one section (3) *Gammarus* spp. are present in all sections and *Asellus* spp. are rare or absent in all sections (4) Fresh signs of *L. lutra* activity are present in each section (5) Salmonids and *C. gobio* are present in each section (6) Either *Calopteryx virgo* or Calopteryx splendens (or both) is present in each section, and both species should be recorded in at least one section
Site-specific definitions		
Major cover-forming aquatic plants		Batrachian *Ranunculus* spp., *Myriophyllum alterniflorum*, *Callitriche brutia*, and *Fontinalis spp.*
Aquatic mesotrophic indicator plant species		Batrachian *Ranunculus* spp., *Myriophyllum alterniflorum*, *Callitriche brutia*, *Fontinalis squamosa*, *Chiloscyphus polyanthos*, *Lemanea fluviatilis*
Clean-water benthic invertebrate families		*Leuctridae*, *Perlodidae*, *Chloroperlidae*, *Ephemerellidae*, *Heptageniidae*, *Odontoceridae*, *Goeridae*, *Brachycentridae*, *Sericostomatidae*
Fresh signs of *L. lutra* activity		Tracks in silt or mud in the river channel, spraints still oily
Salmonids		*Salmo trutta* or *Salmo salar*
River channel		A gently sloping bed of substrate submerged under water
Rare		Less than five individuals per completed kick sample

3. Enough families of clean-water benthic invertebrates co-occur along the length of the river.
4. The fauna that we expect to be associated with the vegetation is present.

The lower limit for habitat extent focuses on the area of river channel covered by aquatic mesotrophic macrophytes, and takes account of the fish, invertebrates and mammals that we would expect to be associated with the macrophyte beds. The requirement for four or more of the mesotrophic indicator species to be present provides confidence that the trophic status of the river is remaining relatively stable:

all of these species have a Species Trophic Rank rating of 6–9 in the MTR methodology. Five of these six indicator macrophyte species have been shown to have high detection rates in observer variation sampling trials (Chapter 14). The remaining targets were informed by the results of the baseline sampling in 2006.

References

Brooks S, Lewington R (1997) Field guide to the dragonflies and damselflies of Great Britain and Ireland. British Wildlife Publishing, Hook

Davies CE, Shelley J, Harding PT, McClean IFG, Gardiner R, Pierson G (2004) Freshwater fishes in Britain: The species and their distribution. Harley Books, Colchester

European Commission (1999) Interpretation manual of European habitats. Eur 15/2. European Commission DG Environment, Brussels

Holmes NT, Newman JR, Chadd S, Rouen KJ, Saint L, Dawson FH (1999) Mean trophic rank: A user's manual. Environment Agency, Bristol

Hynes HBN (1997) Adults and nymphs of British stoneflies (plecoptera): A key. Scientific Publication No 17. Freshwater Biological Association

Wallace ID, Wallace B, Philipson GN (2003) Keys to the case-bearing Caddis larvae of Britain and Ireland. Scientific Publication No 61. Freshwater Biological Association

Ward D, Holmes N, José P (eds) (1994) The new rivers and wildlife handbook. Sandy, RSPB, NRA, WT

Wright JF (2000) An introduction to RIVPACS. In: Wright JF, Sutcliffe DW, Furse MT (eds) Assessing the biological quality of fresh waters. Freshwater Biological Association, Ambleside, pp 1–24

Wright JF, Moss D, Armitage PD, Furse MT (1984) A preliminary classification of running water sites in Great Britain based on macro-invertebrate species and prediction of community type using environmental data. Freshwater Biol 14:221–256

Chapter 18
Monitoring of *Cryphaea lamyana* on the Afon Teifi SSSI/SAC: A Case Study

Sam Bosanquet

Introduction

Multi-fruited river-moss *Cryphaea lamyana* is a patch-forming pleurocarpous riparian moss with a very restricted global distribution: this includes two rivers in Wales, five in south-west England (Holyoak 2002), and a handful of sites in France, Italy, Switzerland, Spain and Portugal. Its European distribution is Oceanic Southern-temperate (Hill and Preston 1998), although Goulborn and Franco (2006) suggest that its Italian and Swiss colonies indicate a Suboceanic distribution. Until recently, the only known Welsh population was on the Afon Teifi, on the borders between the counties of Carmarthenshire, Ceredigion and Pembrokeshire; a colony on the Afon Tywi was discovered in 2005 (Bosanquet *et al.* 2005). The Teifi colony – the largest in the UK – is a designated species of the Afon Teifi Site of Special Scientific Interest (SSSI). By contrast, the Tywi colony is not listed as a designated SSSI species, although it is protected by lying within a SSSI. The Afon Teifi is one of the largest rivers in west Wales, with a 94,000 ha catchment that includes the northern part of Carmarthenshire and most of southern and eastern Ceredigion. It rises in the Cambrian Mountains above the town of Tregaron and flows south-westwards through the raised bog complex of Cors Caron and past the university town of Lampeter before turning westwards and passing the towns of Llanybydder, Newcastle Emlyn and Cardigan. It is a famously back-to-front river, with slow-flowing, meandering upper reaches and faster-flowing, rockier lower reaches. It discharges into the southern part of Cardigan Bay, more than 120 km from its source.

The Teifi population of *Cryphaea* has been relatively well studied since its discovery by C.D. Preston in April 1978. Holyoak (2002, pp. 10–12) provides a list of previous records, although the true number of populations and visits is clouded

S. Bosanquet (✉)
British Bryological Society and Countryside Council for Wales, Cnwc y Llwyn, Brechfa, Carmarthen, Carms. SA32 7QR, UK
e-mail: Sam.Bosanquet@ccw.gov.uk

C. Hurford et al. (eds.), *Conservation Monitoring in Freshwater Habitats: A Practical Guide and Case Studies*, DOI 10.1007/978-1-4020-9278-7_18,

by duplications in this list, in the Biological Records Centre (BRC) dataset and on the Threatened Bryophytes Database (TBDB).

Several records come from a series of surveys by A. Orange and others in 1981 (Brooker 1982a,b), and these are combined with other more recent records in Orange (1993). Holyoak (2002) revisited most of the original sites, relocating *Cryphaea* each time, as well as searching unsuccessfully at five other localities. He also checked locations on five other rivers in Pembrokeshire and Carmarthenshire, but *Cryphaea* was absent. The first record from one of the Teifi tributaries, the Afon Tyweli, was made by Williams (2002). Finally, general local recording revealed additional colonies at Maesycrugiau and Pont Tyweli (Bosanquet *et al.* 2005) (Map 18.1). Despite this succession of different surveys, the only clear information on population extent is from the Tyweli (Williams 2002)): the other surveys provide more general statements, such as "frequent from the bridge to the falls" or a more useful but still imprecise "totals of 45 patches being examined on rock (only a fraction of those present) and 41 patches on bark". This lack of precise information means that it is almost impossible to be sure of anything but the most dramatic population trends.

The most precise way to gauge how the population of *Cryphaea* on the Afon Teifi is (or is not) changing would be to map and measure every patch. However, the fact that *Cryphaea* is relatively robust, and appears to be resilient to pounding by water, makes it difficult to justify the time and effort that this would involve. Nevertheless, as it is limited to so few rivers in the UK, it is important that conservationists do not 'take their eye off the ball' with this species: the loss of a single colony would have a serious impact on the total population. Holyoak (2002, p. 17) recommends "detailed population monitoring of marked patches", but monitoring plots that comprise several patches are probably more suitable for assessing the wider population.

Both Orange (1993) and Holyoak (2002) found that most patches of *Cryphaea* by the Teifi are on rock, in contrast to its near-constant use of riverside trees in England. Given this, monitoring the species on the Teifi needs to focus on the saxicolous colonies (Fig. 18.1), with epiphytic colonies examined in *lesser* detail. At this site, the large majority of the saxicolous patches are on near vertical or slightly overhanging rock faces, with flatter surfaces occupied either by mosses such as *Grimmia lisae* and *Pterogonium gracile*, or are scoured by walkers (who abound at this popular tourist spot). *Cryphaea* favours the north bank of the river: partly because of the greater availability of near vertical rock and partly because it appears to be tolerant of desiccation (this is supported several of the other bryophytes on the rocks at Cenarth favouring insolated rocks in Britain, including *G. lisae*, *P. gracile* and the liverwort *Porella obtusata*). Orange (1993, pp. 2, 3) described the zonation of mosses and liverworts on these riverside rocks: *Cryphaea* occurs in a zone dominated by *Cinclidotus fontinaloides* and *Schistidium rivulare*, with occasional *Porella pinnata* and *Thamnobryum alopecurum*; above a *P. pinnata*-dominated band in which *C. fontinaloides*, *Fontinalis squamosa* and *Rhynchostegium riparioides* are occasional; and below a very rarely flooded zone where *Brachythecium plumosum*, *T. alopecurum*, *P. gracile* and *G. lisae* are the principal species. This suggests that it is important for the monitoring plots to focus only on the potentially suitable zone,

Map 18.1 The enlarged inset area (*bottom*) shows the distribution of *Cryphaea lamyana* on the Afon Teifi (*black dots*) related to towns & villages (*grey*) and administrative areas. This area in shown in the context of Wales as a whole (*top*).

Fig. 18.1 *Cryphaea lamyana* on rocks at Llandysul. The distinctive curled appearance of the shoots is characteristic of the species when dry: the optimum monitoring period would be when the species is in this state. *Photo by Clive Hurford*

rather than extending upwards into the *P. gracile* zone or downwards into areas with *F. squamosa*.

The ability of *Cryphaea* to cope with aquatic pollution was investigated by Goulborn and Franco (2006). Their analyses of data from Environment Agency monitoring stations included records of pH, ammonia, nitrates, nitrites and orthophosphates from English and Welsh rivers between 1999 and 2003, as well as data on metal ions and water hardness from the five English rivers. One Principal Component Analysis (PCA) showed a negative correlation between the abundance of *Cryphaea* and ammonia, nitrate and Zinc levels, whilst another PCA showed negative correlation between eutrophication and the abundance of the moss. This inability to grow in nutrient-enriched environments makes monitoring *Cryphaea* all the more essential. The presence of a sewage treatment works 400 m upstream of the Llandysul colony may be significant, especially as the stronger Cenarth colony is more than 4 km from the nearest treatment works discharge. The Welsh rivers on which *Cryphaea* grows are not thought to have been subjected to high inputs of chemicals, metals or nutrients because the stretches where it occurs are upstream of all but the smallest of urban centres. Furthermore, the main nineteenth century industry in the Teifi Valley was milling of corn and wool (Lloyd *et al.* 2006): neither of these would have caused much pollution. These mills included a number that overhang the river at Pont Tyweli and overlook the *Cryphaea* colonies on the Llandysul bank. It is possible, however, that other large rivers within the broad geographic area that is suitable for the species, for example the River Neath, may have lost it because of pollution during the Industrial Revolution and intervening decades of heavy industry.

Methods

Baseline Survey

The baseline survey focussed on the two largest populations of *Cryphaea* on the Afon Teifi: on rocky stretches of the river at Cenarth and at Llandysul. Initially the population at Henllan was thought to be of similar stature, but a visit in 2007 indicated that it is relatively small. However, providing that the Cenarth and Llandysul populations survive and produce sporophytes, there is the potential for recolonisation at the minor sub-sites.

The set-up phase of the baseline survey involved an initial site visit to take a series of oblique digital photographs from vantage points on the banks opposite the *Cryphaea* colonies. We printed these images on waterproof paper in preparation for the survey and monitoring exercises. The overall distribution and size of the colonies, at both at Cenarth and Llandysul, was then mapped onto these images. Two surveyors were involved in the data collection: one on the bank with the *Cryphaea*, holding up cards to indicate patch sizes, and the other on the opposite bank, marking the location and patch size onto the photographs. We used four patch size categories: <5 scattered plants; dense patch <25 × 25 cm; dense patch >25 × 25 cm; extensive patch of scattered plants.

Monitoring

After completing the baseline survey, we used the information to select five locations for more detailed monitoring: using plots 200 cm wide × 100 cm high. The monitoring focussed on recording the extent of *Cryphaea* within plots (estimated using a 10 × 10 cm square); the largest patch of *Cryphaea* present within the plot; the abundance of *Thamnobryum* (thought to oust *Cryphaea* on shaded rocks); the extent of scoured rock (% with no bryophytes or lichens at all); and a basic estimate of reproductive success (sporophyte frequency) . Invasive vascular plants, notably ivy *Hedera helix* and Japanese knotweed *Fallopia japonica* (see below) are unable to grow in the inundation zone occupied by saxicolous *Cryphaea*, although *H. helix* can just about survive at the upper edge of this zone, so it was included as a possible future threat.

As trees are less regular hosts of *Cryphaea* on the Teifi than they are on the rivers of south-west England or on the Afon Tywi, monitoring the epiphytic colonies was considered less of a priority than monitoring colonies on rock. If resources were not an issue, then the ideal way to monitor the epiphytic colonies would be to identify all of the riverside trees within major *Cryphaea* zones with numbered tags (see Holyoak 2002, p. 22) and to record the extent and reproductive success of *Cryphaea* on each tree. This would have an additional advantage of precisely locating trees that support this protected species (Schedule 8 of the Wildlife & Countryside Act 1981): particularly as epiphytic colonies are much more easily destroyed than saxicolous ones. Nevertheless, this was considered too costly and time-consuming to be justifiable, so we focussed the monitoring on a selection of 10 trees that currently support *Cryphaea*: these were distributed within the two main monitoring areas. We monitored a similar range of attributes to those used for the saxicolous colonies, but the presence of *Fallopia japonica* was added as a negative attribute because both it and *Hedera helix* pose a threat to epiphytic colonies of *Cryphaea*. All of the host trees were photographed, but poor GPS coverage prevented us recording their precise locations.

Results

The Baseline Survey Results

A baseline survey was carried out on 28th & 29th August 2007 – bright, sunny days after the first dry week of a very wet summer. It was important to choose a bright, dry day because *Cryphaea* is more readily identifiable when dry, and because good light was essential for photographing the host trees in the shady river valley. The baseline survey results are shown in Table 18.1: the information gained during the baseline survey informed the decisions on where to locate the monitoring plots, and the extent attribute in the condition indicator table. The location and size of each patch of Cryphaea was plotted on oblique photographs of thc saxicolous habitat at both Cenarth and Llandysul (Fig. 18.2).

The Monitoring Plot Results

The water levels were sufficiently low to allow access to almost all of the visible populations of *Cryphaea* and were at least 20 cm below its lowest limit. Eight plots were recorded for monitoring the saxicolous colonies: a lower number than the 10 initially planned, due primarily to (a) the extremely sparse nature of the Henllan colony and (b) the inaccessibility of much of the south bank of the river at Llandysul, which prevented oblique photography. Ten trees, five each at Cenarth and Llandysul, were surveyed.

Table 18.1 The results from the baseline surveys carried out in 2007 at Cenarth and Llandysul

	Number and size of *Cryphaea lamyana* patches			
Site	Scattered shoots	<25 × 25 cm	>25 × 25 cm	>25 × 50 cm
Cenarth	15	40	6	9
Llandysul	12	18	1	3
Total	27	58	7	12

Fig. 18.2 *Cryphaea lamyana* patches on one section of rocks at Cenarth. Symbols are: circles (patch <25 × 25 cm); triangles (patch >25 × 25 cm; not present on this section); linked squares (extensive patches of scattered plants). *Photo by Clive Hurford*

Table 18.2 The results from the *Cryphaea lamyana* monitoring plots on rocks at Cenarth (✓ means that attribute passed the target, + = present)

	Plot 1	Plot 2	Plot 3	Plot 4	Plot 5
Extent of *Cryphaea lamyana* (cm^2)	4,300	800	1,100	2,100	3,700
Largest pure patch extent (cm^2)	2,400	400	100	2,000	3,100
Extent of *Thamnobryum alopecurum* (cm^2)	0	0	0	0	0
Extent of scoured (totally bare) rock (cm^2)	300	100	900	200	3,000
Sporophytes frequent (present on >25 shoots)?	✓	✓	✓	✓	✓
Ivy present on rocks?					
Plot metadata					
Plot 1. Tall rock face above river, *C. lamyana* much mixed with *Cinclidotus fontinaloides*					
Plot 2. Dry, lichen-dominated rock face set back from river					
Plot 3. Low *C. fontinaloides*-dominated rock face by river					
Plot 4. Large patches of *C. lamyana* with abundant *C. fontinaloides* above the river					
Plot 5. Extensive patches of *C. lamyana*, with much bare rock, above a narrow ledge					

Saxicolous Colonies: Cenarth

The results from the saxicolous monitoring plots at Cenarth (Table 18.2) show that the population is clearly thriving, and few threats to it are evident. It grows on sufficiently steep, inaccessible rocks that there is very little chance of accidental damage by visitors. Potentially the most damaging factors here are increased shading, change in water levels or aquatic pollution, all of which should result in effects that are visible in monitoring plots. The absence of *Thamnobryum alopecurum* from all five plots suggests that trees on the south bank of the river do not cast enough shade to damage the *Cryphaea*, but they might need to be cut back at some point in the future if they grow too tall.

Cryphaea is clearly able to withstand scouring by the river because the largest patches recorded are in Plot 5, where the extent of scoured rock is at its greatest. There are powerful spates on the Teifi almost every year (Fig. 18.3), and these

Fig. 18.3 Most, if not all, of the saxicolous colonies of *Cryphaea* at Cenarth (on the rocks on the left bank in this photo) would have been under water during this spate in November 2007. *Photo by Clive Hurford*

Table 18.3 The results from the *Cryphaea lamyana* monitoring plots on rocks at Llandysul (✓ means that attribute passed the target, + = present)

Attribute	Plot 6	Plot 7	Plot 8
Extent of *Cryphaea lamyana* (cm^2)	650	750	1,800
Largest pure patch extent (cm^2)	300	200	750
Extent of *Thamnobryum alopecurum* (cm^2)	0	0	700
Extent of scoured (totally bare) rock (cm^2)	600	0	800
Sporophytes frequent (present on >25 shoots)?	✓	✓	✓
Ivy present on rocks?			+
Metadata			
Plot 6. Patchy *Cryphaea* mixed with abundant *Cinclidotus fontinaloides*			
Plot 7. Low face set back from river; patchy *Cryphaea* and abundant *C. fontinaloides*			
Plot 8. Tall rock faces overhung by Bramble *R. fruticosus* and *H. helix*			

could easily cause some damage to the bryophyte flora. However, riparian mosses of rocky rivers have evolved to deal with pounding by water and rocks, and *Cryphaea* has clearly survived on the Teifi long enough to have an extensive range on the river. Overall, the results from the plots suggest that the situation at Cenarth is healthy.

Saxicolous Colonies: Llandysul

The Teifi is much more tree-lined at Llandysul than at Cenarth, which explains the presence of both *Hedera helix* and *T. alopecurum* in Plot 8 (Table 18.3). Both species are widespread throughout the stretch, whereas *Cryphaea* has rather a patchy distribution here, although access difficulties make it hard to assess its true extent. In general, the *Cryphaea* grows on rock faces that are set further back from the river at Llandysul than at Cenarth, which might make it more vulnerable to competition from other plants. A further threat comes from a canoe slalom course on this stretch of river. It is unclear whether canoeing takes place when river levels would push canoes against *Cryphaea* patches: liaison with the canoeing fraternity might confirm this. Such a robust moss would probably survive a degree of canoe damage anyway, given its ability to live through spates.

Epiphytic Colonies: Cenarth

Epiphytic colonies of *Cryphaea* are scattered upstream of Cenarth Bridge, but play a minor role in the overall population (Table 18.4). The steepness of the valley, which precludes the use of GPS units to relocate trees, makes monitoring the epiphytic colonies difficult. All five of the trees chosen for monitoring can be identified from photographs, but this is unsatisfactory: fixed markers along the path would be preferable.

Table 18.4 The results from the *Cryphaea lamyana* monitoring plots on trees at Cenarth (✓ means that attribute passed the target, + = present)

Attribute	Tree 1	Tree 2	Tree 3	Tree 4	Tree 5
Extent of *Cryphaea lamyana* (cm^2)	400	1,300	90	150	220
Largest pure patch extent (cm^2)	200	1,300	90	150	220
Sporophytes frequent (present on >25 shoots)?	✓	✓	✓	✓	✓
H. helix present on trunk?	Little		Little		
F. japonica present around base?					
Metadata					
Tree 1: Alder *A. glutinosa* cut off at 100 cm in park on south bank of river, just downstream of bridge					
Tree 2: the first *A. glutinosa* on the north bank upstream of the waterfalls					
Tree 3: small patch of *Cryphaea* on lowest part of curving Ash *Fraxinus excelsior* overhanging river					
Tree 4: *F. excelsior* roots 3 m downstream of *F. excelsior* just downstream of 5th revetted section of path					
Tree 5: base of *F. excelsior* by rock shelf just downstream of bend					

Table 18.5 The results from the *Cryphaea lamyana* monitoring plots on trees at Llandysul (✓ means that attribute passed the target, + = present)

Attribute	Tree 6	Tree 7	Tree 8	Tree 9	Tree 10
Extent of *Cryphaea lamyana* (cm^2)	3,300	120	200	550	500
Largest pure patch extent (cm^2)	1,100	110	200	200	250
Sporophytes frequent (present on >25 shoots)?	✓	✓	✓	✓	✓
H. helix present on trunk?	dead	+			
F. japonica present around base?	+	+			
Metadata					
Tree 6: both trunks of a large twin *F. excelsior* tree					
Tree 7: Y-shaped *F. excelsior* tree opposite grassy park					
Tree 8: small twin-trunked *F. excelsior* opposite Tyweli inflow					
Tree 9: tall *F. excelsior* where concrete path starts					
Tree 10: *F. excelsior* among Sycamores *Acer pseudoplatanus* at east end of sandy bay					

Tree 1 was included to highlight the vulnerability of epiphytic colonies to human interference: the host tree was felled at a height that left the *Cryphaea in situ*, but the difficulty of identifying individual trees makes it likely that other less obvious host trees could be felled at ground/river level, destroying the *Cryphaea*.

Epiphytic Colonies: Llandysul

The Llandysul population of *Cryphaea* includes a greater proportion of epiphytic colonies than the Cenarth population. Most trees support just a single patch, as was the case at Cenarth (Table 18.5). Shading of host trees by *H. helix* and *F. japonica* is more of a threat at Llandysul, and the monitoring should promote targeted removal of these species if any colonies are perceived to be threatened.

The Condition Indicators

The condition indicators listed in Table 18.6 have been informed by the data collected during the baseline survey and monitoring plot recording carried out in 2007.

Rationale Underpinning the Condition Indicators

The population of *Cryphaea* on rocks at Cenarth appears to be very strong, and the current survey shows no evidence of a decline since previous documented visits. The species grows on a good number of rock faces, and a total of 70 patches of varied size were recorded. It certainly appears to be in favourable condition. For favourable condition to be maintained, the distribution should continue to extend throughout the rocky section of the river from the waterfalls to downstream of Cenarth Bridge. It should include small patches (<25 × 25 cm), large patches and extensive scattered stands, although placing figures on how many of each is beset with problems as an extensive stand could fragment to produce five large patches and would then appear to

Table 18.6 The condition indicators for the *Cryphaea lamyana* populations on the Afon Teifi

Condition indicator table	The *Cryphaea lamyana* population on the Afon Teifi will be in favourable condition when:	
Extent	Lower limit	The extent of the populations mapped at Cenarth and Llandysul during the 2007 baseline survey ....
Quality	Lower limit	The saxicolous population at Cenarth comprises >7 extensive stands >4 large patches >35 small patches The area of *Cryphaea* is >1,000 cm^2 in at least four of the five monitoring plots *Cryphaea* sporophytes should be frequent in each of five monitoring plots .... The saxicolous population at Llandysul comprises >4 extensive stands, >4 large patches and >20 small patches. *Cryphaea* sporophytes should be frequent in each of the three monitoring plots The area of *Cryphaea* is >1,000 cm^2 in at least two of the three monitoring plots .... >5 trees host epiphytic colonies at each site
Site-specific definitions		
Extensive stands	Continuous patches of Cryphaea >50 × 25 cm	
Large patches	Dense patches of Cryphaea >25 × 25 cm	
Small patches	Dense patches of Cryphaea <25 × 25 cm	
Frequent	Sporophytes present on ≥25 *Cryphaea* shoots within a plot	

represent a population increase. Even allowing for this, the targets set out in the condition indicator table should suffice to provide an early warning of a decline in condition.

The Llandysul population is smaller than that at Cenarth, and does not appear quite as strong as Orange (1993) and Holyoak (2002) suggest. It may have declined, perhaps because of nutrient enrichment of the water or canoeing, although the imprecise nature of previous data makes it impossible to be certain. *Cryphaea* is currently absent from several apparently suitable rock faces, which combines with the low number of patches present to make the population here in unfavourable condition. As such, we set similar targets as for Cenarth, though taking the shadier situation at Llandysul into consideration. Future monitoring in this section may necessitate downgrading of these targets if they are thought to be unrealistic.

Setting targets for epiphytic colonies was relatively arbitrary, and we settled for maintaining the status quo. Therefore, if *Cryphaea* is lost from any tree, or if a host tree is destroyed, then we would need to consider a more detailed documentation of the epiphytic populations at Cenarth and Llandysul.

Discussion

The monitoring protocol for *Cryphaea lamyana* on the Afon Teifi in south Wales is a good example of how a distinctive lower plant species that grows in the riparian zone can be monitored using a combination of oblique photographs and monitoring plots. Monitoring more obscure species, or those that grow submerged throughout the year, would require some modification of the method used. The extent of the species on two stretches of the Teifi was documented more accurately than before; despite a good number of previous visits by conservationists and bryologists. Base-line mapping of all the saxicolous patches within a population allows future changes to be considered precisely, whereas anything other than a severe change to a colony might otherwise have gone unnoticed. Precise monitoring using well-localised plots, which are easily relocated by comparing rock shapes on the fixed-point photographs, should also reveal subtle changes, giving an opportunity for management changes to be effected before the species suffers a catastrophic decline. At present, the population of *Cryphaea lamyana* on the Afon Teifi as a whole appears to be robust, with few threats and with systems in place to identify most potential problems. However, there is a slight query over the strength of the population at Llandysul, which might have declined because of enrichment of the river.

Future work needs to examine how robust individual patches of *Cryphaea lamyana* are, especially after spate conditions bring tree trunks and rocks down the river. Its growth performance in dry and wet years could also be examined, as could its sporophyte production in different years. The apparent restriction of *Cryphaea* to rocky stretches of the river needs to be examined critically to establish whether its apparent absence from wide floodplains (Goulborn and Franco 2006) is genuine or is an artefact of biased bryological recording and the lack of footpaths. There is little doubt that further small colonies await detection.

Given the vagaries of the Welsh climate and uncertainties over the effects of climate change on river levels, it would be desirable to monitor fluctuations in water levels on the river: to identify any long-term trends, and to establish whether the vertical extent of *Cryphaea lamyana* is varying. Comparison of the height of the top and bottom of bands of *Cryphaea* with mean summer water levels, as suggested by Holyoak (2002), is fraught with difficulties, and measurements relative to a fixed point on the bank would be preferable.

Furthermore, it would be useful to investigate and model humidity levels in the Cenarth section of the Afon Teifi to establish whether this is critical to the growth and survival of *Cryphaea*. Changes in humidity through climate change or water abstraction could then be modelled in order to work out how vulnerable the species is to future drops in river level. Mills were used to harness the power of the Teifi in the past, and scientific evidence may be needed at some point in the future to show whether hydroelectric generators would impact on the *Cryphaea*.

Finally, we should keep an open mind as to whether *Cryphaea* occurs on other Welsh rivers. For a long time, it was assumed that this species was genuinely restricted to the Teifi: a feeling backed up by Holyoak's (2002) brief surveys of the Afonydd Gwaun, Cothi, Tâf and Eastern & Western Cleddaus. The discovery of *Cryphaea lamyana* on a stretch of the Tywi in 2005 indicates that it may have been overlooked elsewhere, not least because the Tywi colony is fairly small (35 patches/tufts) and is restricted to a short rocky stretch of the river in an otherwise open flood-plain. Further work on the Eastern and Western Cleddaus by S.D.S. Bosanquet (pers. obs.) has failed to reveal any more colonies, neither has recording on the Gwaun, but there is a distinct possibility that parts of the Cothi may be suitable, as may the Aeron in Ceredigion or the Dyfi on the Ceredigion/Montgomery/Meirionydd borders, although the latter would represent a considerable northward range extension.

References

Bosanquet SDS, Graham JJ, Motley GS (2005) The mosses and liverworts of Carmarthenshire. Privately published, Dingestow

Brooker MP (ed) (1982a) Conservation of wildlife in river corridors. Part I. Methods of survey and classification. Joint report by UWIST, Welsh Water Authority, RSPB, Otter Haven Project and NCC. Welsh Water Authority

Brooker MP (ed) (1982b) Conservation of wildlife in river corridors. Part II. Scientific assessment. Joint report by UWIST, Welsh Water Authority, RSPB, Otter Haven Project and NCC. Welsh Water Authority

Goulborn A, Franco M (2006) Factors affecting the presence and abundance of the multi-fruited river moss, *Cryphaea lamyana*, in Devon. Plymouth University report, Cornwall and SW Wales

Hill MO, Preston CD (1998) The geographical relationships of British and Irish bryophytes. J Bryol 20:127–226

Holyoak DT (2002) Multi-fruited river moss (Cryphaea lamyana): Report on work carried out in England during 2001 and south Wales during 2002. Plantlife Report 204

Lloyd T, Orbach J, Scourfield R (2006) The buildings of Wales. Yale University Press, Carmarthenshire and Ceredigion. New Haven & London

Orange A (1993) The ecology and distribution of *Cryphaea lamyana* (Bryophyta) on the Afon Teifi. CCW contract report no. 10/92

Williams RJ (2002) *Cryphaea lamyana* on the lower Afon Tyweli, Carmarthenshire (2002). Contract report

Chapter 19
Monitoring Sea Lamprey *Petromyzon marinus* Ammocoetes in SAC Rivers: A Case Study on the River Wye

Jonathan P. Harvey, Richard A. Noble, Andrew D. Nunn, Ryan J. Taylor, and Ian G. Cowx

Introduction

The EC Habitats Directive (92/43/EEC) on Conservation of Natural Habitats and of Wild Fauna and Flora stipulates that Member States maintain or restore habitats and species in a condition that ensures their favourable conservation status in the community. To comply with this Directive, a number of UK rivers were designated as Special Areas of Conservation (SACs) because they support important populations of designated species. Three such species in UK rivers are *Petromyzon marinus* L. (sea lamprey), *Lampetra fluviatilis* (L.) (river lamprey) and *Lampetra planeri* (Bloch) (brook lamprey). All three are widely distributed throughout the British Isles, although *P. marinus* and *L. fluviatilis* are reported north of the Scottish Great Glen only occasionally (Maitland and Campbell 1992, R. Gardiner, pers. comm.). *Petromyzon marinus* and *L. fluviatilis* are anadromous and parasitic, with adults typically inhabiting coastal and offshore waters (Maitland *et al.* 1994). *Lampetra planeri* is a non-parasitic species and lives exclusively in fresh water. All three species spawn in fresh water and the larvae, known as ammocoetes, of all three species use similar microhabitats, but with varying geographical distribution; *P. marinus* is typically found in the lower reaches of rivers, whilst *L. fluviatilis* and *L. planeri* are more closely associated with the middle and upper reaches, where their distributions often overlap.

Lamprey habitat preferences change with life-history stage (APEM 2001; Maitland 2003), with gravel-dominated substratum preferred for spawning, and mainly silt and sand-dominated substratum for nursery habitat. Optimal nursery habitat is defined as extensive areas of stable, fine sediment or sand ≥15 cm deep, with low water velocity and detritus present (APEM 2002). However, small pockets of suitable habitat are often found in open, comparatively high-velocity reaches, interspersed among coarser substrata (APEM 2002).

J.P. Harvey (✉), R.A. Noble, A.D. Nunn, R.J. Taylor and I.G. Cowx
Hull International Fisheries Institute, Department of Biological Sciences, University of Hull, Cottingham Road, Hull, East Yorkshire, HU6 7RX, UK
e-mail: J.P.Harvey@hull.ac.uk

C. Hurford et al. (eds.), *Conservation Monitoring in Freshwater Habitats: A Practical Guide and Case Studies*, DOI 10.1007/978-1-4020-9278-7_19,

Other suitable habitats include detritus overlying coarse substratum, submerged tree roots and silt banks, silt-dominated cattle drinks and bankside vegetation rooted in silt (Harvey and Cowx 2003). These latter areas, designated as sub-optimal habitat, often contain very high densities of ammocoetes and are important when assessing the abundance and distribution of lampreys in rivers. Spate rivers, with high flow velocities, tend to support fewer ammocoetes because they contain smaller areas of suitable habitat.

Harvey and Cowx (2003) described a protocol for assessing the conservation status of lampreys in UK SACs based on the Joint Nature Conservancy Council (JNCC) common standards monitoring guidance. The protocol compares the abundance and size structure of lamprey ammocoete populations against a predetermined set of criteria. This process is known as condition assessment, and is considered appropriate for testing for compliance in other European conservation areas. This chapter tests the protocol using a case study on sea lamprey in the River Wye, and examines the effectiveness of the protocol for condition assessment.

Study Area and Development of Monitoring Programme

The River Wye in Wales is approximately 250 km long from its source at Plynlimon to its confluence with the Severn Estuary at Chepstow, draining an area of over 4,000 km^2. In Wales the River Wye (Fig. 19.1), and many of its tributaries, is designated an SAC based on a number of features, including seven fish species. Populations of each lamprey species are among the primary reasons for the designation of the River Wye as an SAC. The Wye catchment provides habitat for lampreys and supports populations of *P. marinus*, *L. fluviatilis* and *L. planeri*.

Historical data on the occurrence and distribution of lamprey species in the Wye catchment were examined to provide an indication of the distribution of the target species and a basis for the selection of sampling sites. These data were collated from three main sources: (i) the Database and Atlas of Freshwater Fishes (DAFF) (Davies *et al.* 2002; Davies *et al.* 2004), which collates records of fish species occurrence across Great Britain; (ii) Environment Agency of Wales (EAW) fish population monitoring data, which comprised records of the occurrence of lampreys at sites surveyed as part of their standard electric fishing programme; and (iii) data collected by EAW in targeted lamprey surveys. Final site selection was based on the known distribution of lampreys in the River Wye, identified by interrogation of historical data, and by consultation with local EAW and Countryside Council for Wales (CCW) staff. Sites were selected to provide a wide coverage of both optimal and sub-optimal habitats that may support lamprey ammocoetes and metamorphosing ammocoetes ('transformers').

Monitoring Targets

The protocol developed for the LIFE in UK Rivers project, and adopted as the monitoring standard in the UK (Harvey and Cowx 2003), considered that information on the abundance and population demographic structure of ammocoetes is the best

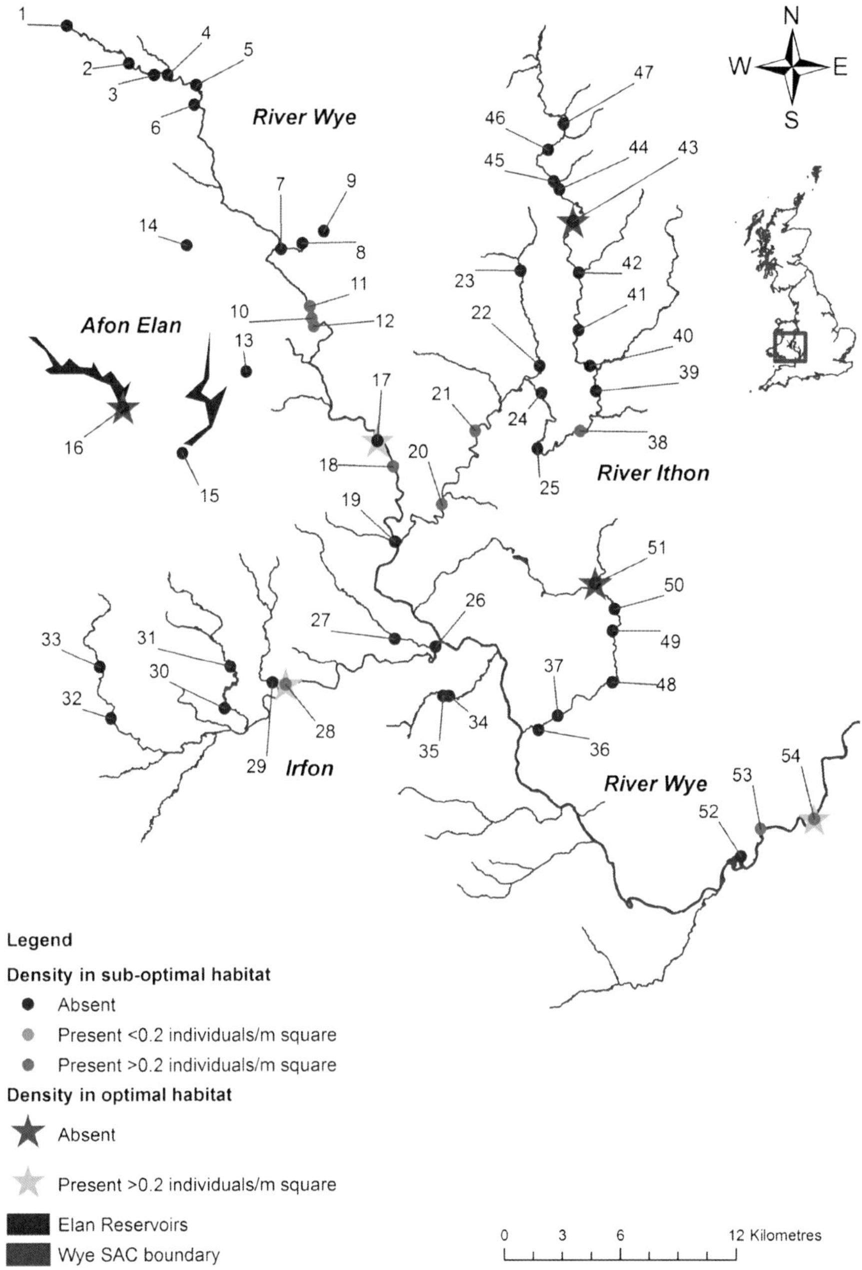

Fig. 19.1 Distribution of sampling sites and density estimates of *P. marinus* in optimal and sub-optimal habitat in the River Wye catchment.

measure for assessing condition status of lamprey populations in SAC rivers. The lamprey ammocoete abundance classification is based on two measures: (i) density in optimal habitat; and (ii) density based on catchment-wide surveys that include a diversity of habitats. Compliance with favourable status is based upon sea lamprey ammocoetes achieving threshold densities of ≥0.2 individuals m^{-2} in areas of optimal habitat and ≥0.1 individuals m^{-2} on a catchment perspective (Harvey and Cowx 2003; Cowx *et al.* 2009). The population demographic structure criterion requires at least two age classes to be present in ammocoete populations from optimal habitats (Harvey and Cowx 2003).

Materials and Methods

Fifty-four sites on the River Wye and its tributaries (Fig. 19.1) were sampled for lamprey ammocoetes in October and November 2005. The sampling strategy followed the LIFE in UK Rivers protocol (Harvey and Cowx 2003), with a minimum of three quantitative or semi-quantitative samples taken at each site. Lamprey larvae were sampled by electric fishing (2 kVA generator, 220 V, 50 Hz pulsed DC). For quantitative surveys, a delimiting framework (equivalent to a quadrat base area 1 m^2) was used (Harvey and Cowx 2003). The framework was placed at the selected sampling point and left to allow settlement of any disturbed sediment. A single anode (40-cm diameter) was immersed approximately 10–15 cm above the substratum, energized for 20 s, then turned off for 5 s. This process was repeated for approximately 2 min. This technique draws lamprey out of the sediment and into the water column. Immobilized lamprey were removed using a fine-meshed (280 μm) net, and transferred to a water-filled container. The sampling process was conducted three times in total, with a resting period of 5 min between each sample. Samples were kept separate for analysis. Where deployment of the framework was not possible (e.g. narrow marginal areas, near overhanging trees, and deep or fast-flowing areas), semi-quantitative electric fishing to catch as many ammocoetes as possible was used, in this case sampling points were fished only once, rather than three times. In both strategies, the microhabitat at each sample point was classified as either optimal or sub-optimal.

Lampreys were identified, measured (total length, mm) and returned to their place of capture. *Petromyzon marinus* larvae were differentiated from *L. fluviatilis* and *L. planeri* in the field by examination of the pigmentation in the oral hood and caudal fin (Fig 19.2; Gardiner 2003). *Petromyzon marinus* were classified into the following categories: (1) ammocoete; and (2) transformer.

Petromyzon marinus ammocoete and transformer densities (number m^{-2}) were calculated for each sampling point. Absolute density was estimated at quantitative sampling points using depletion methods (Zippin 1956; Carle and Strub 1978), while gear calibration was used for semi-quantitative sampling points (Cowx 1995). This involved calculating the mean probability of capture (p) at quantitative sampling sites

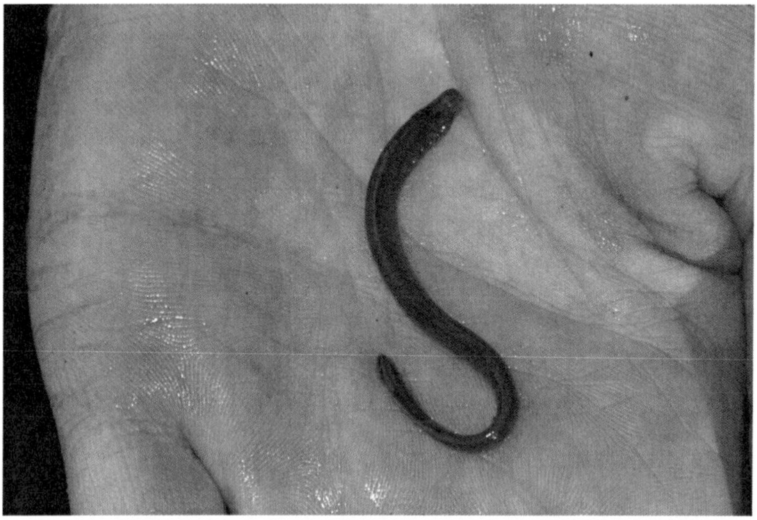

Fig. 19.2 *P. marinus* ammocoete captured in surveys in the River Wye. Note, the pigmentation extending into the caudal fin differentiates *P. marinus* ammocoetes from *L. fluviatilis* and *L. planeri* ammocoetes. *Photo by Clive Hurford*

($p = \sum_{n}^{1} C / \hat{N}$), and using this value to estimate relative abundance ($N = (C/p)/A$), where C is the total number of ammocoetes caught in the first sample at each sampling point, $\hat{N}$ the estimated absolute abundance from quantitative sampling, and A the sampling area (Cowx 1995). Mean density estimates were calculated for optimal and sub-optimal microhabitats at each site by summing the individual sample densities (quantitative and semi-quantitative samples combined) and dividing by the number of samples. Length distributions of *P. marinus* ammocoetes were derived for each site to facilitate interpretation of age-structure of the populations.

Results

A total of 453 *P. marinus* ammocoetes was caught during the study period. The majority of *P. marinus* ammocoetes (68%) were caught at site 54 (Hay-on-Wye) in the immediate vicinity of known spawning areas. *Petromyzon marinus* transformers were not captured at any site, although some individuals were of a size (>140 mm) considered likely to transform in winter 2005/2006. *Petromyzon marinus* distribution was mainly restricted to below the waterfall in Rhayader (Site 10), with only one individual captured upstream (Fig. 19.1).

Density Estimates

Mean densities of *P. marinus* ammocoetes at individual sites ranged from 0.00 to 30.98 individuals m^{-2} in optimal microhabitats; mean river densities ranged from 0.00 to 3.61 individuals m^{-2} (Table 19.1, Fig. 19.1). The overall Wye catchment mean density of *P. marinus* ammocoetes in optimal microhabitats was 9.04 individuals m^{-2}.

Mean densities of *P. marinus* ammocoetes at individual sites ranged from 0.00 to 74.39 individuals m^{-2} in sub-optimal microhabitats; mean river densities ranged from 0.00 to 5.89 individuals m^{-2} (Table 19.1, Fig. 19.1). The overall Wye catchment mean density of *P. marinus* ammocoetes in sub-optimal microhabitats was 2.20 individuals m^{-2}. These densities were above the thresholds established by Harvey and Cowx (2003) (mean ≥0.2 individuals m^{-2} in optimal and ≥0.1 individuals m^{-2} in sub-optimal microhabitats), suggesting the populations were in favourable condition. Assessment of the distribution of densities of ammocoetes at individual sampling sites (Fig. 19.3) showed that the mean density in 54 sites (optimal and sub-optimal combined) was 2.58 individuals m^{-2}, again supporting favourable condition. These densities compared favourably with those reported from elsewhere in UK rivers (APEM 2001).

Size Structure

Demographic structure varied between sites, reflecting variability in inter-annual recruitment success and ability to access certain reaches. Individual *P. marinus*

Table 19.1 Mean density estimates of *P. marinus* ammocoetes captured from optimal and sub-optimal microhabitats on the River Wye. Columns 3 and 4 give the mean density estimates for all sites sampled, column 5 gives the mean density estimates in optimal habitat (n/a = not applicable)

		Mean density of *P. marinus* ammocoetes m^{-2} (area sampled m^2)		Mean density of *P. marinus* ammocoetes m^{-2}
River	Number of sites (total)	Optimal habitat	Sub-optimal habitat	Sites with optimal habitat only
Afon Cammarch	2	n/a	0.00 (24.0)	No optimal habitat
Afon Elan	4	0.00 (1.00)	0.00 (8.40)	0.00
Afon Ithon	14	0.00 (2.20)	1.38 (34.60)	0.00
Afon Marteg	3	n/a	0.00 (3.00)	No optimal habitat
Chwerfri	2	n/a	0.00 (7.50)	No optimal habitat
Clywedog Brook	2	n/a	0.00 (4.00)	No optimal habitat
Duhonw	2	n/a	0.00 (9.50)	No optimal habitat
Dulas	1	n/a	0.00 (6.50)	No optimal habitat
Edw	6	0.00 (1.00)	0.00 (15.70)	0.00
Irfon	3	0.81 (1.00)	2.23 (5.00)	2.44
Wye	15	3.61 (6.00)	5.89 (44.20)	27.10

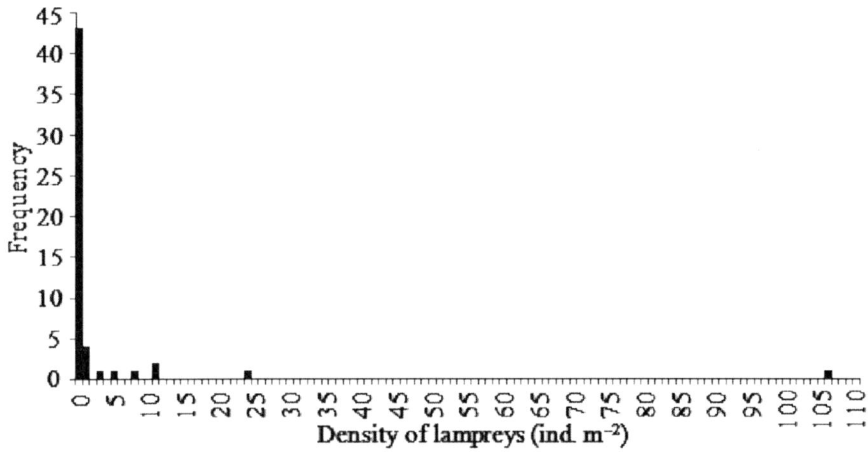

Fig. 19.3 Distribution of densities of *P. marinus* ammocoetes at individual sampling sites in the River Wye catchment

ammocoetes, with lengths between 60 and 75 mm, belonging to the ≥1+ age group (2004 year class), were caught at sites 10, 11 and 12. No evidence of recruitment in 2005 (discriminated by capture of ammocoetes <40 mm in length) was found at these sites. Further downstream, abundance of larger ammocoetes (>40 mm) increased.

At site 17, *P. marinus* ammocoetes ranged from 32–127 mm; the presence of individuals <40 mm indicated recruitment in 2005, and the larger-sized individuals (>80 mm) were >1+ age groups. A number of age groups were also captured at sites 18, 20 and 21, but all were ≥1+ individuals; there was no evidence of recruitment in 2005 (Fig. 19.4). Ammocoetes <40 mm (0+ individuals) were captured at sites 28, 53 and 54, indicating recruitment in these areas in 2005; several other age groups were also evident at these sites (Fig. 19.4). 0+ ammocoetes dominated catches at site 54, suggesting the site is an important spawning and nursery area for *P. marinus* (Fig. 19.4). The presence of more than two ammocoete age classes at optimal sites indicates that *P. marinus* is in favourable condition.

Discussion

Petromyzon marinus populations in the River Wye catchment in Wales achieved favourable conservation status, and provide a baseline for comparison with future surveys. The following sections discuss the significance of the findings and key issues associated with monitoring *P. marinus* populations; the issues are also considered pertinent for monitoring *L. fluviatilis* and *L. planeri* populations.

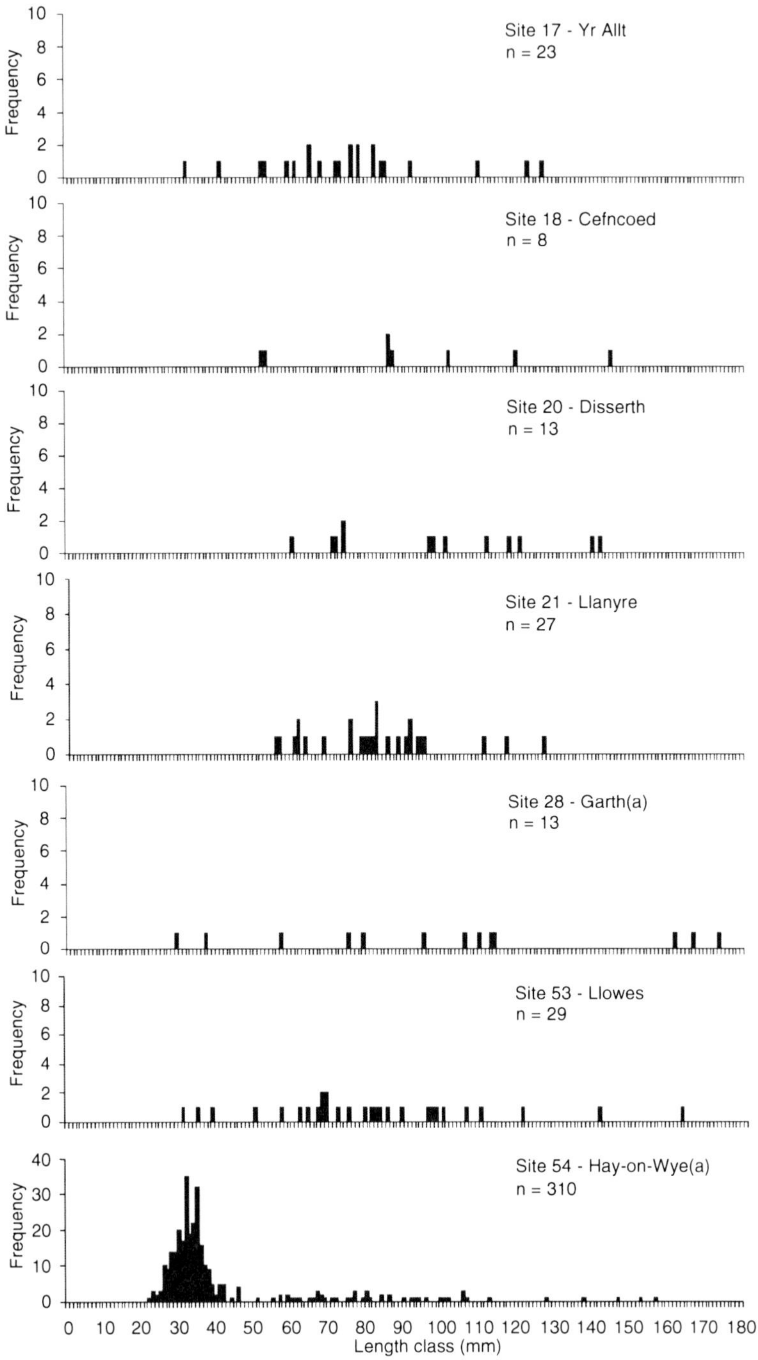

Fig. 19.4 Length distributions of *P. marinus* ammocoetes in the River Wye catchment

Habitat Requirements

A key factor determining the distribution and abundance of lamprey ammocoetes appears to be the availability of suitable substratum, typically fine particulate (organic) matter. Lamprey larvae reside in the river bed, and filter food particles from feeding currents that pass through their burrows. The substratum must therefore be fine enough to burrow into, but stable enough for respiration and feeding purposes. Slow-flowing stretches of rivers or backwaters and eddies, where deposition of sand and silt occurs, provide such conditions (Maitland 2003). *Petromyzon marinus, L. fluviatilis* and *L. planeri* ammocoetes all use similar microhabitats (Harvey and Cowx 2003). The substratum used is predominantly mud, silt or sand, and may vary in depth from a few centimetres to more than 30 cm (Hardisty and Potter 1971).

Availability of optimal and sub-optimal habitats varied greatly in the catchment. Optimal habitat was generally restricted to the lower reaches of the main stem and tributaries, reflecting the gentler gradient, slower flow regime and increased silt loading; this was reflected in high densities of *P. marinus* and also *Lampetra* ammocoetes at these sites (Harvey *et al.* 2006).

The upper reaches of the main stem and tributaries of the River Wye typically only contained sub-optimal ammocoete habitat, with very few areas of deep stable sediment. This reflects the dynamic nature of the river and tributaries in this area. Substratum at the majority of sites was gravel, cobble and bedrock, and ammocoete habitat was generally restricted to areas sheltered from the main flow, where patchy and shallow silt accumulations occurred. Such areas were considered highly mobile in times of high flow, which could cause downstream displacement of ammocoetes. Notwithstanding, sub-optimal habitat appears critical to *P. marinus* populations in the Wye catchment. Applegate (1950) noted that the depth to which lamprey larvae burrow is positively related to their size. Thus, the availability of optimal and sub-optimal habitat may influence the size structure of lamprey populations, with larger ammocoetes occluded from shallow sediments. The issues surrounding the contribution of optimal and sub-optimal habitats to rivers, or reaches, attaining favourable conservation status need further consideration as more information becomes available. For example, exactly what constitutes 'optimal' habitat has not been validated, and may not accurately reflect the requirements and preferences of lamprey ammocoetes. Future research into the factors influencing the suitability of habitat for lamprey ammocoetes should allow 'optimal' habitat conditions to be quantified and revised conservation assessment criteria to be produced.

Abundance Classification

Very few sites in the River Wye catchment had optimal habitat. As such, comparison of densities with the criterion for favourable condition in optimal habitat (ammocoete density $\geq$0.2 individuals m^{-2}) was only possible at a small number of sites. However, the criterion for compliance with favourable condition in optimal habitat was also used for comparative purposes with densities in sub-optimal habitat, primarily to highlight that sub-optimal habitats often contained high densities of ammocoetes. Therefore, for future monitoring, the ammocoete density thresholds for optimal and sub-optimal habitat may need adjustment, in conjunction with habitat definitions, especially in catchments dominated by sub-optimal habitat. Indeed, the thresholds defined by Harvey and Cowx (2003) for attaining favourable condition status were intended to be modified as more extensive data become available. At many sites, *P. marinus* densities met the criterion for compliance for optimal habitat, even when the habitat was classified as 'sub-optimal'. Those of *P. marinus* ammocoetes in the River Wye compared favourably with densities reported elsewhere (Maitland 2003; Nunn *et al.* 2008), reflecting good spawning habitat and few migration barriers in the catchment. Indeed, one *P. marinus* ammocoete was captured upstream of the waterfall at Rhayader, which was initially considered the upstream limit of sea lamprey, indicating that the waterfall may be negotiable by a small number of adult sea lamprey under favourable flow conditions. Ammocoete densities were high in the main River Wye at Hay-on-Wye indicating that this reach is a key spawning and nursery area. Densities in the rivers Ithon and Irfon, tributaries of the River Wye, were also high indicating that smaller watercourses may be important for sea lamprey spawning. Based on the data collected for the River Wye, *P. marinus* ammocoete populations were considered to be in favourable condition in 2005.

Population Demographic Structure

There were differences in the population demography between upstream and downstream reaches. Ammocoetes were rarely caught in upstream reaches, suggesting limited access to this region for adult *P. marinus*. The waterfall at Rhayader appears to be a barrier to migration and must be considered the upstream limit for assessing condition status. At least two age classes of *P. marinus* ammocoetes were captured at all sites containing optimal habitat downstream of the waterfall. This was also the case at most sites dominated by sub-optimal habitat. Recruitment at the majority of accessible sites with suitable habitat was therefore not compromised, and *P. marinus* populations in the River Wye are in favourable condition for this criterion (Harvey and Cowx 2003).

Monitoring Issues

The long-term viability and conservation status of *P. marinus* populations is an important issue in SAC rivers, and spawning activity and distribution within such systems should be surveyed regularly. Adult *P. marinus* are known to locate spawning grounds using a migratory pheromone that is released by resident larval conspecifics

(Sorensen *et al.* 2003). Similarly, female *P. marinus* locate males via a pheromone released by males on the spawning grounds (Li *et al.* 2003). Thus, there is a risk in some systems that *P. marinus* populations could become too small for chemical communication to be effective, particularly where age classes are missing in the larval populations or where spawning stocks are minimal.

In addition, there is a possibility that the use of common spawning and nursery grounds may increase the potential for competition between *P. marinus* and *L. fluviatilis* larvae. Large spawning aggregations in discrete localities (e.g. site 54 – Hay-on-Wye) are extremely susceptible to interference, habitat degradation or environmental perturbations such as pollution. Attention should be given to maintaining effective habitat protection, and access to and prevention of exploitation at spawning grounds (Jang and Lucas 2005). Therefore, in any monitoring programme for lampreys it is important to identify and monitor the spawning areas, and subsequently assess the status of ammocoete populations in the vicinity of the spawning localities. This would ensure that any degradation in spawning areas is identified quickly and mitigating measures implemented to protect such areas.

In the UK, most historical information on lampreys was gathered opportunistically, either directly by government agencies as part of other monitoring programmes (e.g. EA juvenile salmonid monitoring programmes) or indirectly by members of the public, including anglers. There is a need to undertake independent monitoring surveys that specifically target lamprey ammocoetes and transformers. It is, thus, recommended that independent monitoring of the status of lampreys, following the LIFE in UK Rivers protocol (Harvey and Cowx 2003; Cowx *et al.* 2009), is undertaken to comply with reporting needs under the Habitats Directive. Unfortunately, the reporting frequency under the Directive is every 6 years, which may be too long to identify and prevent any catastrophic changes in populations. Thus, more strategic monitoring at key locations is recommended annually or biennially (Harvey *et al.* 2006; Nunn *et al.* 2008; Cowx *et al.* 2009). Such monitoring should assess the distribution, abundance and demographic structure of lamprey populations so that mitigation measures can be initiated should any deleterious changes occur.

Site selection is an important consideration when monitoring *P. marinus*. In this study, it was crucial that sites on tributaries outside the boundaries of the SAC were included in the sampling programme as these probably contributed significantly to the sustainability of lamprey populations in the SAC as a whole. Furthermore, assessing sites outside SACs may identify factors affecting the status of lamprey populations within SACs (Cowx *et al.* 2009). In this study, particular attention was paid to the lower reaches of the main river (below Builth Wells) to account for the main distribution and lower abundance of *P. marinus* ammocoetes compared with *Lampetra* species in most rivers (Gardiner 2003). This is especially important where barriers, either natural or man-made, may prevent or limit upstream migration of *P. marinus* and *L. fluviatilis* to spawning grounds (Harvey and Cowx 2003; Nunn *et al.* 2008).

Critical to the success of any monitoring programme is the review of historical data on lamprey populations, and habitat and water quality. These may provide information on lamprey distribution and abundance, and also identify spawning areas, facilitating the selection of survey sites. Furthermore, habitat data (e.g. River Habitat Survey – RHS) and chemical and biological water quality data (e.g. EA General Quality Assessments – GQA) can be linked with fisheries data to determine if physical or chemical characteristics compromise population viability. This allows assessment of lamprey populations in relation to the full range of JNCC common standards monitoring guidance criteria, which gives targets for a number of water quality attributes that contribute to favourable condition assessment (JNCC 2005).

It appears that one of the key factors dictating whether lampreys succeed in reaching their spawning grounds, particularly in regulated rivers, is the river level during their spawning migrations. This may be the case for *P. marinus* (Fig. 19.5), which tends to migrate in spring and early summer, when river levels are lower and more stable than in the autumn and winter (Nunn *et al.* 2008). There is a potential, tangible link between flow characteristics and lamprey spawning success (Waterstraat and Krappe 1998; Oliveria *et al.* 2004). Indeed, Nunn *et al.* (2008) identified migration barriers and low flows as factors potentially contributing to the poor recruitment of *L. fluviatilis* in the River Ure in 2003 compared with less regulated rivers in the Yorkshire Ouse catchment. Therefore, there is a need to understand the relationships between flow characteristics, migration barriers and the potential impacts that climate change, and concomitant shifts in flow regime, may have on lamprey population viability. Assessment and future regulation of flow in many rivers is currently being undertaken by the Environment Agency within the Review of Consents procedure. The findings of these reviews will be important to the assessment of conservation status of lamprey populations in SAC rivers.

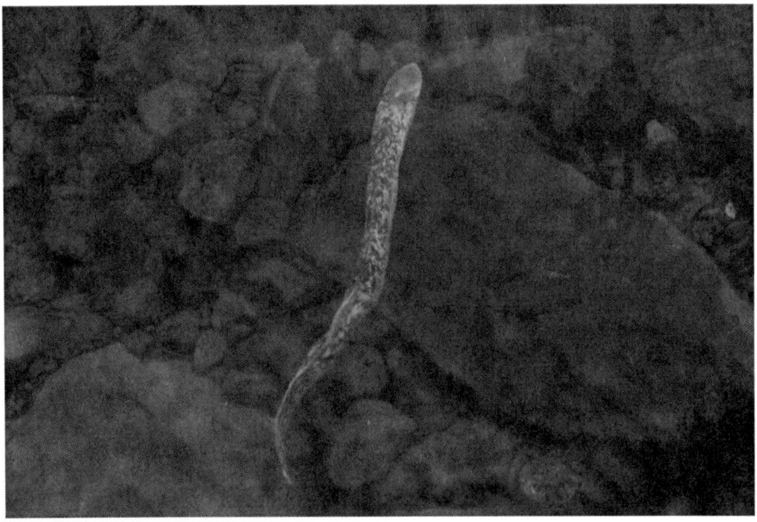

Fig. 19.5 Adult sea lamprey swimming upstream to spawning grounds. *Photo by Clive Hurford*

Acknowledgements The authors acknowledge the Countryside Council for Wales for provision of funds to carry out the survey work. The authors wish to thank Clive Hurford, Tristan Hatton-Ellis, Rhian Thomas, Chris Dyson and Karen Wilkinson (CCW) for organising access arrangements to the survey sites and Bill Purvis, Ray Dobbins, George Woodward, Keith Jenkins and Steve Barker (Environment Agency Wales) for provision of electric fishing survey consents and valuable advice on sampling sites. Many thanks are extended to the various landowners and fishery owners for permission to survey the sampling sites. The authors gratefully acknowledge the assistance of Darren Rollins and David Hunter in electric fishing surveys.

References

APEM (2001) Standardised sampling strategies and methodologies for condition assessment within SAC rivers for sea, river and brook lamprey and bullhead – Phase I Report. Life in UK Rivers Project, December 2001, 48 pp

APEM (2002) Standardised sampling strategies and methodologies for condition assessment within SAC rivers for sea, river and brook lamprey and bullhead – Phase 2a Final Report. Life in UK Rivers Project, Report to English Nature 29 pp

Applegate VC (1950) Natural history of the sea lamprey (*Petromyzon marinus*) in Michigan. USA. Fish Wildlife Service Special Sci Report Fish 55:1–237

Carle FL, Strub MR (1978) A new method for estimating population size from removal data. Biometrics 34:621–830

Cowx IG (1995) Fish stock assessment – Biological basis for sound ecological management. In: Harper D, Ferguson A (eds) Biological basis for river management. Wiley, London, pp 375–388

Cowx IG, Harvey JP, Noble RA, Nunn AD (2009) Establishing survey and monitoring protocols for the assessment of conservation status of fish populations in river Special Areas of Conservation in the UK. Aquat Conserv Marine Freshwater Ecosyst 19:96–103

Davies CE, Harding PT, Arnold HR (2002) Database and atlas of freshwater fishes in Great Britain: Final report. Environment Agency R&D Report, 88 pp

Davies C, Shelley J, Harding P, McLean I, Gardiner R, Perison G (2004) Freshwater fishes in Britain: The species and their distribution. Harley Books, Colchester, UK 176 pp

Gardiner R (2003) Identifying lamprey. A field key for sea, river and brook lamprey. Conserving Natura 2000 Rivers Conservation Techniques Series No. 4. English Nature, Peterborough, 27 pp

Hardisty MW, Potter IC (eds) (1971) The biology of lampreys. London, Academic 423 pp

Harvey JP, Cowx IG (2003) Monitoring the river, brook and sea lamprey, *Lampetra fluviatilis*, *L. planeri* and *Petromyzon marinus*. Conserving Natura 2000 rivers monitoring series No. 5. English Nature, Peterborough, 32 pp

Harvey JP, Noble RA, Cowx IG, Nunn AD, Taylor RJ (2006) Monitoring of lamprey in the Rivers Wye and Usk SACs 2005–2006. Report to Countryside Council of Wales, 120 pp

Jang MH, Lucas MC (2005) Reproductive ecology of the river lamprey. J Fish Biol 66:499–512

JNCC (2005) Common standards monitoring guidance for freshwater fauna. Joint Nature Conservancy Council, 39 pp

Li WM, Scott AP, Siefkes MJ, Yun SS, Zielinski B (2003) A male pheromone in the sea lamprey (*Petromyzon marinus*): An overview. Fish Physiol Biochem 28:259–262

Maitland PS (2003) Ecology of the river, brook and sea lamprey. Conserving Natura 2000 rivers ecology series No. 5. English Nature, Peterborough, 52 pp

Maitland PS, Campbell RN (1992) Freshwater fishes of the British Isles. HarperCollins, London, 368 pp

Maitland PS, Morris KH, East K (1994) The ecology of lampreys (Petromyzonidae) in the Loch Lomond area. Hydrobiologia 290:105–120

Nunn AD, Harvey JP, Noble RA, Cowx IG (2008) Condition assessment of lamprey populations in the Yorkshire Ouse catchment, north-east England, and the potential influence of physical migration barriers. Aquat Conserv Marine Freshwater Ecosyst 18:175–189

Oliveria JM, Ferreira MT, Pinheiro AN, Bochechas JH (2004) A simple method for assessing minimum flows in regulated rivers: The case of sea lamprey reproduction. Aquat Conserv Marine Freshwater Ecosyst 14:481–489

Sorensen PW, Vrieze LA, Fine JM (2003) A multi-component migratory pheromone in the sea lamprey. Fish Physiol Biochem 28:253–257

Waterstraat A, Krappe M (1998) Distribution and abundance of *Lampetra planeri* populations in the Peene drainage (NE Germany) in relation to isolation and habitat conditions. Ital J Zool 65:Suppl 137–143

Zippin C (1956) An evaluation of the removal method for estimating animal populations. Biometrics 8:163–189

Chapter 20
Monitoring Juvenile Atlantic Salmon and Sea Trout in the River Sävarån, Northern Sweden

Hans Lundqvist, Kjell Leonardsson, Ulf Carlsson, Stefan Larsson, Jan Nilsson, Johan Östergren, Lars Karlsson, Peter Rivinoja, Ignacio Serrano, Daniel Palm, and John Ferguson

Introduction

Wild salmon stocks have declined worldwide (NRC 1996). In many Baltic Sea rivers most wild populations of Atlantic salmon (*Salmo salar* L.) and anadromous trout (sea trout, *Salmo trutta* L.) have been destroyed, with the remaining stocks found primarily in rivers within northern Sweden and Finland. Here they suffer high rates of fishery exploitation, while hydropower regulation and the re-engineering of rivers for floating timber has led to the loss of spawning and rearing habitat and to a loss of connectivity among habitats (McKinnell 1998).

To remain viable in the face of demographic and environmental stochasticity, salmonid populations require a certain level of abundance, positive growth rates, adequate spatial structure, and access to (connectivity among) habitats of sufficient quantity and quality to express their life history and genetic diversity (McElhany *et al.* 2000). To understand what is limiting their productivity and viability and develop conservation actions for these threatened populations, we need information on both the freshwater and marine phases of the salmon and sea trout life cycles.

The Salmon Action Plan (SAP) 1997–2010 was adopted by IBSFC (International Baltic Sea Fishery Commission), and states that by 2010 natural production in

H. Lundqvist(✉), K. Leonardsson, J. Nilsson, J. Östergren, P. Rivinoja, D. Palm and J. Ferguson
Department of Wildlife, Fish and Environmental Sciences, SLU, SE-901 83, Umeå, Sweden
e-mail: hans.lundqvist@vfm.slu.se

U. Carlsson
The County Administrative Board of Västerbotten, SE-901 86, Umeå, Sweden

L. Karlsson
National board of Fisheries, SE-814 94, Älvkarleby, Sweden

K. Leonardsson, S. Larsson and I. Serrano
Department of Ecology and Environmental Sciences, Umeå universitet, SE-901 87, Umeå, Sweden

J. Ferguson
Northwest Fisheries Science Center National Marine Fisheries Service, 2725 Montlake Boulevard East, Seattle, Washington, USA

C. Hurford et al. (eds.), *Conservation Monitoring in Freshwater Habitats: A Practical Guide and Case Studies*, DOI 10.1007/978-1-4020-9278-7_20,

Baltic rivers should be >50% of the maximum production potential. To date, maximum natural production levels have primarily been based on expert knowledge rather than empirical estimates (e.g. WGBAST 2008). The Swedish Government now recognises the need for index rivers to obtain reliable estimates of abundance, productivity, population structure, and to collect the information on life-history diversity needed to manage salmonid stocks.

From 2005 to 2008, a pilot study was implemented in the River Sävarån (a small, unregulated forest river in northern Sweden), to monitor the downstream migrations of salmon and trout, and explore its suitability as an index river. Rotary screw traps were used to investigate the abundance of smolts as well as their timing, size and age, and to obtain samples to analyse the genetic composition of the stock. Parr densities from electro-fishing surveys were compared with screw-trap data to determine whether the two approaches produced similar smolt production estimates.

The Study Site

The Sävarån is 140 km long, empties in the Bothnian Bay north of Umeå in northern Sweden, and drains a 1,160 km^2 watershed (Fig. 20.1). Salmon, resident brown trout (*Salmo trutta* L.), and sea trout inhabit the river. Water temperatures and flow regimes are typical of small, coastal rivers that drain the lowland forests of north-eastern Sweden (Fig. 20.2). The river has been fished for salmon since the sixteenth century (Johansson 2004), and recently, efforts have been undertaken to restore spawning and rearing habitats (Norberg and Ahlström 2007). Between 18 and 94 salmon have been counted annually since 1998 at a fish ladder located at river kilometer 50. The potential production from an estimated 20 ha of spawning habitat is estimated at 4,000 salmon smolts per year (Rappe *et al.* 1999).

Methods

Smolt Migration: Abundance and Timing

Rotary screw traps (EG Solutions, Oregon, USA) (Thedinga *et al.* 1994) were located at river kilometer 15 to monitor smolt abundance and migration timing (Fig. 20.3). Two traps were used in 2005 and 2006, but high recapture rates allowed a single trap to be used in 2007 and 2008. Traps were checked twice daily; to record the number of smolts collected, species, date, time of day, and smolt length (mm). All fish not needed for acoustic telemetry, or previously marked for trap efficiency studies, were immediately released downstream from the trap. Water temperature and level (flow) were noted daily, except in 2008 when data loggers were used to record water temperature and level every 30 min.

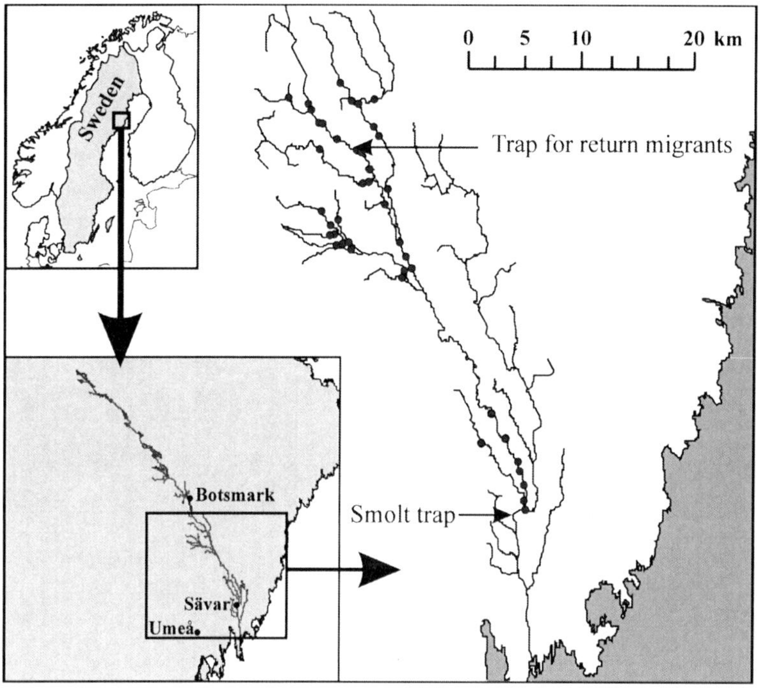

Fig. 20.1 Map showing the location of the smolt trap (installed 15 km upstream from the river mouth; Lat: N63° 54′ 38.6″, Long: E20° 33′ 56.5″) and electro-fishing sites (*black circles*) on the River Sävarån

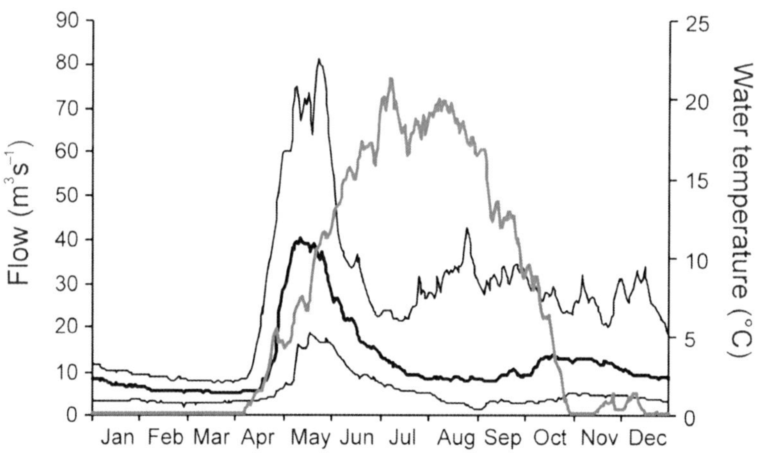

Fig. 20.2 Annual flow and temperature for the River Sävarån near the rotary-screw trap site. *Black lines* denote daily flow (thick = median, thin = percentiles; 2.5 and 97.5). *Grey line* denotes average daily water temperatures from 2006 to 2007

Fig. 20.3 River Sävarån smolt traps with their lead-arms that guide downstream migrating fish toward(s) the traps, as seen from downstream

Trapping efficiency was evaluated through daily mark-recapture experiments. Captured smolts were tagged with external Carlin tags, transported 2–4 km upstream, and released. Tags from recaptured fish and date and time of recapture were recorded. Trap efficiencies and population estimates were calculated based on a standard, single mark-recapture event (Seber 1982; Carlsson *et al.* 1998).

To improve the estimates, stratification was tested in 2007 by dividing the out-migration period into three periods: initial (few smolts, high flows and low recapture probabilities), intermediate (high numbers of smolts and high recapture rates), and final (low smolt numbers, low flow, and low recapture rates).

Genetic and Age Structure Monitoring

Tissue and scale samples were collected from a subsample of trapped fish to analyse genetic structure and age composition, to initiate a long-term dataset for assessing the genetic composition and stability of salmon and trout populations in the region, and to assess the effects of non-indigenous hatchery fish straying into the River Sävarån.

Microsatellite and mitochondrial DNA markers were selected for the genetic monitoring, based on a database being available for these marker systems that encompass most wild and cultivated Baltic Sea stocks. DNA was extracted from adipose fin tissues from 192 trout and 483 salmon collected throughout the migration period each year, 2005–2008. Standard procedures for analysing DNA were used (see Nilsson *et al.* 2001; Säisä *et al.* 2005), and reference DNA samples were used to calibrate allele sizes with existing data. Analysis of the amount of variability (heterozygosity) and measures of genetic differentiation (Fst) were performed using GENPOP (Raymond and Rousset 1995) software. The software STRUCTURE (Pritchard and Wen 2004) was used to estimate the number of populations contributing to a sample and search for non-indigenous individuals. The age of each sampled individual was determined from scale reading.

Parr Density

Electro-fishing has been performed in the River Sävarån since 1989 to estimate parr densities using the standardized methodology for Swedish rivers (Degerman and Sers 1999; Rivinoja and Carlsson 2008). Initially, salmon stock abundance in the river was low, and to increase the number of salmon observed nine sites were selected for monitoring from the best habitats, rather than being randomly assigned or from a stratified sample. Using this approach, population trends could be detected with higher statistical power. For a variety of reasons, including inter-annual and seasonal variability in flow, all nine salmon-monitoring sites could not be sampled each year and there were no sites with samples from all years. The missing observations from different years complicated the data analysis; therefore data from a geographically expanded electro-fishing programme of 15 sites were used to estimate 0+ parr densities for comparison with the smolt abundances.

Results

Smolt Migration Abundance and Timing

The downstream migrations of salmon and sea trout occurred in a relatively narrow window from mid-May to mid-June (Fig. 20.4). In general, smolts appeared in the trap when the water temperature reached 8°C for salmon and from 5 to 6°C for sea trout. While the number of trapped smolts varied between years, the migration period was relatively fixed. Trap collection efficiencies, measured as recapture rate of tagged fish released upstream, ranged from 15% to 27% for salmon and from 11% to 27% for trout in 2005 and 2006, and were high for salmon (21–31%) and trout (8–15%) when a single trap was used in 2007 and 2008 (Table 20.1).

Estimated salmon smolt population size ranged from 2,600 to 3,900 fish (Table 20.1). Uncertainty in the estimates, expressed as percent deviation between the 95% confidence limit and the mean was less than 15%. Estimated trout population size ranged from 500 to 1,500 fish, but uncertainty in the estimates ranged from 30% to 65%. The stratified method of estimating population size used in 2007 did not change the estimate markedly.

Genetic and Age Structure Monitoring

In 2005 and 2006, salmon smolts in the River Sävarån were genetically distinct from those in other Gulf of Bothnia rivers but were similar to smaller, forest-origin rivers in the region. Genetic variability was similar to other Gulf of Bothnia populations. Analyses of stock composition revealed that 11% of the gene pool originated from salmon stocked from the nearby River Byskeälven. Sea trout in the River Sävarån were also genetically differentiated from other nearby rivers, even though these populations have served as brood stock for fish stocked into the River Sävarån.

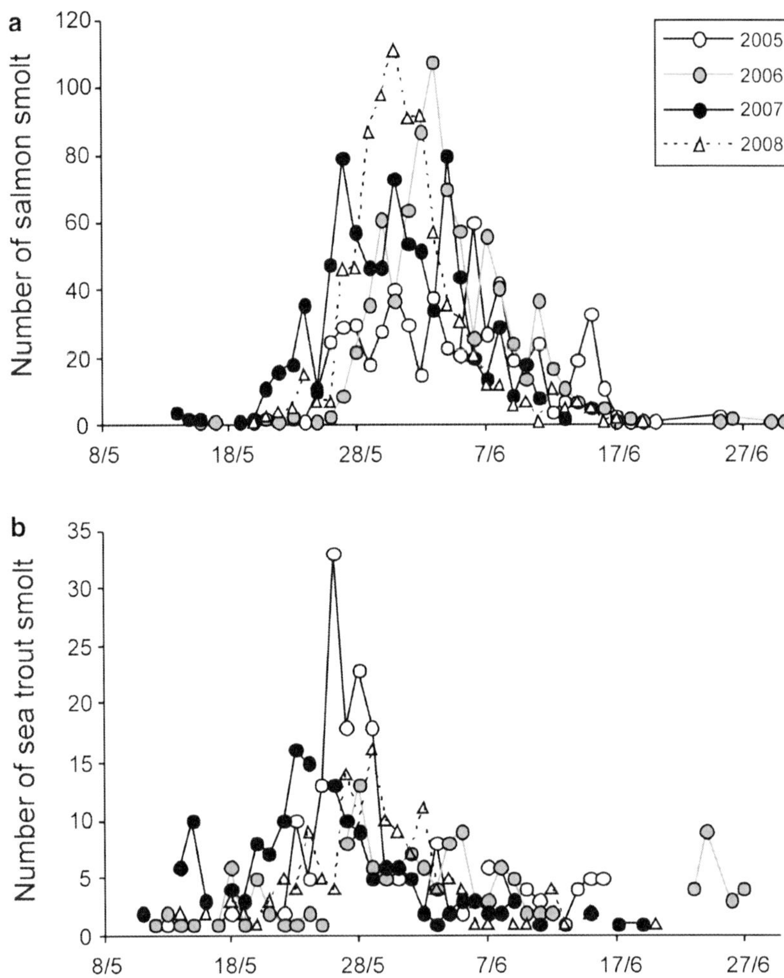

Fig. 20.4 Seasonal (day/month) variation in number of trapped salmon and sea trout smolt during 2005–2008

Table 20.1 Data from smolt trapping in the River Sävarån

Species	Year	Tagged	Recapts	Untagged	$Prob_C$	LCI95	Mean	UCI95
Salmon	2005	289	42	294	0.15	2,780	3,930	5,080
Salmon	2006	763	205	49	0.27	2,620	3,010	3,410
Salmon	2007	689	213	134	0.31	2,320	2,650	2,990
	*2007**					*2,650*	*2,850*	*3,000*
Salmon	2008	723	151	106	0.21	3,340	3,950	4,560
Trout	2005	35	4	160	0.11	290	1,404	2,520
Trout	2006	112	30	28	0.27	340	510	680
Trout	2007	122	18	39	0.15	590	1,040	1,490
Trout	2008	85	7	56	0.08	530	1,520	2,500

$Prob_C$ denotes recapture efficiency, Mean, LCI95 and UCI95 denote the mean, lower and the upper 95% confidence limits of the mark-recapture population estimates. The asterisk denotes estimates using three time frames of the migration period (initial, intermediate and final).

Estimated introgression from stocked fish was higher in sea trout than salmon, and genetic differentiation among rivers was less pronounced for sea trout, suggesting that gene flow was higher in this species.

In 2007 and 2008, salmon smolts originated from two different sources in roughly equal proportions, Sävarån and Byskeälven salmon (Fig. 20.5). Whether the introgression of the Byskeälven stock into the previously unique Sävarån stock will change the life-history patterns of future offspring in the River Sävarån is unknown, and requires additional monitoring.

Smolt sizes were similar between years; salmon smolts averaged 14.1 ± 0.6 (±1SD) cm, and trout averaged 16.7 ± 0.9 cm. Salmon and trout smolt age varied from 2 to 4 years, and averaged 2.7 and 2.9 years, respectively. In general, salmon smolt age in the 13–17 cm size range overlapped considerably, and the dominant age at smoltification varied among years (Table 20.2). This information, along with the expected time at sea, is crucial for assessing salmon and sea trout populations.

Parr Density

Salmon parr densities were low at the beginning of the sampling period and increased slowly to a density of about eight 0+ parr/100 m^2 at the end of the time series (Fig. 20.6). During the last three years, parr density ranged from 0 at some sites to average of 30 individuals/100 m^2 at others. To reduce the influence of the spatial variation between sites, data from the years when all sites were sampled were averaged for each site. These averages were then subtracted from each site's time series as a standardization of all sites. To standardize the time series between sites, the level of the normalized time series was then shifted to have the same densities in the initial phase as the original time series sites.

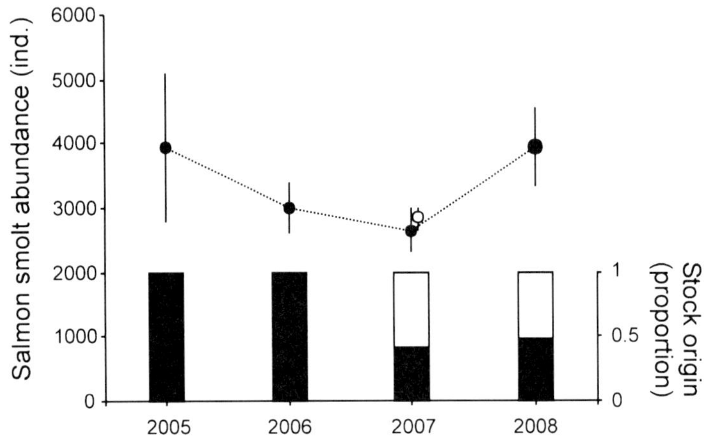

Fig. 20.5 Estimated salmon smolt abundance (*individuals*) over the study period. *Vertical bars* show proportion of smolts belonging to the wild Sävarån stock (*black*) and the introduced Byskeälven stock (*white*) assessed using DNA-samples from trapped smolts. *Thin vertical lines* show 95% CL. The *open circle* denotes the abundance from the stratified method

Table 20.2 Proportion of individuals in each of the three age classes recorded from scale reading

Species	Year	2 years	3 years	4 years	N
Salmon	2006	0.14	0.81	0.05	104
	2007	0.47	0.36	0.16	165
	2008	0.46	0.51	0.03	127
Sea trout	2006	0.15	0.64	0.22	74
	2007	0.30	0.66	0.04	122
	2008	0.32	0.61	0.07	90

Variation in river flow between years resulted in different amounts of area being sampled at each site and available for the fish (which directly influences fish density). This variability within the time series data was impossible to separate from temporal population fluctuations. To address this, a three-year moving average was used to smooth the time series (Fig. 20.6). Based on this approach, the population was recovering during the monitoring period and increased almost 10% each year.

Parr densities ranged from 2.6 to 6.6 individuals/100 m^2 from 2005 to 2008, which is an increase of 150%. From 2002 to 2006, the corresponding increase for the 17 sites sampled was 260%. By contrast, trapping efficiency data estimated salmon smolt populations to be between 2,700 and 4,000 fish from 2005 to 2008, and increased only 50% throughout the period.

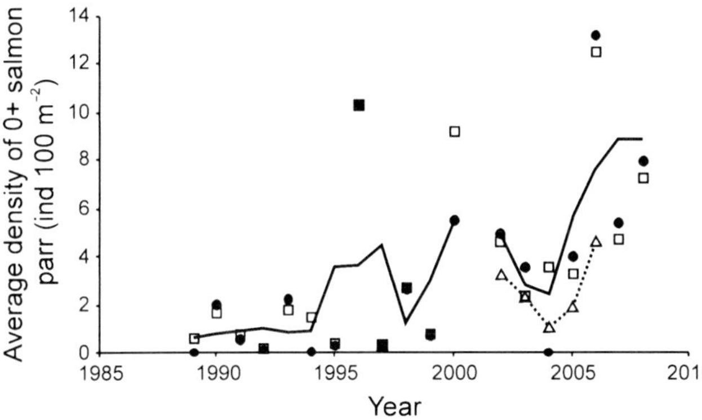

Fig. 20.6 Average densities of one-summer old salmon from electro-fishing at 9 and 17 sites in Sävarån upstream of the trapping site. *White squares* denote annual averages and *black dots* denote adjusted annual averages from the 9 salmon monitoring sites. The *thick black line* denotes the three-year moving average for the adjusted means. The dotted line and the unfilled triangles denote the mean densities at the 17 monitoring sites (including tributaries) that were sampled each year between 2002 and 2006

Discussion

Several important observations were forthcoming from the smolt migration monitoring in the River Sävarån. The estimated number of salmon smolts produced in the R. Sävarån based on smolt trapping was similar to the estimated production potential of 4,000 (Rappe *et al.* 1999). If accurate, this would mean that 75–100% of the potential production (and international conservation limits) of salmon smolts is being achieved. However, in a separate research project, acoustic telemetry was used to estimate sea trout smolt survival from the trap (personal communication, Ignacio Serrano, Umeå University, Sweden) and indicated that 86%, 84%, and 74% reached the lower river, estuary and coastal area, respectively. If salmon survival rates are similar to survival rates observed in sea trout, then we should reduce our estimates of smolt abundance (which are based on smolt-traps) by about 20% in order to obtain estimates of the production of smolt leaving the river.

The time window for downstream migrating smolts in the R. Sävarån was generally consistent among years, and trout smolts migrated 5–10 days earlier than salmon on average. Thus, the trap was in place before the onset of the migration and the population estimates should not be biased by missing fish before the trapping started. The highest frequency in downstream migrating salmon smolts occurred between 10–12°C, which coincided with published reports that found increased downstream migration at temperatures of 8–10°C (Österdahl 1969;

Jonsson and Ruud-Hansen 1985). This information can be used to prepare for the start of trapping each year.

Supplemental stocking of salmon from non-indigenous stocks was undertaken to build up a new population in the belief that the original population of Sävarån salmon was lost. However, this study revealed that this was not the case, and that a unique population exists without signs typically associated with population bottlenecks, such as low heterozygosity. Genetic studies in 2005 and 2006 found that salmon smolts were unique to the Sävarån (Nilsson *et al.* 2008), while in 2007 and 2008 a large fraction of smolts originated from hatchery released fish. The origin of these smolts must come from supplemental stockings done before 2006 since no stockings were performed after 2005. The observation that a considerable component of the original salmon stock exists suggests that management of Sävarån salmon should be redirected toward preserving the indigenous population. Managers need accurate estimates of annual smolt production to predict the number of adult returns and to safeguard stocks. To evaluate the goal of the Salmon Action Plan, it is also important to have reliable measures on the smolt production. In most rivers no explicit monitoring of smolt occurs, but the estimate is based on electro-fishing. The smolt production is then calculated using the number of recruiters from adults that spawned 2–4 years earlier. For Sävarån, annual electro-fishing data were compared with the observed smolt production but no relationship was found between parr densities based on electro-fishing and estimated smolt abundance based on trapping. Thus, electro-fishing did not provide a good estimate of smolt production. This may have been a result of several factors.

- Electro-fishing is sensitive to flow level and currents at the time of sampling (e.g. the high densities observed in 2006 for both 0+ and older fish probably resulted from extremely low flows that year) and can lead to biased estimates during extreme years that are difficult to separate from temporal variation in recruitment success.
- The electro-fishing sites were not randomly selected or stratified initially, so up-scaling the recruitment success to the watershed level without bias was not guaranteed.
- The method focused on seemingly high quality 0+ parr habitats, but other critical rearing habitats may have been missed.
- The method is not applicable for estimating trout smolt production. More trout were captured during electro-fishing than salmon. However, sea trout smolt counts at the trap were significantly lower than salmon, implying that the majority of the trout parr electro-fished were resident brown trout, not anadromous sea trout.

The ultimate goal of conservation efforts in the Baltic region is to preserve viable populations of anadromous salmonids, and the conservation of multiple populations is considered vital for the conservation of a species (Allendorf and Luikart 2007). However, before stock conservation plans can be developed, the basic unit of management effort has to be identified. For example, evolutionary significant units (ESUs), based on populations being reproductively isolated or genetically or ecologically unique (Allendorf and Luikart 2007), are used in the Pacific North-western USA to manage salmon populations. In addition, estimates of interaction and dispersal rates between populations are crucial to the development of adequate management strategies. To develop conservation plans and track their success, monitoring is needed on populations or sub-populations to document levels of abundance, and life-history and genetic diversity over time. This should include recruitment success, downstream migrations, harvest, upstream migrations, and the amount and quality of spawning and rearing habitat available. In the Sävarån case study, data on recruitment success and downstream migrations were collected, but data on the additional monitoring components need to be collected for the River Sävarån in the future. We conclude that the smolt trapping, genetic and tagging methods described provided critical information on wild salmon and trout population status in the River Sävarån, and that the river is a suitable index site for monitoring salmonid populations and trends in the forest rivers of northern Sweden.

Acknowledgements We thank B. S. Wiklund, H. Rönnholm, R. Grönlund (field work and technical support), H. Köningsson (genetical analysis), Å. Forssen (Norrfors hatchery), I. Holmgren (scale readings) and finally Sävaråns FMU for support. This multi-project was financed by governmental funding for fishery management within the SAP (Salmon Action Plan) under ICES, by the Swedish Board of Agriculture for the development of environmental monitoring at the Swedish University of Agricultural Sciences (SLU) (FoMa-*fisköververvakning*), by the Swedish National Board of Fisheries and by the Swedish Environmental Protection Agency. The Swedish Animal Welfare Agency (CFN, Umeå, Vb-län, Dnr. A35-05 and CFN Dnr. A5-06) approved these studies.

References

Allendorf FW, Luikart G (2007) Conservation and the genetics of populations. Blackwell, Oxford

Carlsson SR, Coggins LG Jr, Swanton CO (1998) A simple stratified design for mark-recapture estimation of salmon smolt abundance. Alaska Fish Res Bull 5:88–102

Degerman E, Sers B (1999) Elfiske. Standardiserat elfiske och praktiska tips med betoning på säkerhet såväl för fisk som fiskare. Fiskeriverket Information 1999:3. Fiskeriverket, Sötvattenslaboratoriet, Örebro (in Swedish)

Johansson S (2004) Laxfisket i Sävarån i forna tider. Redovisning av uppdrag 020927 från FiV försöksstation (Älvkarleby). Appendix in U. Carlsson (2004) Förstudie: Sävarån som indexvattendrag för lax. Rapport 2004-06-10. Länsstyrelsen i Västerbottens län, Umeå (in Swedish)

Jonsson B, Ruud-Hansen J (1985) Water temperature as the primary influence on timing of seaward migrations of Atlantic salmon (*Salmo salar*) smolts. Can J Fish Aquat Sci 42:593–595

McElhany P, Ruckelhaus M, Ford M, Wainwright T, Bjorkstedt E (2000) Viable salmonid populations and the recovery of evolutionary significant units. NOAA Technical Memorandum NMFS-NWFSC-42

McKinnell SM (1998) *Atlantic salmon* (*Salmo salar* L.) life history variation: Implications for the Baltic Sea fishery. Doctoral thesis, Swedish University of Agricultural Sciences. SLU, Umeå

Nilsson J, Gross R, Asplund T, Dove O, Jansson H, Kelloniemi J, Kohlmann K, Löytynoja A, Nielsen EE, Paaver T, Primmer CR, Titov S, Vasemägi A, Veselov A, Öst T, Lumme J (2001) Matrilinear phylogeography of Atlantic salmon (*Salmo salar* L.) in Europe and postglacial colonisation of the Baltic Sea area. Mol Ecol 10:89–102

Nilsson J, Östergren J, Lundqvist H, Carlsson U (2008) Genetic assessment of Atlantic salmon *Salmo salar* and sea trout *Salmo trutta* stocking in a Baltic Sea river. J Fish Biol 73:1201–1215

Norberg M, Ahlström J (2007) Plan för biologisk återställning i kalkade vatten i Västerbottens län 2007–2010. Umeå, Länsstyrelsen i Västerbottens län (in Swedish)

NRC (1996) Upstream: Salmon and society in the Pacific Northwest. Report of the committee on protection and management Pacific Northwest anadromous salmonids. National Academy Press, Washington, DC

Österdahl L (1969) The smolt run of a small Swedish river. In: Northcote TG (ed) Salmon and trout in streams. H.R. MacMillan Lectures in Fisheries. University of British Columbia Press, Vancouver, pp 205–215

Pritchard J, Wen W (2004) Documentation for STRUCTURE software: Version 2. Available at: http://pritch.bsd.uchicago.edu/structure.html (accessed on 2 May 2009)

Rappe C, Ranke W, Soler T, Funegård P, Karlsson L, Thorell L (1999) Baltic salmon rivers – Status in the late 1990s as reported by the countries in the Baltic Region. International Baltic Sea Fishery Commission (IBSFC) and Baltic Marine Environment Commission – Helsinki Commission (HELCOM) in co-operation with the Swedish Environmental Protection Agency and the Swedish National Bord of Fisheries. Göteborgs Länstryckeri, Göteborg

Raymond M, Rousset F (1995) GENEPOP (version 1.2): Population genetics software for exact tests and ecumenicism. J Hered 86:248–249

Rivinoja P, Carlsson U (2008) Miljöövervakning av lax i Västerbotten: Effekter av laxutplantering i Öre- och Lögdeälven under femton år. Rapport nr 59, Institutionen för Vilt, Fisk och Miljö, SLU. SLU, Umeå (in Swedish)

Säisä M, Koljonen M-L, Gross R, Nilsson J, Tähtinen J, Koskiniemi J, Vasemägi A (2005) Population genetic structure and postglacial colonisation of Atlantic salmon (*Salmo salar*) in the Baltic Sea area based on microsatellite DNA variation. Can J Fish Aquat Sci 62:1887–1904

Seber GAF (1982) The estimation of animal abundance and related parameters, 2nd edn. Macmillan, New York

Thedinga JF, Murphy ML, Johnson SW, Lorentz JM, Koski KV (1994) Determination of salmonid smolt yield with rotary-screw traps in the Situk River, Alaska, to predict effects of glacial flooding. North Am J Fish Manage 14:837–851

WGBAST (2008) Report of the Baltic Salmon and Trout Assessment Working Group (WGBAST), 1–10 April 2008. ICES, Copenhagen

Chapter 21
Shad Monitoring in the Afon Tywi SAC: A Case Study

Richard West

Introduction

Shad species are members of the herring family (Clupaeidae) and have a broad global distribution, with approximately 200 different species found in all seas around the world, except for the Antarctic. Shad primarily inhabit marine ecosystems, but some species are anadromous, i.e. adults return to fresh water to spawn and juvenile fish spend part of their life in fresh water before returning to the sea to continue their development (Maitland and Hatton-Ellis 2003).

Shad have a generally westerly distribution across Europe and have been recorded primarily in major rivers discharging into the Atlantic. The allis shad (*Alosa alosa* L.) and twaite shad (*Alosa fallax* L.) are the only members of the herring family found in freshwater habitats in the UK (Maitland and Hatton-Ellis 2003). They are similar in appearance but can be identified to species level through the number, length and spacing of their gill rakers. Abundances of shad species have declined in recent years throughout their geographic ranges, primarily as a result of migration barriers such as dams and weirs, pollution and habitat destruction (Aprahamian *et al.* 2003). As a result, both allis and twaite shad are protected under Appendix III of the Bern Convention and Annexes II and V of the EC Habitats Directive 92/43/EEC.

Allis and twaite shad populations are protected as designated features under the EC Habitats Directive within the Afon Tywi Special Area of Conservation (SAC) in South-west Wales (Fig. 21.1). Although allis shad are extremely rare, with no confirmed records of the species within recent years, twaite shad have regularly spawned in the river (Aprahamian and Aprahamian 1990; CCW 2005).

The freshwater phase of shad spawning migration is temperature dependent, generally taking place when water temperature ranges from 12°C to 20°C. Tidal cycles and river flow levels also influence migration, which can be halted during strong spate conditions (Aprahamian *et al.* 2003).

R. West (✉)
Environmental Monitoring Officer (Marine & Water Quality), Environment Agency, Rivers House, St Mellons Business Park, Fortran Road, St Mellons, Cardiff, CF3 0EY, UK
e-mail: Richard.West@environment-agency.gov.uk

C. Hurford et al. (eds.), *Conservation Monitoring in Freshwater Habitats: A Practical Guide and Case Studies*, DOI 10.1007/978-1-4020-9278-7_21,

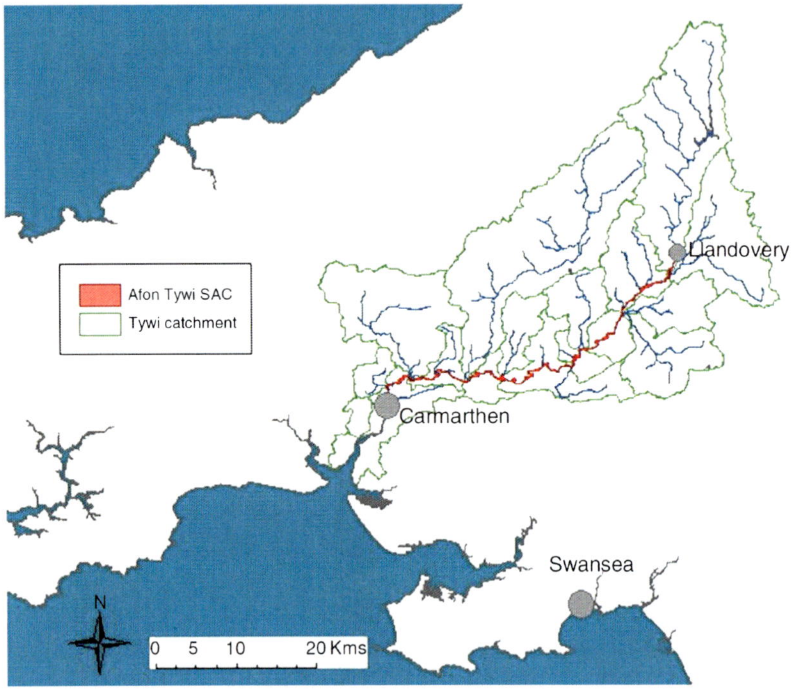

Fig. 21.1 The location of the Afon Tywi Special Area of Conservation in South Wales, UK.

Shad demonstrate a preference for upstream movement at temperatures between 11–13°C and spawning generally takes place once water temperatures rise above 15°C. Shad eggs are particularly sensitive to water temperatures below 16°C and the optimal temperature regime for egg incubation is temperatures exceeding 18°C throughout June and July (Maitland and Hatton-Ellis 2003; McIlquham, 2006).

Water temperatures in the Tywi River have been impacted by the Llyn Brianne dam, which was constructed in the upper reaches of the catchment in 1971 for public water supply and subsequently modified for hydropower production. Water has historically been released from the lower levels of Llyn Brianne below a thermocline in the lake, which develops over summer months, introducing cold water into the upper Tywi at the same time as the freshwater phase of the shad life cycle. We had to take this factor into account when developing the methods for shad monitoring in the Tywi.

For monitoring purposes, Hillman *et al.* (2003) recommended that, where sufficient baseline data were available, the adult shad run should be assessed using hydroacoustic fish counters and video footage – and that the target should be based on the 6-year average of the total shoal count. Shad egg surveys should then commence following the sighting of the first adult shoal run of the season and the distribution investigated through a comprehensive kick-sampling survey, the results from the first year providing the baseline target for subsequent monitoring events. Reductions of more than 50% spawning distribution would represent an adverse change in the distribution of the species. In addition, three juvenile seine nettings surveys should be carried out between July and October each year, with annual comparisons of catch per unit effort data used against population targets to inform condition assessments.

The primary aim of this case study is to outline an approach developed for monitoring the spawning distribution of shad in rivers, as a contribution to assessing the conservation status of the species.

Pre-existing Survey Information

Aprahamian and Aprahamian (1990) identified a spawning population of shad within the Afon Tywi catchment and Environment Agency staff carried out the first attempt at shad egg surveys within the catchment in 1998. Thereafter, the initial Habitats Directive Review of Consents funded shad egg monitoring work took place in 2004 (Smith 2005a). The sites surveyed for shad eggs in 1998 and 2004 informed the selection of sites for subsequent surveys, with the number of sites later extended using aerial photography images and salmonid survey site data to identify other suitable locations on the river.

In addition, an underwater camera array was deployed in 2005 in the lower reaches of the catchment, downstream of one of the major abstraction points under review. Camera array video recordings were primarily used to identify the timing of adult shad runs up into the Tywi catchment to spawn.

An Environment Agency temperature monitoring study on the Afon Tywi was carried out in 2004 by Smith (2005b) using spot samples at a limited number of sites. Subsequently, an extensive network of continuous temperature data loggers was established within the catchment to gather time series temperature data. Monitoring sites were also located for comparative purposes on the Doethie tributary in the Upper Tywi catchment, which flows to the west of Llyn Brianne and joins the upper Tywi river some 3 km downstream of the dam. In addition, a passive thermistor array was deployed under contract in Llyn Brianne to record water temperatures at a range of depths throughout the reservoir water column.

The 1998 shad egg survey recorded spawning activity at three locations, upstream of White Mill (NGR SN468215), Llanegwad (NGR SN515212) and downstream of Llandeilo Bridge (NGR SN627220) (Fig. 21.2). Smith (2005a) recorded shad eggs at six locations from the tidal limit at Llangunnor (NGR SN424203) to Nantgaredig Bridge (NGR SN493203) on the Afon Tywi and on the Afon Cothi just upstream of its confluence with the Tywi (NGR SN501201). Smith (2005b) noted that river flows were low during May and early June 2004 and subsequently a freshwater release from Llyn Brianne was authorised on 10–13 June 2004 to encourage salmonid migration into the river system. An ensuing reduction in water temperature was recorded up to 70 km downstream of the reservoir and

Smith (2005b) noted that shad eggs were not observed above the point where temperatures on the Tywi were impacted by releases from Llyn Brianne, upstream of its confluence with the Afon Cothi.

The Generic Conservation Targets

As the EU Habitats Directive requires member states to report on the conservation status of habitats and species listed in Annex I and Annex II of the Directive on a 6-yearly cycle, CCW (2008) specified indicators for the condition of both shad species in the Afon Tywi (under the heading of 'Performance Indicators' in the management plan). These indicators of condition, which had to be met in order to achieve favourable conservation status, were stated as

1. No decline in the adult annual run size greater than would be expected from variations in natural mortality alone
2. No decline in the spawning distribution of both species

These same 'generic' indicators of condition were to be applied to all rivers in the UK where either twaite or allis shad had been designated a Natura 2000 species. The methods described below were developed to address the second of these targets.

Methods

Onset Hobo Water Temp Pro passive temperature loggers were installed by the Environment Agency at a number of locations throughout the Afon Tywi catchment from March – October 2005 and April – November 2006 to monitor the thermal impact of regulated releases from Llyn Brianne (Fig. 21.2). Loggers were located at accessible sites to give a wide distribution of locations throughout the catchment. The loggers were attached using re-useable cable ties to the lowest available point on trees, roots or any other available point as low in the water column as possible to try and maintain continuous submersion. Data were downloaded monthly, using an infrared Onset base station plugged into a laptop computer.

Hydro-acoustic surveys were carried out at Ty Castell near Nantgaredig (NGR SN491203) using a 200 kHz HTI counter in conjunction with a 420 kHz system, installed in 2002 with the aim of recording shad shoal data, that detects shad without disturbing their migration. Adult shad are known to avoid a standard 200 kHz frequency counter of the type normally used to monitor salmonids (McIlquham 2006). Video cameras were also installed at White Mill (NGR SN466214) to assist with species identification of migrating fish shoals. Nine submersible multiplexed cameras were placed across a 40-m channel in three arrays of three cameras. The river width was split into three 13 m sections and an array was placed randomly

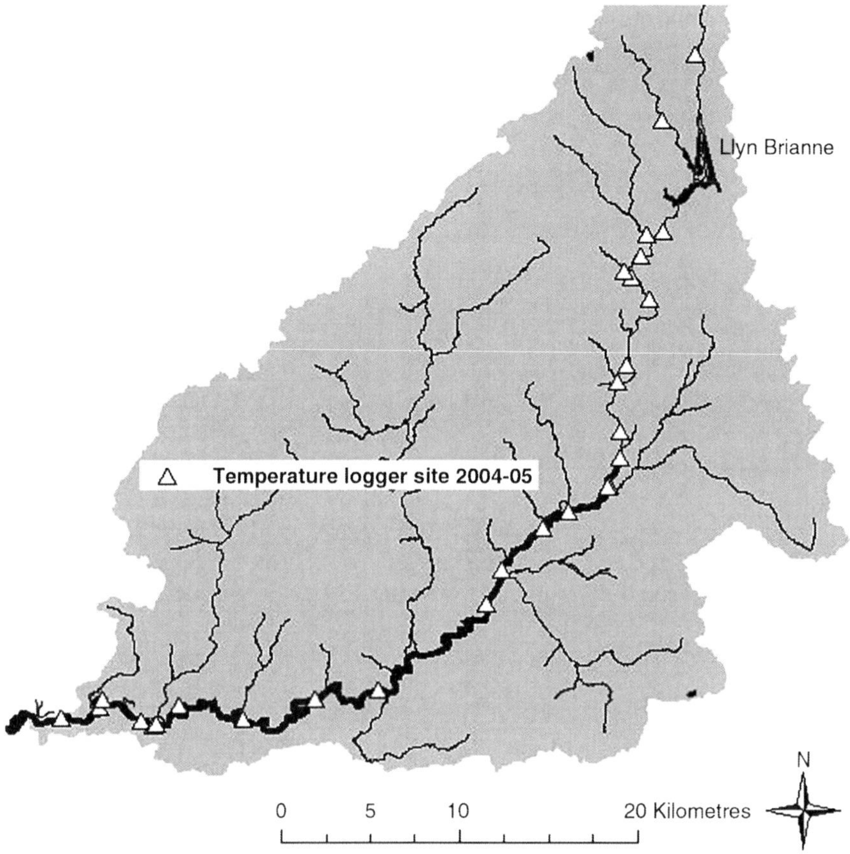

Fig. 21.2 Afon Tywi temperature logger locations 2005–2006. © Crown Copyright and/or database right. All rights reserved. Licence number 100043571

within each section. Each camera view was perpendicular to the direction of flow to enable species identification. Time-lapse recording was used to maximise the data stored on each videotape, recording at 10 frames per second. Data were recorded from 5 April 2005 to the 29 June 2005 to ensure the whole shad migration period was surveyed (McIlquham 2006).

Shad egg surveys were carried out at 10 sites between 10 May 2005 and 18 July 2005 and 14 sites between 12 May 2006 and 4 July 2006, between Llangunnor (NGR SN424205) and Llandovery Bridge (NGR SN761347) on the Tywi, and from just upriver of the Tywi/Cothi confluence (NGR SN500201) to Pont-ar-Gothi on the Cothi (NGR SN504219). Sites were selected using aerial photographs and previous Environment Agency salmonid survey records to identify areas of accessible riffle habitat typically associated with shad spawning. The 2005 and 2006 shad egg surveys were conducted in accordance with the methodology set out in Hillman *et al.* (2003). A standard 250-μm macroinvertebrate pond net was used to complete 25 fifteen-second kick samples at each site in the river stretch downstream of the riffle to within the riffle area (Fig. 21.3). Particular care was taken to include any areas of eddying water where shad eggs could be expected to congregate. Each kick sample was examined in a white tray with a little river water to identify whether any neutrally buoyant and non-aggregating shad eggs were present (Fig. 21.4) and the number of shad eggs found was recorded. Data for 1998 and 2004 shad egg surveys were obtained from Environment Agency records.

Results

Hydroacoustic and video data found shad migration into the Tywi occurred at temperatures between 11°C and 20°C between 5 April and 29 June 2005, with the highest numbers of shad at temperatures between 12°C and 15°C and a flow regime of 8–19 cumecs (m^3 sec^{-1}) (McIlquham 2006). The first migrating shoals of shad were detected on 10 May 2005 and the final shoal was detected on 29 June 2005.

Shad egg surveys were carried out between 10 May and 18 July 2005 at ten locations on the Tywi and Cothi rivers. Shad eggs were confirmed at only one site in the Tywi catchment in 2005, at White Mill (NGR SN468215) on 23 June 2005.There was a marked increase in the distribution of shad eggs observed in the 2006 survey period between 12 May and 4 July 2006. Shad eggs were recorded for the first time since 1998 at Llandeilo Bridge (NGR SN626220) on 8 June 2006 and the first confirmed observations of shad eggs higher in the catchment were recorded at Manordeilo (NGR SN687268) also on 8 June 2006, Llanwrda (NGR SN717308) on 20 June 2006

Fig. 21.3 The author sampling for shad eggs downstream of a riffle in the Afon Tywi in June 2006. *Photo by Clive Hurford*

and close to Lwynjack Farm (NGR SN754331) on 21 June 2006. This last location is 1.8 km downstream from Llandovery and 56 km above the tidal limit (Fig. 21.5).

Average daily water temperatures in the 5 km of the Tywi immediately upstream of the tidal limit between Carmarthen and White Mill did not consistently exceed the 11°C shad migration threshold until 27 April 2005. Temperatures between Carmarthen and White Mill fluctuated throughout June between 14–20°C and only consistently exceeded 18°C from 9 to 19 July 2005 before declining below the threshold survival limit of 16°C for shad eggs and larvae on 28 July 2005.

The weather conditions over the Tywi catchment were more favourable for shad species in 2006 with persistent rainfall throughout May, resulting in Llyn Brianne overtopping the dam for the entire month, causing higher flows throughout the Afon Tywi and warmer water temperatures on the main river. When temperature monitoring commenced on 28 April 2006, average daily temperatures were already in excess of the 11 °C migration threshold 47 km above the tidal limit at Llangadog (NGR SN513210). Although average daily water temperatures fell back below 11°C between 30 April and 2 May and from 22 to 24 May, the 15°C spawning threshold was exceeded from 3 June 2006 until the end of the study period (2 August 2006) between Llanegwad and the tidal limit, representing a 22-km stretch of the lower Tywi river. Average daily temperatures exceeded the 16°C egg and larvae survival threshold from the end of June until early August up to Llandeilo (NGR SN626220) and average daily temperatures of >16°C were recorded at Llangadog (NGR SN513210), 47 km above the tidal limit, between 15 and 27 July 2006.

The results of comparative temperature monitoring work on the lower 75 km stretches of the rivers Wye and Tywi between spring and autumn 2005, showed that average daily water temperatures were consistently higher in the River Wye than in the

Fig. 21.4 A shad egg (the spherical object below the '20' on the 20 pence piece) in a white sampling tray. *Photo by Clive Hurford*

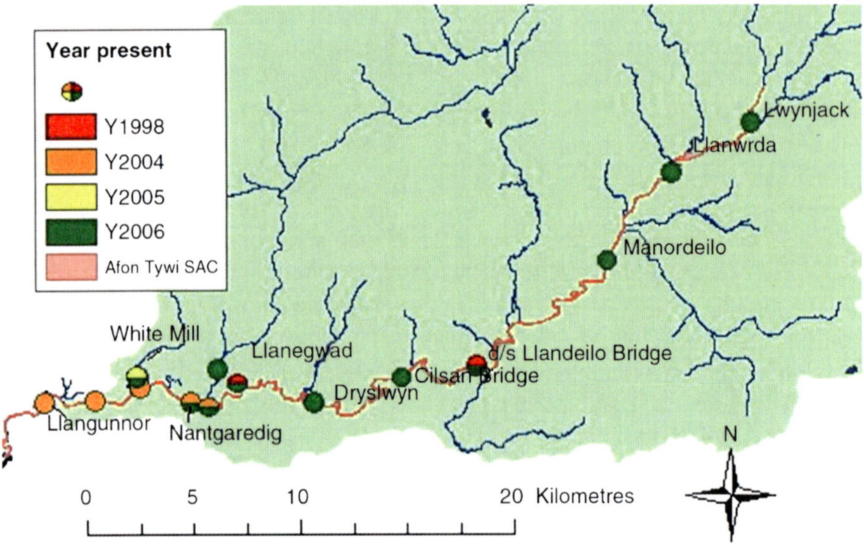

Fig. 21.5 Afon Tywi shad egg survey results, 1998–2006. © Crown Copyright and/or database right. All rights reserved. Licence number 100043571

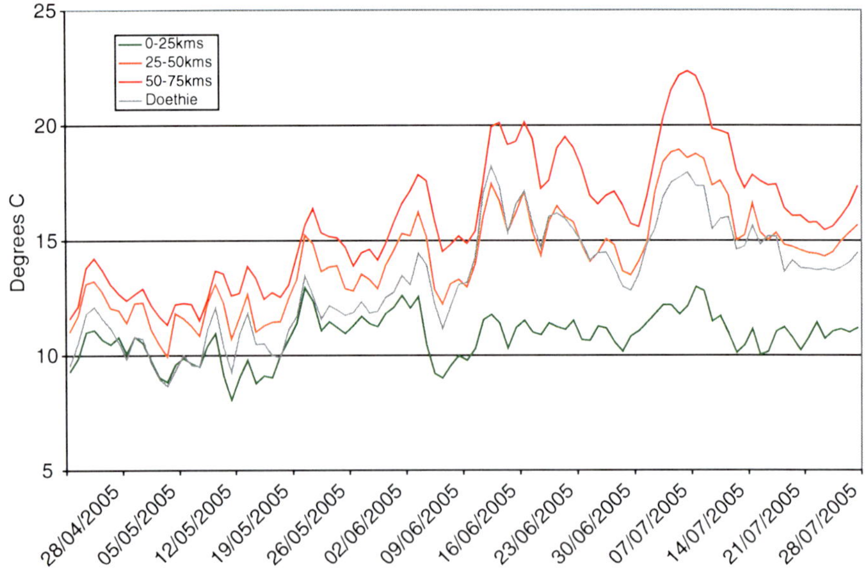

Fig. 21.6 Afon Tywi average daily water temperature (°C) in different 25 km stretches downstream of Llyn Brianne and in the Doethie tributary, 2005

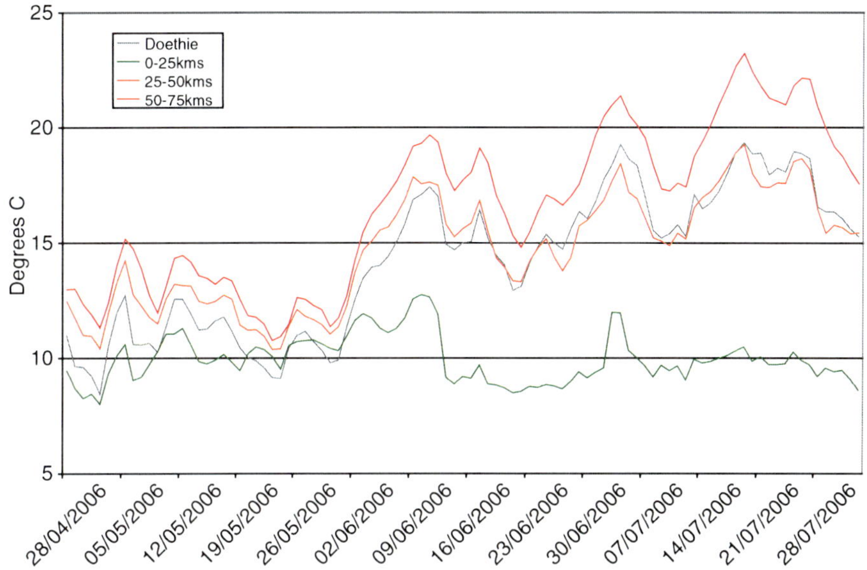

Fig. 21.7 Afon Tywi average daily water temperature (°C) in 25 km stretches downstream of Llyn Brianne and in the Doethie tributary, 2006

comparable stretch of the River Tywi. Temperatures on the River Wye exceeded the shad spawning threshold of 15°C from the end of May until mid-August 2005.

Monitoring showed that average daily water temperatures in the Afon Tywi were impacted by regulatory releases of water from below the thermocline in Llyn Brianne between spring and late summer in 2005 and 2006 (Figs. 21.6 and 21.7). There was a marked difference between average daily water temperatures in the Doethie tributary, which joins the main Afon Tywi 2.8 km downstream of Llyn Brianne, and temperatures in the 25 km of main river immediately below the dam.

Discussion

At the time of our survey, the shad monitoring and condition assessment targets (see Cowx *et al.* this volume) were primarily informed by the Afon Tywi Special Area of Conservation Site Issue Briefing published by the Countryside Council for Wales (Version 1.5: CCW 2005). This document aimed to provide a general description of the site and its designated features, a future vision for the site, a detailed breakdown of the factors affecting the individual designated features (including some targets and priorities), an indication of feature condition and a table of issues and actions.

The targets for maintaining the presence of allis and twaite shad and enhancing their conservation status within the Afon Tywi Special Area of Conservation focused on achieving favourable water quality, ensuring that in-channel features were conducive to spawning and migration, maintaining adequate flow regimes to support migration, providing sufficient riparian vegetation for the requirements of the species and preventing any potentially detrimental developments impacting on the designated protected area. Allis shad, which were considered to be rare within the Tywi catchment, were provisionally described as being in unfavourable condition, whereas Twaite shad, which have been sighted in large shoals within the river system at certain times of year, were provisionally said to be in favourable condition. These were amended to unfavourable (unclassified) condition for both species in the subsequent Core Management Plan.

These performance indicators highlight the main problem with setting monitoring targets for the Tywi River and indeed many other rivers. The Joint Nature Conservancy Council (JNCC) recommended that targets for adult run size, spawning distribution and juvenile density should be set using baseline data gathered from surveys in six of the 7 years comprising the monitoring cycle for SACs (Hillman *et al.* 2003). To date, however, no such comprehensive survey effort has been made to assess allis and twaite shad populations within the Tywi catchment and the information available is based on ad hoc survey effort primarily aimed at particular life stages of the fish. A more comprehensive and sustained survey programme is therefore required before informed monitoring targets can be set for the Afon Tywi Special Area of Conservation.

The shad monitoring work carried out by the Environment Agency demonstrated that regulation of the Afon Tywi through freshwater releases from Llyn Brianne impacted on temperatures in the upstream reaches of the river. Shad migration, spawning and egg survival are known to be temperature dependent (Maitland and Hatton-Ellis 2003; Aprahamian *et al.* 2003). Although the temperature differences between the Tywi and Wye Rivers could be caused by a number of additional factors, including the different geomorphology of river substrata within the two rivers, the most upstream evidence of shad spawning activity in the Tywi prior to 2006 was 33 km upstream of the tidal limit at Llandeilo Bridge (NGR SN626220), whereas shad eggs were found 190 km upstream of the tidal limit on the River Wye at Builth Wells (NGR SO045512) (McIlquham, 2006). The shad egg survey carried out on the Afon Tywi in 2006 showed that it was possible for shad to spawn higher into the catchment when temperature and flow conditions were favourable and that the spawning distribution of twaite shad within the catchment could reach almost to the upstream boundary of the Special Area of Conservation under such circumstances.

Environment Agency research on shad and other species and habitats protected under the EU Habitats Directive within the Afon Tywi SAC has input into the Review of Consents process, which is to this date ongoing. Discussions are continuing with licence holders over amendments to existing water abstraction licenses to effectively limit their impacts upon protected species and habitats, including the regulatory regime of the river through releases from the reservoir in the headwaters of the Afon Tywi.

Additional research on other shad life stages is adding to the foundation of knowledge on the species within the Afon Tywi catchment. This includes juvenile netting surveys, which were carried out in 2005–2006 in the Tywi and a number of other South Wales estuaries to investigate the diets and parasites of larval and juvenile shad (Nunn *et al.* 2008). There remains, however, an incomplete picture of both the population density and the distribution of shad within the catchment, making it impossible to provide well-informed conservation targets for the Tywi River at the present time.

The monitoring protocols described by Hillman *et al.* (2003) provided details of an effective survey procedure for investigating juvenile, adult and spawning shad life stages. The importance of measuring temperature, flow and other water quality variables was emphasised by this case study. It was also useful to consider historical salmonid monitoring sites, where suitable riverine conditions exist to carry out shad egg surveys, in identifying the shad egg survey site locations in conjunction with River Habitat Survey sites mentioned by Hillman *et al.* (2003). Additional research by Nunn *et al.* (2008) highlighted the importance of further investigation into shad diet and parasite loading as useful indicators of population stress and species ecology of larval and juvenile shad life stages.

A comprehensive and sustained survey effort following the methodology set out by Hillman *et al.* (2003), with adult spawning run, shad egg and juvenile netting research carried out simultaneously represents a significant task in terms of staff and resources for any one publicly funded agency. Consequently, collaborative work using existing expertise in educational institutions, consultancies and public conservation bodies is needed to gain sufficient data on shad populations in all freshwater life stages. These data could then be used to develop informed 'site-specific' conservation targets for assessing the condition of the shad populations within a river catchment.

The shad egg surveys carried out in the Afon Tywi during the summers of 2005 and 2006 gave an insight into the distribution and levels of shad breeding we can expect to find in the river under both sub-optimal and optimal water temperature conditions. These data should be used to inform site-specific targets for shad on the Afon Tywi.

Acknowledgements Thanks are due to Gemma Dickson and Gareth Davies who helped enormously with the shad monitoring survey work.

References

Aprahamian MW, Aprahamian CD (1990) Status of the genus *Alosa* in the British Isles; past and present. J Fish Biol 37:257–258

Aprahamian MW, Baglinière JL, Sabatié R, Alexandrino P, Aprahamian CD (2003) *Alosa alosa* and *Alosa fallax* spp. literature review and bibliography. Environment Agency Research and Development Report W1-014, Bristol

CCW (2005) Afon Tywi special area of conservation site issue briefing, Version 1.5. Countryside Council for Wales, Bangor, p 38

CCW (2008) Core management plan including conservation objectives for Afon Tywi/River Tywi SAC (Special Area of Conservation). Countryside Council for Wales, Bangor, p 42

Hillman RJ, Cowx IG, Harvey JP (2003) Monitoring the Allis and Twaite Shad. Conserving Natura 2000 rivers ecology series No. 3. English Nature, Peterborough

Maitland PS, Hatton-Ellis TW (2003) Ecology of the Allis and Twaite Shad. Conserving Natura 2000 rivers ecology series No. 3. English Nature, Peterborough

McIlquham G (2006) Analysis of shad migration in the River Tywi during 2005: Data collected by video monitoring. Environment Agency Technical Report EASE/REP/06/02, p 23

Nunn AD, Noble RAA, Harvey JP, Cowx IG (2008) The diets and parasites of larval and juvenile twaite shad in the lower reaches and estuaries of the rivers Wye, Usk and Towy, UK. Hydrobiologia 614:209–218

Smith M (2005a) River Tywi cSAC: Potential impacts of abstraction and river regulation on shad, *Alosa* spp. Environment Agency Wales Technical Report EATW/05/01.1, 40 pp

Smith M (2005b) Llyn Brianne reservoir: Temperature effects in the river Tywi and their effects on shad, *Alosa* spp. Environment Agency Wales Technical Report EATW/05/01.3, 40 pp

Chapter 22
Monitoring the Effects of Acidification and Liming on Water Quality in a Boreal Stream: The River Stridbäcken in Northern Sweden

Johan Ahlström and Mats Johansson

Introduction

Acidification of lakes and streams has been, and still is, one of the most serious environmental problems of our time. Due to its base-poor bedrock, Scandinavia is one of the worst affected areas in Europe. Although the northern parts of Sweden are far from the main point sources in central Europe, the biota of both lakes and streams has been extensively damaged (Degerman *et al.* 1987; Degerman *et al.* 1992; Ahlström *et al.* 1995).

Acidification in Sweden has primarily been caused by the deposition of sulphur, mostly culminating during the latter half of the 1970s. The sulphur originates primarily from the combustion of fossil fuels. Dissolved in the atmosphere, the sulphur may travel hundreds of kilometres from the source. The main part of the sulphur deposition in Sweden originates from foreign coal power plants.

The liming of lakes and rivers is carried out extensively in Sweden and Norway, primarily financed by state subsidies. In Sweden, liming started in 1977 and the total cost to date is 415 million €. Presently, 21 million € in state subsidies is spent annually on liming. In 2007, 143,000 t of lime were spread in Sweden, with around 7,000 lakes and 15,000 km of rivers treated.

The River Stridbäcken

The Stridbäcken is a small river draining directly into the Bothnian Sea (Fig. 22.1). The catchment area is 16 km^2 and is characterised by a steep topography. During the last Ice Age, the area was covered by an almost 3 km thick ice-sheet, which pressed the earth crust down considerably. After the deglaciation, the sea extended

J. Ahlström (✉) and M. Johansson
County Administration, Västerbotten County, SE-901 86, Umeå, Sweden
e-mail: Johan.Ahlstrom@lansstyrelsen.se

C. Hurford et al. (eds.), *Conservation Monitoring in Freshwater Habitats: A Practical Guide and Case Studies*, DOI 10.1007/978-1-4020-9278-7_22,

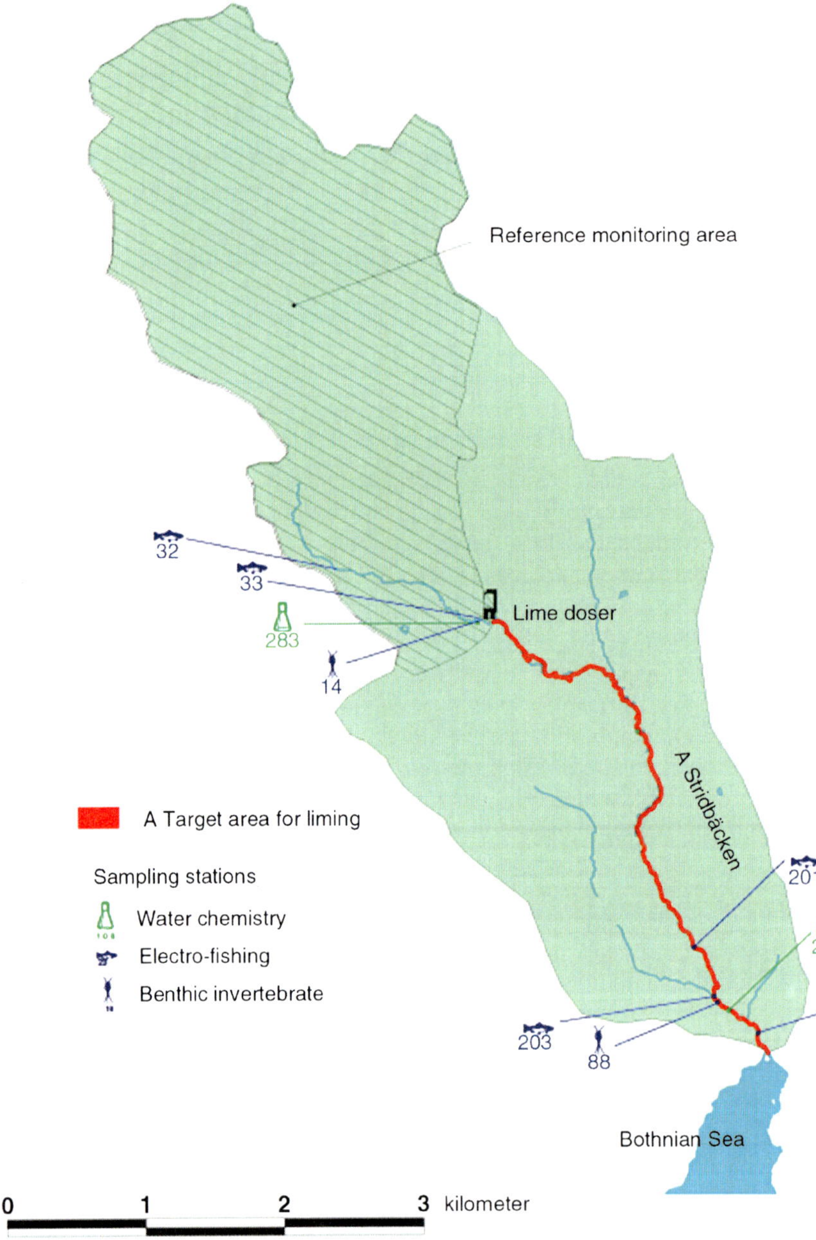

Fig. 22.1 Map of the River Stridbäcken with catchment area and monitoring stations

much further inland than it does today. Since then, the sea has slowly and gradually receded because of the crustal rebound. The land upheaval in the region is today 8 mm per year.

The Stridbäcken catchment area is completely below the former highest coastline and has been exposed to intensive wave action since the last Ice Age. Bare bedrock constitutes half of the catchment area. The rest is largely covered with washed moraine (till) and sandy, washed sediments. These soil types are characteristically sensitive to acidification because of a low content of fine particles and thereby a low capacity for ion exchange and a low degree of weathering. The topography and soils also accelerate run-off, which allows precipitation to reach the river rapidly with no time for accompanying acids to be neutralised.

Acidification in this part of Sweden is primarily caused by deposition of sulphur. This deposition culminated during the latter half the 1970s: pH in the precipitation then reached a minimum of 4.2 measured as an annual mean. This has increased to 4.8 today. Acidification has caused considerable damage to the fauna in the R. Stridbäcken. The river is a spawning and nursery area for sea trout, and in the 1950s, sea trout of 2–3 kg in size were captured in the river mouth. Before the liming started in the early 1990s, however, these had virtually disappeared from the river, and the benthic invertebrate community was entirely dominated by acid tolerant species.

Liming

Liming started in spring 1993, when a lime doser was installed 4.7 km from the river mouth. This first doser, however, had a primitive technical design, which led to recurring breakdowns in the operation. The amount of lime disseminated was regulated manually, which made it impractical to align the lime dose with the current discharge of water in the river. Improvements were made in later years, however, and in autumn 1999, the doser underwent a radical renovation: this enabled the automatic regulation of lime dosing through using an echo sounder to continuously record the water levels. Before the spring flood in 2003, the technique was again upgraded with a discharge-regulated dose system.

Liming Methods

Three methods of liming are used in Sweden:

- Lake liming – finely ground lime (∅ 0–0.2 mm) spread by boat or helicopter directly on the lake surface.
- Wetland liming – coarsely ground lime (∅ 0.2–1.0 mm) or granulated limestone powder spread by helicopter on the ground in areas with upwelling groundwater (fens).
- Dose liming – limestone powder contained in silos applied dose-wise directly into rivers. The lime dosers are often regulated by remote control.

From a catchment perspective, several methods are usually combined to achieve an optimal result, especially when the whole river constitutes the target area for improvement.

Chemical and Biological Objectives

The overall objective of liming is to create the chemical requirements for viable populations of the native fauna to re-establish or flourish; pH is used as an operative target parameter (Naturvårdsverket 2002). Depending on the present (or previously present) fauna, the pH-target is set at 5.6, 6.0 or 6.3. The 6.3 target is used mainly for salmon rivers, with 6.0 used for rivers with freshwater pearl mussels and/or sea trout and for lakes with roach and/or crayfish. The objective requires the pH not to fall below the relevant target.

The water chemical objective in the R. Stridbäcken was initially set at pH 5.6. In 2004, this was increased to 6.0, following new national guidelines (Naturvårdsverket 2002).

There is also an objective to prevent and counteract over-liming. This is defined as a specified level of alkalinity that should not be exceeded at high river discharges. At a pH-target of 5.6, alkalinity should not exceed 0.07 mekv/l and at a target of pH 6.0 the threshold is 0.10 mekv/l.

The biological objectives are defined as the occurrence of benthic species sensitive to acidity or the occurrence of young individuals of acid-sensitive fish species, e.g. salmon *Salmo salar*, L. roach *Rutilus rutilus* (L.) and trout *Salmo trutta* L.

Design of the Monitoring Programme

The monitoring is designed to verify changes in water chemistry and biology after the introduction of the liming in the system and to maintain the water chemical responses at an optimal level. Sampling stations to monitor the effects of liming were placed downstream from the lime doser, while the area upstream from the lime doser served as an untreated reference area (Fig. 22.1).

Monitoring the Water Chemistry

Sampling Frequency

The purpose of the water chemistry monitoring is twofold, firstly to establish whether the liming objectives are being met, but also to serve as a basis for optimising the liming effort. As a consequence of how the objective is formulated, sampling is carried out when the risk of failing the objective is highest. In rivers, this means that samples are taken mainly during high flows caused by snowmelt or heavy rains. Samples are also taken if the liming effect declines or if the lime doser malfunctions. To meet these requirements, a flexible discharge-based sampling scheme is used. The recommended standard is to take one sample during the winter

Fig. 22.2 Lime doser. In front Mats Näslund and Jorun Storeng from the Nordmaling Municipality administration. *Photo by Johan Ahlström*

base-flow, six samples during the spring flood period, and four samples during high autumn flows.

In rivers limed by dosers, a more intensive sampling scheme is advised, i.e. with one sample in winter, 10 samples during spring, and 10 during the autumn. In reality, however, the sampling frequency is determined by the variation of flow throughout the year and may deviate considerably from the recommended schedule.

Sampling Stations

The sampling stations for water chemistry are located where the objective is considered to be most difficult to achieve. This is based on the pH-levels measured before liming and on the variation in the efficiency of the lime dose along the river. Usually, one sampling station per river is sufficient, but some rivers may require up to four stations. In rivers with lime dosers, an additional reference sampling station is located upstream of the doser. Accessibility is an important factor when choosing the location of the sampling stations: often they are placed near roads that can be accessed at all times of the year.

Analyses

All samples are analysed for pH, alkalinity, conductivity, colour, calcium, and magnesium. During spring flood, inorganic aluminium is also analysed on two occasions per station. The results from the analyses are available from the laboratory through a web service within 24 h of the samples arriving.

Biological Monitoring

The benthic invertebrate fauna is sampled once per year, directly after the spring flood has declined, i.e. when the critical effects of the spring flood on the composition of the benthic community can be observed. A selective method is used (the M42-method; Naturvårdsverket 2008a). This involves sub-sampling at 30 locations along a 50 m stretch of river. The sub-samples are taken from all habitat types represented along the stretch. Each sub-sample plot is about 0.2 m^2 and is sampled during 5 s by stirring up the substrate with the landing net or by kicking. Since the sampling is made in a larger and more variable area, this method is more effective in maximizing the number of species and recording indicator species than the standard non-selective method used in time-series monitoring (Naturvårdsverket 2008b). The non-selective method involves five sub-samples from a smaller area with homogenous bottom structure. The disadvantage of the selective M42-method is that it is more sensitive to observer variation (in individual sampling skills and experience) than other methods. After collection, the samples are sorted under a microscope; all specimens are picked out and determined to species level (or the lowest taxonomic level possible). The taxa are then classified according to a five-degree scale in acidic tolerance, based on Degerman *et al.* (1994):

Index 0 – Unknown tolerance
Index 1 – Tolerates pH < 4.5
Index 2 – Does not tolerate pH < 4.5
Index 3 – Does not tolerate pH < 4.9
Index 4 – Does not tolerate pH < 5.5

The fish fauna is monitored by electro-fishing according to the national standard method (Naturvårdsverket 2008c). This is carried out in the summer when the yearlings have reached a sufficient size (about 30 mm) to be captured effectively. The primary objective is to estimate the density of juvenile trout and salmon. The electro-fishing stations are therefore distributed in areas believed to be juvenile salmonid nursery habitat. The sampling areas are typically 50 m long and generally include the entire width of the river. Quantitative electro-fishing is practiced, which means three consecutive, exhaustive fishing efforts. All fish specimens are determined to species, weighed (g) and measured (fork length, nearest mm). Trout and salmon were separated into young-of-the-year (YOY) and older age classes using length frequency analysis. Population sizes of both groups were calculated according to Zippin (1956).

Results

Water Chemistry

Before the onset of liming, 27 water samples were collected at the station 282. The lowest recorded pH-value was 4.45 (Fig. 22.3). Since liming commenced in 1993, 286 water samples have been taken, which corresponds to an average of 19.1 samples per year. A maximum of 30 samples were taken in some years (2001 and 2007) and as few as 10 in 1996, when no spring flood occurred (Fig. 22.4). In 2006, flow variation required 16 samples to be taken, two in winter, nine during the spring flood, and five during the autumn high flows (Fig. 22.5).

The lowest recorded pH-value after liming started was 5.25, noted during spring 1995 and on two occasions during autumn 1997. The highest pH-value 7.5 was recorded during spring 2003. On eight occasions, pH was lower than the target value, which corresponds to less than 3 % of the sampling occasions.

Benthic Fauna

The first benthic invertebrate sample was taken in spring 1993, and although the lime doser was already in operation during the spring flood, it is unlikely that any new species had colonised the river as a consequence of changes in water chemistry at this time. Thus, this sample may be regarded as representing the pre-liming benthic community composition. In total, 18 taxa were found in this sample, of which 17 belonged to the most acid tolerant class (Index 1 in Fig. 22.6). In the samples from 1995, two mayflies *Baetis rhodani* and *Nigrobaetis niger* were found. These species have been noted every year since, except for 1996, when *B. rhodani* was absent, and 1999 when *N. niger* was absent. From 2000, we observed an increase

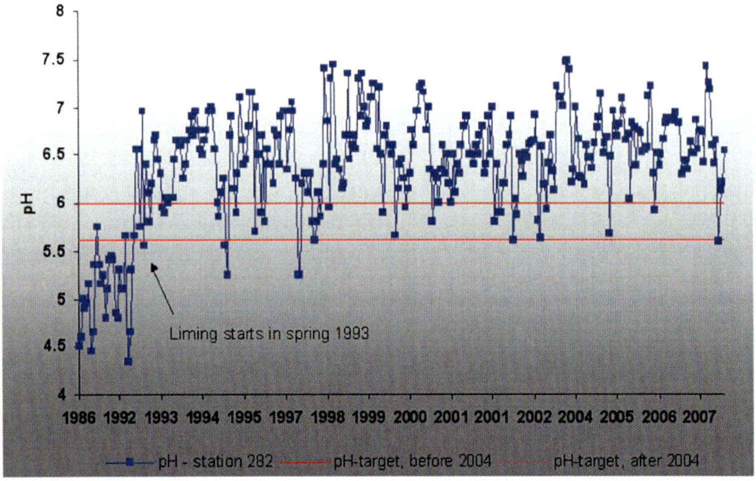

Fig. 22.3 pH at station 282. This station, which is downstream of the doser, has been influenced by liming since spring 1993

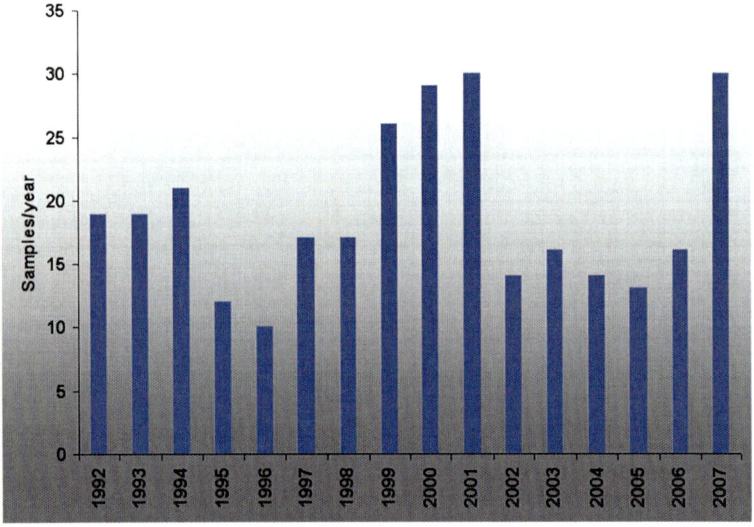

Fig. 22.4 The number of water samples at station 282

in the number of taxa. In 2002, a species from the most acid-sensitive class (Index 4), the stonefly *Capnopsis schilleri*, was found for the first time, and this has been observed annually since.

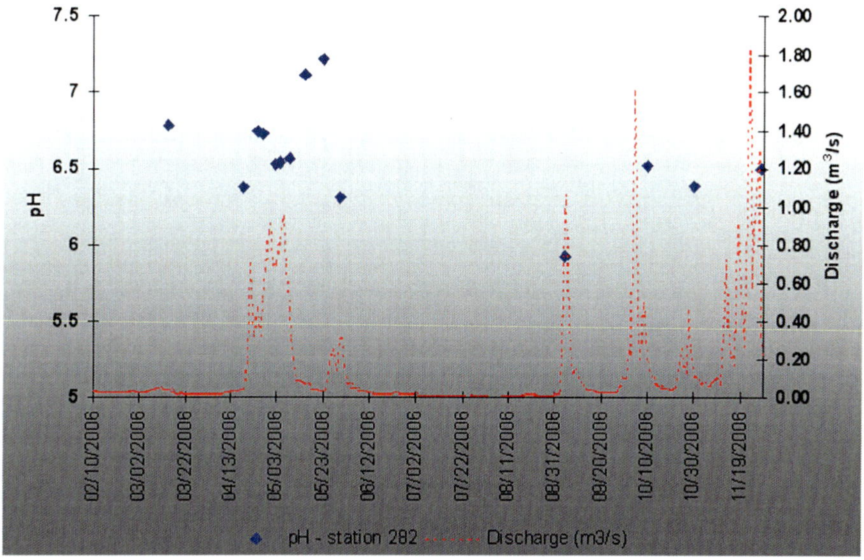

Fig. 22.5 Water samples and discharge at station 282 in 2006

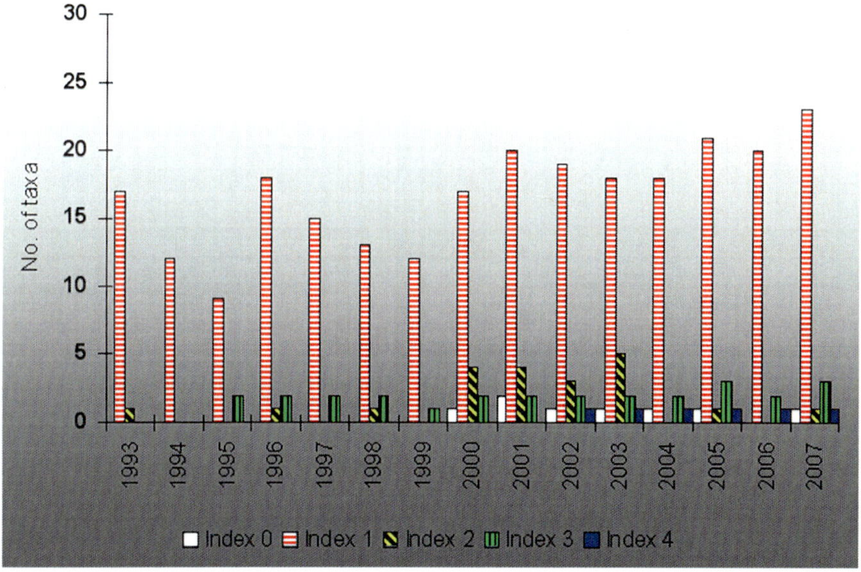

Fig. 22.6 Benthic fauna at station 88 between 1993 and 2007. The station has been influenced by liming since 1993. The sample from 1993 was regarded as the pre-liming situation. The index refers to acidic tolerance. Index 0 – unknown tolerance, index 1 – tolerates pH < 4.5, index 2 – does not tolerate pH < 4.5, index 3 – does not tolerate pH < 4.9 and index 4 – does not tolerate pH < 5.5

Fish Fauna

No trout reproduction was observed in the year before liming started, or during the first summer after the lime doser was installed. In 1994, however, YOY were captured for the first time at the station furthest downstream. Thereafter, YOY were captured in every year except 1996. At the two upstream stations, YOY were first observed in 1998 and 2000, respectively (Fig. 22.7). At all stations, the density of older trout increased progressively.

The Situation Upstream of the Lime Doser

The upstream part of the River Stridbäcken serves as a regional reference monitoring station, with water chemistry data available for the period 1993–2007 and biological data available for 1999–2007.

Since spring 1993, 297 samples have been collected from station 283 – which is upstream from the doser. The lowest pH was at 3.85 in November 1993 (Fig. 22.8) and highest pH 5.75 in October 1996.

The benthic invertebrate community was dominated by acid-tolerant species. On three occasions (1999, 2001 and 2006), a more acid-sensitive taxon (the Diptera family Limoniidae, Index 3) was found (Fig. 22.9). No changes in the number of taxa or proportion of sensitive taxa were observed during the period. In comparison

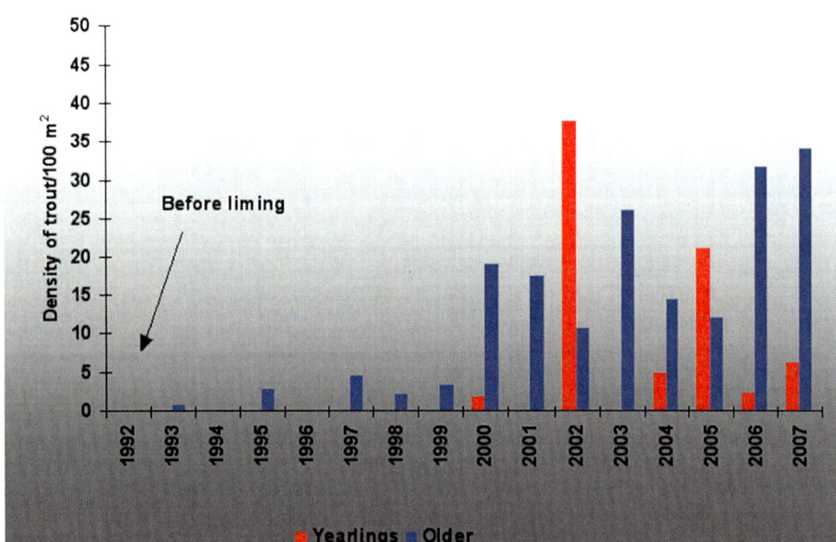

Fig. 22.7 Density of YOY (grey) and older (black) trout at station 203. The station, which is downstream of the lime doser, has been influenced by liming since 1993

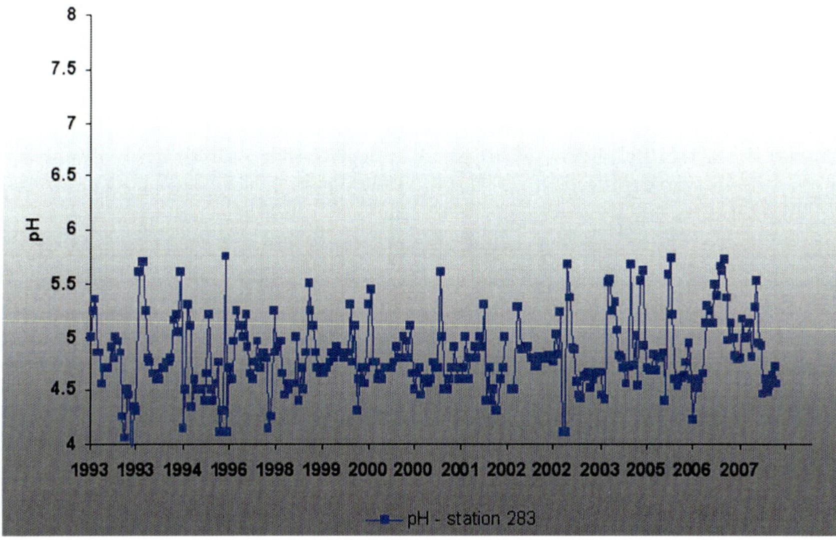

Fig. 22.8 pH at station 283, which is situated upstream from the lime doser

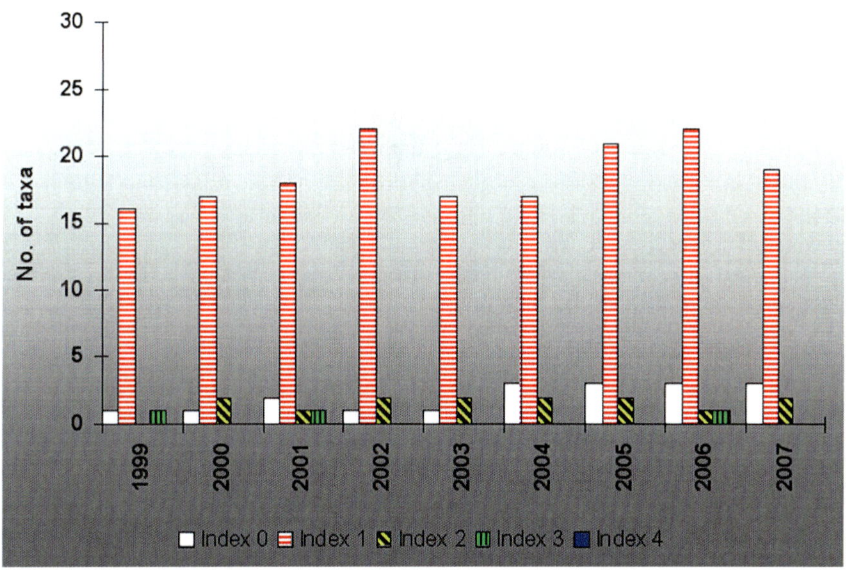

Fig. 22.9 Benthic fauna 1999–2007 at station 14. The station is situated upstream from the lime doser. The index refers to acidic tolerance. Index 0 – unknown tolerance, index 1 – tolerates pH < 4.5, index 2 – does not tolerate pH < 4.5, index 3 – does not tolerate pH < 4.9 and index 4 – does not tolerate pH < 5.5

with the limed part of the river, the average number of taxa was lower, 22.7 compared with 24.3 (1999–2007). The proportion of acid-sensitive species (Index 3 & 4) varied between 0% and 6 % at the unlimed station compared with 7–15% at the limed station.

A few older trout, but no YOY, were recorded at the monitoring station 150 m upstream from the lime doser. At the station further upstream, only one older trout (length 19 cm) was captured in 2004, with another there (length 22 cm) in 2005.

Discussion

In smaller rivers, discharge may increase rapidly. In the River Stridbäcken, there are examples of threefold increases in discharge in 1 day. Changes in discharge are generally accompanied by dramatic changes in water chemistry. This relationship constitutes a challenge to the practice of liming a river. In a small river with a maximum discharge of 5 $m^3 s^{-1}$, the required lime dose may increase from zero to 20–40 kg h^{-1} in just 1 day.

In 1993–1999, the lime doser in the R. Stridbäcken was primitive and unreliable. The river failed to meet the water chemistry target on seven occasions during this period, on five of these mostly as a consequence of technical malfunction. The doser also had a manual lime release, and this resulted in frequent over-liming and large fluctuations in water chemistry (with occasional high values of calcium, alkalinity and pH).

The technical improvements in 2000 involved a change in the lime-discharge method, from vibration to feeding screws. Furthermore, an automatic system was installed to prevent lime from clogging up in the silo. These changes led to a more robust and reliable operation. Since 2000, the water chemistry target has been failed only three times and on no occasion was this due to technical malfunctioning. The change to automatically controlled lime release in 2000 also had positive effects on the water chemistry. After 2000, no extreme high alkalinity events were observed. The change to a new control system in 2003, however, created some disturbance and resulted in some cases of high alkalinity.

The purpose of liming is to raise the pH of a waterbody to a set target. Liming of lakes and rivers always involves a higher release of lime than is theoretically required. For example, to raise the pH from 4.4 to 6.0 in a river, the *Handbook of Liming* (Naturvårdsverket 2002) recommends a lime dose of 12 g m^{-3} for lime dosers, 15 g m^{-3} for wetland liming, and 18 g m^{-3} for liming in lakes in the catchment area of a river. These recommendations are based on practical experience. The theoretical amount of lime required to achieve a rise from pH 4.4 to 6.0 is only about 5 g m^{-3}.

There are several reasons for these large differences in practiced and theoretical lime requirements. Most importantly, lime is normally distributed only once a year and the dose is thereby highest directly after the lime release, then gradually decreasing until the next liming event. Therefore, for the amount of lime to suffice for the period between lime releases, the dose needs to exceed the theoretical needs throughout the period, except for a short period of time before the release. This results in a pH much higher than the pH-target directly after liming and an excessive release of calcium. The availability of suitable lakes and wetlands also influences the lime requirement. Generally, systems with a short turnover time should be avoided, particularly lakes with rapid turnover and periodically flooded wetlands. As the release of lime from these types of systems will be rapid, the initial lime dose needs to be very high to have a lasting effect until the next liming.

Consequently, liming by dosers is the most effective method, because overliming is avoided through continuous output of lime. If the lime dose is controlled automatically by the discharge, the lime output is continually adjusted and appropriate amounts applied throughout. A modern lime doser may also be regulated so that the lime dose is adjusted in relation to changes in water volumes at different discharges. Generally, a higher dose is required at high discharges, as pH drops when the discharge increases. Many limed rivers have higher pH-values than the target pH at low discharge. By setting a low-discharge limit for lime release, this can be automatically interrupted during low-flow periods with high pH-values.

In the R. Stridbäcken, the lime use in recent years corresponded to a dose of about 6 g m^{-3}: only half of the recommended dose in the *Handbook of Liming*. The reason for this is that the recommendations in the Handbook were based on experience gained from using lime dosers with a less sophisticated regulating system than the one used in the R. Stridbäcken. Lime use in the R. Stridbäcken is probably only 20% higher than the theoretically required amount. We therefore argue that the total lime usage in Sweden could be reduced by roughly half if all liming was carried out with the same efficiency as in the R. Stridbäcken.

Successful reproduction of sea trout was observed within 1.5 years of the first liming, at the monitoring station closest to the sea (which is easily accessible for migrating sea trout). This quick re-establishment probably involved trout straying from the nearby population in the River Saluån. Upstream from this monitoring station, migration is hindered by a bedrock formation, which is passable only under rarely occurring flow conditions. This probably delayed re-colonisation in the stretch containing the upper monitoring stations. Although the liming carried out initially was not optimal, it still created conditions conducive to the re-establishment of a viable sea trout population. By comparison, no trout reproduction was observed in the unlimed stretch of the river. The absence of YOY suggests that the older individuals captured were the result of successful reproduction downstream of the lime doser.

During the 2004 spring flood, a comparative experimental study of trout was made upstream from the lime doser and at the river mouth (Ahlström 2005). At the unlimed upstream location, aluminium content in the fish gills increased during the study period: this led to an impaired gill functioning that affected ion balance and respiration. Some of the fish died during the study period. In the limed location, no accumulation of aluminium occurred in the gills and the fish showed no signs of impaired gill functioning. These results showed that the water chemistry upstream of the lime doser was unfavourable for trout and that liming with the doser improved conditions.

The response in the benthic invertebrate community was first observed 2 years after liming, with the appearance of the mayflies *Baetis rhodani* and *Nigrobaetis niger*.

These species are known to colonise limed rivers rapidly if the water quality is satisfactory (Fjellheim and Raddum 1992; Lingdell and Engblom 1995; Ahlström *et al.* 1997). After 2000, the number of species continued to increase, and in 2003, the stonefly *Capnopsis schilleri* colonised the river. Upstream from the doser, the benthic invertebrate community is still dominated by acid-tolerant species. The liming has therefore created conditions suitable for benthic invertebrates that are sensitive to a low pH.

Acknowledgements Jorun Storeng and Mats Näslund at the Nordmaling Municipality administration have dutifully managed the lime doser in Stridbäcken and also collected water samples. Thanks to Iwan Zetterqvist, Kent Holm, Lars Argelius, Terje Lysnes and Ulf Olofsson for contributing to the technical development of the lime doser. Thanks to all personnel at the County Administration in Västerbotten who have performed the biological monitoring and analysed water samples throughout the years, and to ALcontrol, Pelagia and Hushållningssällskapet, who have analysed water samples and examined benthic fauna during the last years. Thanks to Erik Degerman for comments on the manuscript.

References

Ahlström J (2005) Försurning och kalkning av sjöar och vattendrag i Västerbottens län – Årsrapport 2004. Länsstyrelsen i Västerbottens län (in Swedish), Umeå

Ahlström J, Degerman E, Lindgren G, Lingdell P-E (1995) Försurning av små vattendrag i Norrland. Naturvårdsverket, Rapport 4343 (in Swedish)

Ahlström J, Bondestad L, Norberg M (1997) Kalkning av vattendrag i Västerbottens län. Metoder och effekter 1985–1995. Meddelande 5:1997, Länsstyrelsen i Västerbottens län, Umeå (in Swedish)

Degerman E, Lindgren G, Lingdell P-E, Nyberg P (1987) Kartering av strömfauna och fisk i mindre vattendrag i Norrlands inland och fjälltrakter i relation till försurning. Information från Sötvattenslaboratoriet 1987/2. Sötvattenslaboratoriet, Drottningholm (in Swedish with English summary)

Degerman E, Engblom E, Lingdell P-E, Melin E, Olofsson E (1992) Försurning i fjällen? Information från Sötvattenslaboratoriet 1992/1. Sötvattenslaboratoriet (in Swedish with English summary), Drottningholm

Degerman E, Fernholm B, Lingdell P-E (1994) Bottenfauna och fisk i sjöar och vattendrag, utbredning i Sverige. Naturvårdsverket, Rapport 4345 (in Swedish)

Fjellheim A, Raddum GG (1992) Recovery of acid-sensitive species of Ephemeroptera, Plecoptera and Trichoptera in River Audna after liming. Environ Pollut 78:173–178

Lingdell P-E, Engblom E (1995) Liming restores the benthic invertebrate community to "pristine" state. Water Air Soil Pollut 85:955–960

Naturvårdsverket (2002) Kalkning av sjöar och vattendrag. Handbok 2002:1, Naturvårdsverket (in Swedish)

Naturvårdsverket (2008a) Bottenfauna i sjöars litoral och vattendrag. M42-inventering med riktat urval. Handledning för miljöövervakning, undersökningstyp. Naturvårdsverket (in Swedish)

Naturvårdsverket (2008b) Bottenfauna i sjöars litoral och vattendrag. Tidsserier. Handledning för miljöövervakning, undersökningstyp. Naturvårdsverket (in Swedish)

Naturvårdsverket (2008c) Elfiske i rinnande vatten. Handledning för miljöövervakning, undersökningstyp. Naturvårdsverket (in Swedish)

Zippin C (1956) An evaluation of the removal method of estimating animal populations. Biometrics 12:163–169

Section 5
Lake and Wetland Case Studies

Chapter 23
A Baseline for Monitoring the Freshwater Natura 2000 Vegetation of the Teifi Pools (Afon Teifi SAC), Wales

David Broughton and Jane Southey

Introduction

The European Council Directive 92/43/EEC (the Species and Habitats Directive) provides for the protection of animal and plant species of European importance and the habitats that support them, particularly through the establishment of a network of protected sites.

In the United Kingdom these protected sites are referred to as Special Areas of Conservation (SACs), or in the case of birds, Special Protection Areas (SPAs), and these are jointly termed Natura 2000 sites. European member states have a duty to demonstrate through site monitoring and reporting that steps are being taken to ensure that the designated features within Natura 2000 sites are being adequately protected and managed. This process will ensure the maintenance of, or where appropriate the restoration of, the 'favourable conservation status' of habitats and species features for which SACs and SPAs are designated.

The Teifi Pools form part of the Afon (River) Teifi SAC (SAC EU Code UK 0012670). The Afon Teifi, located in the Welsh county of Ceredigion (Fig. 23.1), is one of the longest rivers in Wales and one of the most pristine and least modified river catchments in lowland Britain. It is 122 km from the source in the Cambrian Mountains to the outfall in Cardigan Bay. The source originates at an altitude of 455 m above Ordnance Datum in Llyn (Lake) Teifi, one of five clear water upland lakes that form the series known as the Teifi Pools. Llyn Teifi is the largest of the five lakes, the others being, in descending order of surface area, Llyn Egnant, Llyn y Gorlan, Llyn Hir and Llyn Bach (Fig 23.1 inset).

The solid geology in the upper reaches of the Teifi catchment is Silurian mudstones and shales of the Llandovery series and the landscape is largely rural in character.

D. Broughton
Scott Wilson Ltd., West One, Leeds, LS1 1BA, UK

J. Southey (✉)
Scott Wilson Ltd., Unit 33, Mansfield i-Centre, Oakham Business Park, Hamilton Way, Mansfield, NG18 5BR, UK
e-mail: jane.southey@scottwilson.com (Lead scientist)

C. Hurford et al. (eds.), *Conservation Monitoring in Freshwater Habitats: A Practical Guide and Case Studies*, DOI 10.1007/978-1-4020-9278-7_23,

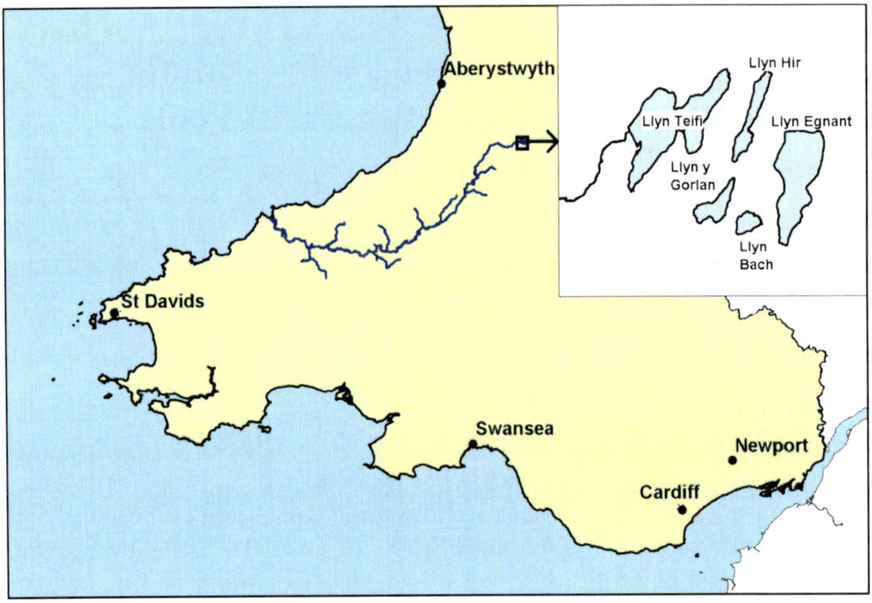

Fig. 23.1 The location of the Teifi Pools in south Wales. The inset box shows the distribution of the Pools at the source of the Afon Teifi. © Crown Copyright and/or database right. All rights reserved. Licence number 100043571

The majority of the catchment area is used for livestock-based agriculture and approximately 15% of the catchment area is used for forestry. The habitats immediately around the Teifi Pools comprise sheep-grazed moorland and acid grassland.

The Teifi Pools are notable for their freshwater plant communities that are of the Natura 2000 'oligotrophic to mesotrophic standing waters with vegetation of the *Littorelletea uniflorae* and/or *Isoëto-Nanojuncetea* standing water' habitat type (EU Habitat Code: 3130) and for an internationally important population of the aquatic plant floating water-plantain (*Luronium natans* (L.) Raf.) (EU Species Code: 1831).

The *Littorelletea uniflorae* and/or *Isoëto-Nanojuncetea* standing water habitat type (subsequently referred to as '*Littorelletea* habitat') is characteristic of clear water lakes of naturally low fertility levels and with stoney substrates. In Britain the habitat is typically found in the uplands of western and northern Britain (Rodwell 1995).

Luronium natans is endemic to an area that extends from southern Norway and Sweden in the north, through the Republic of Ireland and France to northern Spain, east to Poland and it is most frequent in the western part of its range (Britain, France, northern Germany and the Netherlands) and rare elsewhere. It can be a difficult species to identify and so it is probably under-recorded, but is still considered to be in decline throughout its known range. In Wales, *Luronium natans* is extant in 77% of its known historic sites and eleven populations have only been known since the 1980s, suggesting that the Welsh population is relatively stable (Lansdown and Wade 2003).

Luronium natans can be found in a wide-range of aquatic habitats, both upland and lowland, but is intolerant of competition (Willby and Eaton 1993; Greulich *et al.* 2000).

As such, it requires some form of mechanism to retard competition and succession if it is to persist at a site for any length of time (Lansdown and Wade 2003). In the Teifi Pools the authors have observed that the best stands are associated with accumulations of deep, unconsolidated silty and peaty substrates where other plant species perform less well. It is able to exploit a wide variation in water depths and is tolerant of periodic drawdown and exposure, which has the benefit of promoting flowering and seed set.

Luronium natans has a number of apparently discrete reproductive strategies, defined as: annual flowering; perennial flowering; and perennial vegetative. It occurs as dynamic metapopulations that contain different populations with different reproductive strategies. These are explored further in Lansdown and Wade (2003). Perennial vegetative is the principal life strategy associated with this species in the UK in oligotrophic lakes (Willby *et al.* 2000).

While acknowledging the lack of available information on the management of upland lakes for *Luronium natans*, Lansdown and Wade (2003), quoting Welsh data, suggest that upland lake populations are amongst the most stable, and that management is unnecessary unless there is a change in water chemistry or the processes suppressing succession.

Methods

Survey and Monitoring Needs

In 2003, when this survey was commissioned, there was no agreed standard methodology for the monitoring and condition assessment of freshwater Natura 2000 features. In the case of the Teifi Pools only a limited macrophyte (aquatic plant) dataset was available to inform the development of a baseline monitoring strategy and site management plan.

As part of the requirement for monitoring freshwater Natura 2000 features, Scott Wilson Ltd was tasked by the Countryside Council for Wales (CCW), the Welsh Government agency responsible for the management of the countryside and nature conservation, to develop efficient and reliable monitoring methods for the vegetation features of the Teifi Pools.

The objectives of this study were to:

- Determine through survey the extent, distribution and condition of the *Littorelletea* habitat of the Teifi Pools.
- Determine through survey the extent of the *Luronium natans* population of the Teifi Pools.
- Develop with CCW a concise definition of the *Littorelletea* habitat and its component communities, and define how to recognise when the habitat is in 'good' and 'poor' condition.
- Develop and implement a field monitoring methodology for the *Littorelletea* habitat (against targets agreed with CCW).

Development of a Survey Approach

Based on prior experience, consideration of existing standing water survey methodologies and discussions with CCW, a survey approach was developed to meet the objectives of this contract.

The key requirements of the survey approach taken were that it should be:

- Time and cost efficient as a long-term, baseline monitoring approach.
- Easily implemented and applicable in a range of survey conditions, a key consideration for a physically varied upland site subject to rapidly changing, local weather conditions.
- Produce reliable and repeatable results.
- Provide a detailed but manageable baseline dataset that could contribute to a range of management, reporting and decision-making outputs.

A diving survey was not considered necessary by CCW and such an approach was constrained by budgetary limitations.

Given the lack of systematic baseline macrophyte data for the Teifi Pools and because this project came ahead of the development of a defined UK condition assessment methodology, it was determined at the outset that the 'condition' of the target features would largely be assessed in terms of presence or absence. It was also necessary that the survey provide data to allow the production of distribution maps showing where features occurred and to indicate the relative distributional abundance of target species such as shoreweed (*Littorella uniflora* (L.) Asch.), quillworts (*Isoetes lacustris* L. and/or *I. echinospora* Durieu) and *Luronium natans*.

Survey Methodology

Over a 1-week period in October 2003, two experienced freshwater botanists, working as a pair, surveyed the five lakes. The survey of each lake took, depending on their size and complexity, between half a day and a day to complete.

The shoreline of each of the five lakes was walked in full enabling subsequent division of the shoreline into easily identifiable survey compartments. For each compartment, the communities (largely based on those given in Rodwell 1995) present were mapped and the macrophyte flora recorded along with an indication of their local population status. Potentially pertinent background data was also recorded for the benefit of informing future management decisions.

Many of the approaches to lake monitoring and condition assessment in the UK, both shore and boat methodologies, involve the sampling of only a proportion of the total lake perimeter. The simplicity of the baseline survey techniques, deployed by two experienced field surveyors, enabled data collection around the entire perimeter of each lake to be achieved in a comparable timeframe and at a competitive price.

Survey equipment used comprised a small hand held water viewer (bathyscope), a standard (Palmer *et al.* 1992) double-headed grapnel, ranging poles and a digital camera.

The range of the grapnel from the shore was 4 to 5 m, depending on wind conditions.

The survey methodology applied varied slightly, depending upon the physical nature of each lake, with a grapnel being used more heavily for those lakes or sections of lakes where it was not practical to wade the margins. Conditions during the survey period were too windy for the use of a boat to be an advantage and it is demonstrated later that the use of a boat was not a prerequisite for the collection of a robust dataset for this site.

The general survey approach was that, where safe to do so, the shallow marginal waters (depths up to 70 cm) were waded in a zigzag route covering a band of between 3 to 5 m from the shore. Where possible, recognisable land features were used to delimit the survey compartments; otherwise they were temporarily marked out using ranging poles. A viewer was used to observe the submerged macrophytes present. The grapnel was used up to three times every 15 m when wading was not possible, for example where the steep rocky shore prevailed over deep water, where the bottom was unstable and potentially overly deep, or where macrophytes were obscured by wave action. All plants species observed *in situ*, along with material caught by the grapnel, floating to the surface after the grapnel had been thrown or washed-up on the shore of the lake, were recorded. Quantitative values for plant cover were attempted in local areas where it was considered accurate to do so. Representative photographs were also taken.

Where possible local estimates of plant cover were made using the DAFOR scale, where:

D = dominant (total vegetative cover of greater than 70%)
A = abundant (total vegetative cover of 30–70%)
F = frequent (total vegetative cover of 10–30%)
O = occasional (total vegetative cover of 3–10%)
R = rare (total vegetative cover of 0–3%)

The data was collected in a form that would allow it to be easily input into a Geographic Information System (GIS) and the software standard currently adopted by CCW is Mapinfo Professional (Mapinfo Corporation). Incorporation of data into GIS has many advantages. It allows data to be presented in a visual format that allows it to be more easily understood, by both specialist and non-specialist staff and stakeholders, and disseminated. It also ensures a standardisation of visual output so that data can be compared over a time series. Data can also be manipulated to generate summary statistics on an area, although such statistics always require due consideration of the objectives and limitations of the survey techniques applied. In this instance, the data collected provided a detailed picture of the distribution of species and communities around the surveyable margins of the lakes, but could not be used to define the extent of vegetation and species in areas of deeper water, or in areas with deep, unconsolidated substrates.

It should be noted that *Luronium natans*, as a European Protected Species, benefits from strict legal protection within the UK and it is an offence to deliberately pick, collect, cut, uproot or destroy a wild plant of this species. As such, when undertaking survey work at a location where this species is known to occur it is necessary to first obtain the appropriate derogation licence from the relevant national statutory nature conservation agency. In Wales this is CCW.

Results

The survey identified that the five lakes within the Teifi Pools series are all quite different in terms of their physical characteristics and, as a consequence, the vegetation that they support. A total flora of 30 macrophyte species was recorded from the series (Table 23.1). Stands of the European target *Littorelletea* vegetation and *Luronium natans* were recorded in all of the five lakes. For a more detailed account of the characteristics and plant communities of the Teifi Pools the reader should consult the source report (Southey and Broughton 2004).

Table 23.1 The aquatic flora recorded in 2003 from each of the five Teifi Pools

	Lake				
Plant species	1	2	3	4	5
Agrostis stolonifera			*		
Callitriche hamulata	*	*			
Carex rostrata			*		*
Equisetum fluviatile				*	
Glyceria fluitans			*		
Hydrocotyle vulgaris	*				
Isoetes echinospora	*				
Isoetes lacustris			*	*	*
Juncus acutiflorus			*		
Juncus bulbosus	*	*	*	*	*
Juncus conglomeratus				*	
Juncus effusus			*	*	*
Littorella uniflora	*	*	*	*	*
Lobelia dortmanna			*		
Luronium natans	*	*	*	*	*
Lythrum portula	*				
Menyanthes trifoliata			*		
Myriophyllum alterniflorum	*	*	*		
Nuphar lutea				*	
Pellia epiphylla		*			
Potamogeton polygonifolius	*	*	*	*	
Racomitrium aciculare	*	*	*		
Ranunculus flammula			*	*	
Ranunculus omiophyllus			*		
Scapania undulata				*	
Schoenoplectus lacustris				*	

(continued)

Table 23.1 (continued)

	Lake				
Sparganium angustifolium	*	*	*	*	
Sphagnum denticulatum			*	*	*
Subularia aquatica			*		
Utricularia minor			*		
Species total	11	9	20	14	7
Code to lake numbers					
Lake 1 = Llyn Teifi					
Lake 2 = Llyn Egnant					
Lake 3 = Llyn Hir					
Lake 4 = Llyn y Gorlan					
Lake 5 = Llyn Bach					

Llyn Hir is a long and narrow oligotrophic lake that covers 5.1 ha and is situated in a relatively sheltered valley. It proved to be the richest lake for macrophytes, with twenty species recorded within the largely boulder and cobble dominated shallows with occasional bedrock and fines. It is the only lake where water lobelia (*Lobelia dortmanna* L.), was proved to still persist. The presence of extensive beds of *Lobelia dortmanna* growing in combination with *Littorella uniflora* and *Isoetes* species has produced a subtype of the *Littorelletea* habitat that is unique in the context of the five Teifi Pools (although historically it may have been more widespread). Llyn Hir is also the only lake where awlwort (*Subularia aquatica* L.), a species that is rare in Wales and in the wider UK (except Scotland), was recorded.

Llyn Teifi is a controlled reservoir and the largest lake in the series, covering an area of 24.1 ha. It is oligotrophic, and is notable for its large sheltered bays with extensive draw-down zones. These bays support excellent populations, the most extensive of any of the five lakes, of *Luronium natans* (Fig. 23.2), with this species showing a strong affinity for accumulations of thick silty and peaty sediments. The maintenance of these deposits will be important for the ongoing persistence of such a strong population of this species. Llyn Teifi also supports good populations of *Littorella uniflora* and *Isoetes* species and a total of eleven macrophytes were recorded. Along with Llyn Hir, this lake supports the best-developed examples of the *Littorelletea* habitat present in the Teifi Pools. Substrate and degree of shelter, in particular, seem to be the primary factors determining the distribution of the *Littorelletea* habitat within the lake and the assemblage of plants it contained at any given location.

Llyn Egnant is oligotrophic and the other lake in the series that is managed as a controlled reservoir. It covers an area of 17.4 ha and has a varied shoreline that includes steep rocky headlands, gently shelving cobble and shingle deposits and areas of eroding peat deposits. The flora of this lake was observed to be relatively species-poor, supporting a total of nine aquatic plant species. The flora was found to be largely restricted to the more sheltered eastern shores of the lake, where a

Fig. 23.2 Floating water-plantain *Luronium natans* occurs at internationally important levels in the Teifi Pools. *Photo by Clive Hurford*

more stable substrate combined with much reduced wave action provided conditions more favourable for the development of the *Littorelletea* habitat.

Llyn y Gorlan is a small oligotrophic lake that covers an area of 3.8 ha. It has a relatively diverse flora, second only to Llyn Hir, with fourteen species of aquatic plant recorded. The lake was notable for its large deep-water beds of bogmoss (*Sphagnum denticulatum* Brid.) and, unusually for an upland lake, a large stand of yellow water-lily (*Nuphar lutea* (L.) Sm.) growing as a canopy over an otherwise typical example of the *Littorelletea* habitat.

Llyn Bach is, with a surface area of 2 ha, the smallest of the five lakes and the only truly dystrophic waterbody present in the series. The water within the lake was strongly peat-stained and the flora observed was sparsely distributed and species-poor, with only seven aquatic plant species recorded.

Populations of *Luronium natans* were identified from all five of the Teifi Pools with greatest concentrations associated with Llyn Teifi (Fig. 23.3) and a locally abundant presence in Llyn Egnant and Llyn y Gorlan. *Luronium natans* was found to be rare in both Llyn Hir and Llyn Bach, but the reasons for this are unclear. In Llyn Hir, it is possible to speculate that the sheltered position of the lake along with the nature of the substrate types present may allow a more stable plant community to persist with reduced opportunities for a poorly competitive species such as *Luronium natans*. As Llyn Bach was the only dystrophic lake in the series, it is possible that water chemistry may be a factor behind the apparent rarity of *Luronium natans*, as well as the general scarcity of the rest of the flora.

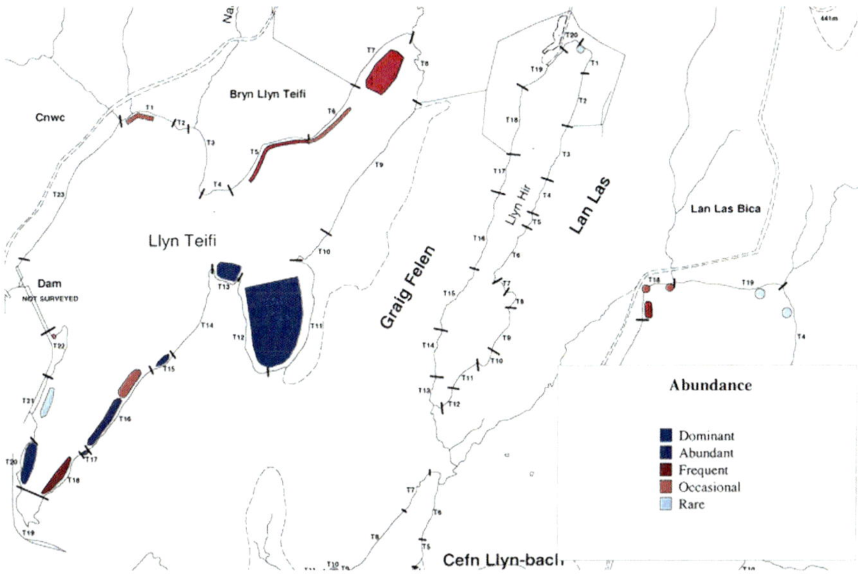

Fig. 23.3 Map showing the distribution and abundance of floating water-plantain *Luronium natans* at Llyn Teifi, and the location of survey transects.

Discussion

Applicability of the Survey Approach Taken

The simple survey techniques and equipment utilised by two experienced freshwater surveyors proved to be highly effective in detecting the presence of macrophytes, even locally scarce species, and in determining their relative abundances at any one site. As well as being technically appropriate, these techniques have also shown to be time and cost efficient. These conclusions have been further substantiated as a result of a detailed diving survey undertaken for Welsh Water and reported in Pickard (2005). The data and recommendations made as part of this contract have allowed CCW to develop well-informed and site-specific conservation objectives for the management and monitoring of the Afon Teifi SAC (CCW 2008).

It is reasonable to expect that these techniques could be effectively applied in other similar situations, particularly where there is a need to find an efficient and effective means to address a lack of detailed baseline data.

The limitations of this survey approach, in terms of the depth range that could be covered, are appreciated. *Luronium natans* grows most abundantly at 1–2.5 m water depth in the three largest Teifi Pools (CCW 2008). *Littorella uniflora* and communities dominated by *Littorella uniflora* can occur over a wide depth range, from seasonally exposed areas of drawdown to areas of relatively deep water. In contrast, *Isoetes* species and *Isoetes* dominated communities are generally thought to predominate in deeper waters beyond the depth limit of communities dominated by *Littorella uniflora* (Rodwell 1995). However, it has been acknowledged that the Rodwell classification and some of its conclusions reflect the limitations of the dataset available for analysis (Duigan *et al.* 2006). While Pickard (2005) found stands of *Isoetes* species growing in the deeper waters of the Teifi Pools, the survey data collected for this study has demonstrated that *Isoetes* species can be found growing in shallower waters as a mixed community with *Littorella uniflora*. As a consequence, a shore-based shallow water survey can provide informative data representative of the aquatic flora of a lake as a whole, even if there are significant limitations on the range of waters that can be sampled.

Regardless of any inherent bias towards the shallow and inshore waters of the Teifi Pools, the survey work undertaken succeeded in successfully categorising the macrophyte assemblages present and provided baseline distributional data on key species and habitats, allowing each lake to be compared and contrasted. As a consequence, this survey has demonstrated that a boat or diving based survey, while both techniques have their own specific merits, is not always an essential requirement for the effective survey of freshwater lakes. This has also been acknowledged in more recent survey guidance for the condition monitoring of standing waters published by the UK Joint Nature Conservation Committee (JNCC 2005). In our experience, there will always be situations where the use of a boat and/or divers are not a suitable option due to access constraints (upland lakes can often be remote from networks of roads and tracks), the physical characteristics of a site, prevailing weather conditions at the time of survey and related health and safety considerations. Similarly, the costs and relatively labour intensive nature of such techniques may be prohibitive and they rarely entail the sampling of an entire lake perimeter, being instead restricted to sampling a subset of any one water body.

Site Management Issues

A range of issues relevant to the long-term sustainability of the Natura 2000 features within the Teifi Pools have been identified during the various studies implemented over the past 5 years and these have made an important contribution to the rationale behind the core management plan (CCW 2008).

Both *Luronium natans* and the *Littorelletea* habitat require adequate water levels with a suitable water quality. Given this, adverse factors for the standing water habitats may include elevated nutrient levels (particularly phosphate), and in the case of the habitats only, artificial regulation of water levels resulting in extensive or prolonged drawdown (limited natural drawdown during the warmer summer months is to be expected) in the reservoirs of Llyn Teifi and Llyn Egnant (Fig.23.4) and poaching of the resultant exposed lakeshores by livestock. It has been

Fig. 23.4 Llyn Egnant is one of the Teifi Pools used as a reservoir. *Photo by David Broughton*

concluded that drawdown events do not present a negative threat to floating water-plantain and that such events are likely to stimulate flowering of deep water populations, leading to opportunities for seed set. However, drawdown has been cited as a causal factor that has resulted in the restriction of *Lobelia dortmanna* to Llyn Hir (CCW 2008).

The maintenance of suitable water quality for both of the identified features is likely to require catchment-wide measures to control diffuse pollution from agriculture, the principal source of phosphate pollution. Significant livestock reduction measures have recently been implemented in the Teifi Pools catchment under the auspices of the Tir Gofal agri-environment scheme, and these will contribute to reducing nutrient enrichment from these sources (CCW 2008).

An additional potential threat to the vegetation features of the Teifi Pools is considered to be the risk of colonisation by the invasive non-native plant New Zealand pigmyweed (*Crassula helmsii* (Kirk) Cockayne). A source of such introductions could be via the boots, clothing or equipment of anglers visiting the Teifi Pools (CCW 2008) or the less frequent activities of aquatic surveyors.

In relation to the above, it is also of concern that the highest aquatic plant diversity and the best example of the *Littorelletea* habitat are concentrated within just one lake, Llyn Hir. In the event of a pollution incident or other extreme event, a potentially irreplaceable element of the Teifi Pools could be permanently lost from the SAC. This concern is the justification for the requirement that the existing condition of the *Littorelletea* habitat within the other waterbodies, including those subject to water abstraction licences, be maintained, allowing the possibility of future species recolonisation (CCW 2008).

Condition Assessment of the European Features Within the Afon Teifi SAC

As part of the conservation management plan for the Afon Teifi SAC, CCW has identified the following vision for the freshwater vegetation of the Teifi Pools (CCW 2008):

'There will be healthy populations of *Luronium natans* in the Teifi Pools and in the river around Tregaron. The Teifi Pools will continue to contain their current range of distinctive aquatic plants that are characteristic of these clear-water upland lakes.'

The results of this survey, supplemented by the findings of Pickard (2005) and Burgess *et al.* (2006) along with guidance given in JNCC (2005), have allowed CCW to define a final set of site-specific conservation objectives for the *Luronium natans* and the *Littorelletea* features of the Teifi Pools.

The conservation objectives consist of two elements, a vision for each feature and condition indicators. It is considered that a 'vision for the feature' brings meaning and substance to the conservation objectives, while it is the role of the condition indicators to make the conservation objectives measurable and as such, the two elements are considered to be inextricably linked (CCW 2008). The condition indicators are given in Tables 23.2 and 23.3.

A formal condition assessment of the Teifi Pools was successfully undertaken in 2006 (Burgess *et al.* 2006) based on the dataset collected for this study, supported

Table 23.2 The condition indicator table for the *Luronium natans* population at Teifi Pools. The population at Llyn Bach is excluded as the population in this very small water body is not considered critical at present, as there is no obvious threat. The requirement for floating flowers will show that the lake populations have the potential for seed dispersal and genetic exchange. It is important that there is evidence of sexual reproduction, especially in the long-term

Condition indicator table	The *Luronium natans* population at Teifi Pools SAC will be in favourable condition when:	
Population extent	Lower limit	Live vegetative material is present in each of Llyn Teifi, Llyn Egnant, Llyn Hir and Llyn y Gorlan
Quality	Lower limit	Floating flowers are present in at least one of Llyn Teifi, Llyn Egnant, Llyn Hir and Llyn y Gorlan (or in any part of these) in 1 year in six And Epiphytic filamentous green algae indicative of eutrophication should have a cover value of not greater than 50% on the surface of each plant for the first 9 out of any 10 aquatic macrophytes examined in any of the pools And The presence of non-native invasive plant species, including but not limited to *Crassula helmsii*, will not be tolerated in any of the Teifi Pools

Table 23.3 The condition indicator table for the *Littorelletea* habitat at Teifi Pools. This table suggests that it is necessary to maintain a fully developed *Littorelletea* community in Llyn Hir only. *Utricularia minor* is also a key species of the community, but is not considered appropriate for effective monitoring as it is easily overlooked. It is not considered necessary for the remaining four lakes to support a fully developed *Littorelletea* community

Condition indicator table	The Oligotrophic to mesotrophic standing waters with vegetation of the *Littorelletea uniflorae* and/or *Isoëto-Nanojuncetea* habitat at Teifi Pools SAC will be in favourable condition when:	
Habitat extent	Lower limit	All of the following characteristic species are present in Llyn Hir: *Lobelia dortmanna*, *Littorella uniflora*, *Isoetes* spp.*, *Subularia aquatica*, *Sparganium angustifolium*, *Luronium natans*, *Carex rostrata* *Both *Isoetes lacustris* and *I. echinospora* are recorded in the Teifi Pools And Those characteristic species (listed above) recorded as present in Llyn Teifi, Llyn Egnant, Llyn y Gorlan and Llyn Bach, between 1997 and October 2005, should be present
Quality	Lower limit	Epiphytic filamentous green algae indicative of eutrophication should have a cover value of not greater than 50% on the surface of each plant for the first 9 out of any 10 aquatic macrophytes examined in any of the pools And The presence of non-native invasive plant species, including but not limited to *Crassula helmsii*, will not be tolerated in any of the Teifi Pools

by data presented in Pickard (2005), against the draft conservation objectives set by CCW and formalised in CCW (2008). Both the *Luronium natans* and the *Littorelletea* features were found to be in a 'favourable state' meaning that for both of the features the specified limits for each of the attributes given in Tables 23.2 and 23.3 were met (CCW 2008).

We are encouraged that relatively simple survey techniques can be employed by skilled aquatic plant surveyors to obtain robust baseline datasets that can be used to define conservation objectives and condition indicators for *Luronium natans* and the *Littorelletea* habitat features. These can then subsequently be employed effectively to undertake a formal condition assessment of these features and to contribute to the production of a formal management plan for an SAC.

We hope that in undertaking this work for CCW we have contributed in a small way to them succeeding in their vision that 'the Teifi Pools will continue to contain their current range of distinctive aquatic plants that are characteristic of these clear-water upland lakes' and that in so doing the site will remain a wild and special place to visit.

Acknowledgements Thanks go to all the staff at CCW who assisted with, or were involved with, this contract. In particular, we would thank Jon Turner (Senior Conservation Officer, Afon Teifi) for commissioning us to undertake this work, Dan Guest for preparing the location map, and Clive Hurford for encouraging us to contribute a chapter on this project. We are also grateful to landowners and Welsh Water for allowing access to the Teifi Pools.

References

Burgess A, Goldsmith B, Hatton-Ellis T (2006) Site condition assessments of Welsh SAC and SSSI standing waters features. CCW Contract Science Report No. 705. CCW, Bangor

CCW (2008) Core management plan including conservation objectives for Afon Teifi/River Teifi SAC (Special Area of Conservation). CCW, Bangor

Duigan C, Kovach W, Palmer M (2006) Vegetation communities of British lakes: A revised classification. JNCC, Peterborough

Greulich S, Bornette G, Amoros C (2000) Persistence of a rare aquatic species along gradients of disturbance and sediment richness. J Vegetat Sci 11:415–424

JNCC (2005) Common standards monitoring guidance for standing waters. Version March 2005. JNCC, Peterborough

Lansdown RV, Wade PM (2003) Ecology of the floating water-plantain, *Luronium natans*. Conserving Natura 2000 rivers ecology series no. 9. English Nature, Peterborough

Palmer M, Bell SL, Butterfield I (1992) A botanical classification of standing waters in Britain: Applications for conservation and monitoring. Aquat Conserv Marine Freshwater Ecosyst 2:125–143

Pickard M (2005) Botanical investigations supporting appropriate assessment of Teifi Pools. Hyder Consulting Limited/D r Cymru Welsh Water report BM00846/ENV4/8

Rodwell JS (1995) British plant communities, vol 4. Aquatic communities, swamps and tall-herb fens. Cambridge University Press, Cambridge

Southey J, Broughton DA (2004) Development of monitoring methods to assess the condition of the 'oligotrophic to mesotrophic standing waters with vegetation of the *Littorelletea uniflorae* and/or of the Isoëto-Nanojuncetea' Natura 2000 habitats and the *Luronium natans* population at Afon Teifi cSAC. CCW Environmental Monitoring Report No.13. CCW, Bangor

Willby NJ, Eaton JW (1993) The distribution, ecology and conservation of *Luronium natans* (L.) Raf. in Britain. J Aquat Plant Manage 30:70–76

Willby NJ, Eaton JW, Clarke S (2000) Monitoring the floating water-plantain, *Luronium natans*. Conserving Natura 2000 rivers ecology series no. 9. English Nature, Peterborough

Chapter 24
Aquatic Plant Monitoring in the Broads

Andrea Kelly and Jane Harris

Introduction

The Broads is an extensive lowland wetland, incorporating shallow lakes, slow flowing rivers and associated fenland, together they form part of the UK's National Park family. The Broads catchment covers 3,181 km^{2,} or almost two-thirds of Norfolk and much of Suffolk in the South-east of England. The rivers and lakes are alkaline, fed by calcium-rich water from the underlying boulder clay. With over 60 shallow lakes (or broads) and five main slow flowing rivers, The Broads offers ample opportunity to monitor and research water plants and their ecology.

The broads originate from peat cutting for fuel in the Middle Ages. Cutting took place mainly in the extensive peats formed in the upper river valleys and reached a depth of around 3–4 m. These peat pits flooded when sea levels rose, creating shallow lakes. Since then, most broads have been connected to the rivers to create a connected network for transport purposes, though some remain isolated.

Most broads are freshwater apart from those in the lower valleys, which are subject to saline incursion from the estuary and where brackish waters from salt seepage affect the coastal broads of the River Thurne. Extensive drainage of the coastal strip for agriculture has increased the amount of salt entering the drainage ditches over the last century threatening the freshwater ecology.

Excess nutrients, combined with the disturbance from motorised pleasure craft, have resulted in a complete loss of macrophyte communities within the many river-connected broads. By contrast, some water plants have survived in isolated broads less affected by eutrophication. However, investigations of the sediment records (palaeoecology) reveal that although these plant communities are still present, they have also been altered by sedimentation and nutrient change.

A. Kelly (✉)
Broads Authority 2 Gilders Way, Norwich, NR3 1BU, UK
e-mail: andrea.kelly@broads-authority.gov.uk

J. Harris
Kepwick Ecological Services, Kepwick Cottage, Wymondham Road, East Carleton, Norwich, NR14 8JB, UK

C. Hurford et al. (eds.), *Conservation Monitoring in Freshwater Habitats: A Practical Guide and Case Studies*, DOI 10.1007/978-1-4020-9278-7_24,

The Broads is listed by Plantlife as a UK Important Plant Area; with the aquatic focus of this listing on the sites that support Natura 2000 habitats, i.e. the 'hard oligo-mesotrophic waters with benthic vegetation of stonewort species' (Habitat 3140), and the 'natural eutrophic lakes with Magnopotamion or Hydrocharition-type vegetation' (Habitat 3150).

Aquatic plants have a key structuring role in shallow lake systems (Jeppesen *et al.* 1998), with conservation decisions frequently undertaken on the basis of submerged macrophyte assemblages (Pollard and Huxham 1998; Willby *et al.* 2006). This case study outlines some of the monitoring methods that have revealed the slow ecological recovery occurring in the Broads since their decline 40–50 years ago (Phillips *et al.* 2005).

Survey

A range of methods are used to assess submerged aquatic plant populations in the Broads, these include transect, point, SCUBA and hydroacoustic surveys. More detail of the sampling methods used during these surveys can be found in the 'Methods' section below.

Transect Survey – 26 Year Dataset (1983–2008)

Since 1983, the Broads Authority and its predecessor organisations have undertaken an annual macrophyte survey, using a set transect method (Kennison *et al.* 1998). The data from these surveys have proved useful for assessing long-term major changes in the submerged macrophyte flora in the Broads over the last 26 years and have informed the UK classification for the EU Water Framework Directive. This survey method remains the foundation of the current suite of plant surveys undertaken by the Broads Authority.

Prior to this survey, naturalists had visited sites on an ad hoc basis dating back to Victorian times. The data have been stored in a range of places, but mainly as biological records in the local museum.

The transect surveys were set up to enable an overview comparison of turbid lakes, using a simple replicable method to enhance the potential for continuity in data collection and interpretation.

The transect survey routinely covers 80% of the water area, or around 40% of the 62 lakes of the Broads (excluding transitory scrapes and turf ponds). We alter the survey locations each year to ensure that all broads are covered on a 5 or 6-year rotation, except the key broads that are covered annually (the sampling frequency is shown in Fig. 24.1a, b).

The survey sites were selected according to the restoration programmes, prioritising large lakes impacted by improvements to sewage treatment works and small

a

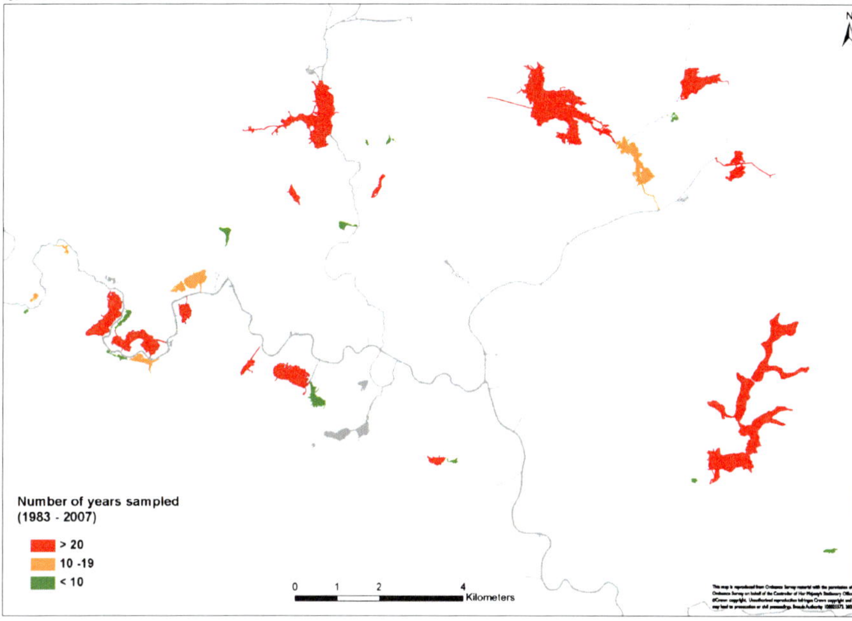

b

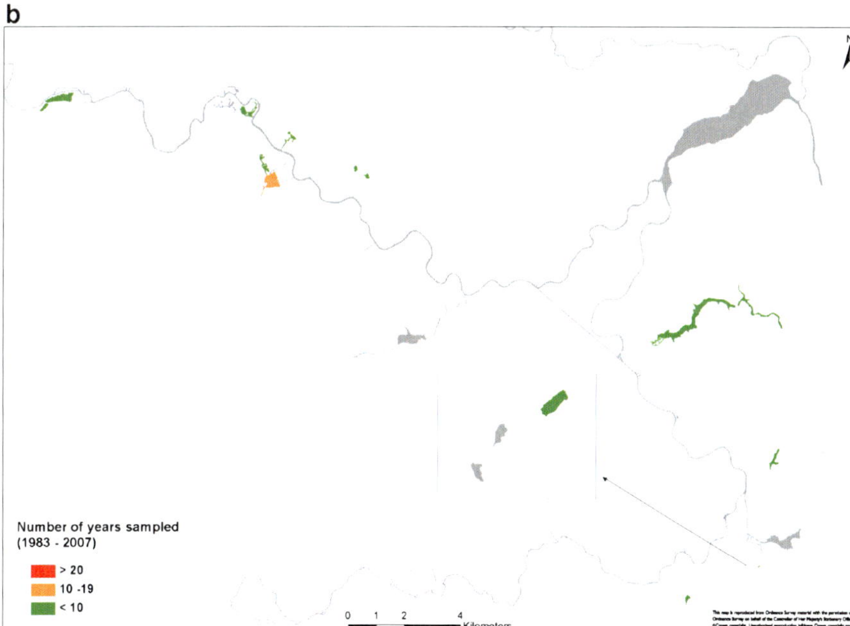

Fig. 24.1 (**a**) Locations of transect macrophyte surveys in the northern broads. © Crown Copyright and/or database right. All rights reserved. Licence number 100043571. (**b**) Locations of transect macrophyte surveys in the southern broads. © Crown Copyright and/or database right. All rights reserved. Licence number 100043571

lakes owned by the Norfolk Wildlife Trust, the Royal Society for the Protection of Birds (RSPB) or sympathetic landowners.

Point Survey – 13 Year Dataset (1996–2008)

Since 1996, a more detailed methodology has been used in the Trinity Broads in the Bure valley. The method was introduced to cover more of the lake area and gain better estimates of the quantity of macrophytes present: this would help us to evaluate the impact of management changes in a series of clear water lakes. Here we weighed the macrophytes, rather than using the more subjective visual assessment applied in the transect method. We also extended the sampling to include more points, as this will better represent the plant conditions across the lake basin.

SCUBA Survey – 13 Year Dataset (1994–2006)

We initiated a SCUBA survey in 1994, in response to macrophyte cutting proposals to facilitate navigation in Hickling Broad in the Thurne valley. This survey gave detailed *in situ* measurements that could not be obtained from the more simple, but destructive, methods of the transect survey (see Methods). The diver determines the height of specific plants from the bed, as well as making assessments of abundance, uprooting and bird grazing.

Hydroacoustic Survey

Whilst the SCUBA survey provides excellent plant height, condition and species data, you cannot use it (within a limited budget at least) to rapidly assess a large lake or several basins. Consequently the Broads Authority has adopted a novel methodology to address these issues. This came in the form of technological advances in hydroacoustics. This new method will replace the routine SCUBA survey to give data on overall plant height and percent volume infested (PVI), although the identification of species will still need to be confirmed using complementary methods.

Monitoring Targets

Monitoring macrophytes over the past 25 years has shown how restoration of The Broads is beginning to have a positive effect on macrophyte communities.

The Broads have a long, and well-documented, history of recording species (George 1992), with Victorian naturalists seeking rarities such as holly-leaved naiad (*Najas marina*) and stonewort (*Chara* spp., *Nitellopsis* spp. and *Nitella* spp.) species. More recently the replicated surveys undertaken by the Broads Authority aim to show, year on year, long-term change, site condition assessment as well as helping to guide site specific management decisions.

Since 1983 the Broads Authority has surveyed around 24–30 broads for aquatic plants annually. These transect surveys are the best records for shallow lakes within the UK. They are important in a local and national context and form an important evidence base for effective restoration management planning. Some examples of how our data contribute to the conservation management of the Broads are listed below:

- By assigning the condition of Sites of Special Scientific Interest (SSSIs): for example, aquatic plant status, derived from the annual transect survey, is used (in combination with water quality status and management actions) to determine if the government's Public Service Agreements for 95% of the area of SSSI to achieve favourable or recovering condition will be met by 2010
- By providing evidence of lake restoration as a result of removing fish and sediments
- By informing targets for lakes under the European Water Framework Directive
- By feeding back on the delivery of Biodiversity Action Plan targets for holly-leaved naiad and five rare stoneworts
- By providing data for research on shallow lake ecology
- By guiding the cutting of aquatic plants for navigational purposes, through determining if thresholds or trigger levels are reached. These thresholds are area, abundance, height and species-related as well as being site-specific (dependent on the type of boats, frequency of summer use, site ownership and conservation designation)
- By informing the development of lake management plans for stakeholders

Our data are also assessed against the National Park indicators to determine the condition and service delivery in the UK National Parks, as well as being used to informing stakeholder lake management plans.

Several waterbody-based stakeholder liaison groups are involved in the decision-making process and these receive regular updates about the plants and water quality in the lakes. The lakes that are navigated and have the potential to, or already, support water plants have well-developed water space management plans. These plans were developed with the stakeholders, ensuring that the many people who enjoy using the lakes for recreation contribute to the decision-making process for their management.

The Broads Authority's water space management plan for Barton Broad includes maps showing the areas used by different craft and the areas designated for wildlife (wintering wildfowl and water plants): these maps define refuge zones for wildfowl and non-intervention areas for protecting plants.

Fig. 24.2 A section of Barton Broad showing the macrophyte flora. *Photo by the Broads Authority*

Methods

Transect Survey

The annual transect survey is carried out in June and August by the Broads Authority and is supported by site partner organisations and volunteers who all receive training in monitoring and plant identification to improve reliability.

The method was set up to rapidly survey turbid lakes where visual assessment is not possible, i.e. where vegetation grows throughout the lake basin and there is no strand line.

The full method used for the plant transect surveys is outlined in Kennison *et al.* (1998). In summary, a series of transects is designed to cover each water body, with the total transect length kept relatively constant and related to the area of open water (1 ha:100 m transect). A double headed rake is dragged along the transect routes behind a boat travelling at around 3–4 mph, ensuring that the rake is on the bed. A Global Positioning System (GPS) is used to record the start and end points of the transects, as well as the locations of regular stopping points along each transect, which are determined by transect length and macrophyte abundance.

When macrophyte abundance is low the rake is pulled up and sampled at approximately every 50–100 m, when abundance is high, shorter distances are travelled between samples, sometimes only a few metres before the rake is full. When the plant beds are very dense and difficult to trawl but the water clarity is good, spot samples can be taken by throwing the rake from the boat and pulling it in to estimate vegetation cover at intervals along the transect.

Abundance scores for each species are estimated on a five-point scale based on the percentage of the rake covered in plant material:

1 = 5% or less (this may also be a fragment of plant)
2 = 6–25%
3 = 26–50%
4 = 51–75%
5= 76% or greater

Point Survey

The point survey is carried out in June and repeated in August each year to collect data for early and late emergence species. Each broad has a number of survey points fixed by GPS which are returned to with a tolerance of 10–20 m of the GPS position.

At each survey point we record plant height and secchi depth. Where the water is too turbid to see plant height, the length of the longest section of each species in the rake is measured, ensuring that only the growing stem is measured (i.e. not rhizomes or stolons). Macro-algal species, although recorded in the survey, are not measured for length.

Percentage overall macrophyte cover is estimated for the point, using 5% intervals, with aid of a bathyscope.

At each of the four points of the compass a double-headed rake is thrown and dragged 5 m along the bottom of the broad. Plant material on the rake is washed to remove silt, transferred to a net with a drawstring and swung several times until no appreciable amounts of water appear to be emerging from the sample. The material is then placed into a white tray for recording.

The plant material is gently pushed into a corner of the tray and percentage of the tray volume taken up by the material estimated, to the nearest 5% with an odd fragment recorded as 1%. This value represents the abundance of plant material in the sample.

Species are identified and an estimate of how much each species contributes to the total cover estimated as a percentage (in 5% intervals, 1% fragments). For example, if the total abundance of plants was 30% of the tray volume, and there were equal proportions of two species in the sample, each species would have a 15% abundance value given to them. The material from each species is then weighed.

SCUBA Survey

By using SCUBA, aquatic plants can be monitored *in situ* and without uprooting or damage. The main challenge for the SCUBA survey in the Upper Thurne catchment was the limited underwater visibility in the broads, which was often as low as 30 cm at the bed.

The method employs circular sample areas in which the species, abundance and maximum height are recorded, together with sediment type. The centre of the sample area is marked with a 2.5 m length pole with square base plate (12 cm × 12 cm) above a pointed spike at the bottom. This pole is pushed into soft sediments until the base plate is firmly located on the bed thus standardising the degree of penetration into the sediment. For hard sediments such as compacted sand and gravel, no spike is used and the base plate is held firmly on the surface of the sediment. The diver swims round the pole at arm's length and records the plants within a circle of approximate area 0.4 m^2. The species present and their maximum height in the water column are recorded by attaching colour-coded tags to a Velcro strip on one side of the pole. The pole is passed back to the cover boat, where species, abundance, height and water depth are recorded. Five samples are recorded for each sample station, all within a radius of 3 m of the cover boat and GPS antenna. In addition, the diver makes an overall estimate if the abundance of individual species on the DAFOR scale for each sample station.

For each waterbody, a stratified random sampling pattern is employed by dividing the broad into a grid and sampling at a random location in each grid square. The grid squares are approximately 100 × 100 m, but size and shape is dependent on the size and shape of the broad, and is designed to give an even coverage. The grid squares are located by eye using a map of the broad with the superimposed grid, marked features in and around the broad and permanent buoys positioned at selected locations. The precise position of the samples is recorded using a GPS accurate to 1 m. Monitoring is carried out at monthly intervals each summer from May to September.

The SCUBA survey monitors seasonal and annual changes in species frequency, abundance, distribution and maximum height. Abundance is also used to map plant cover for the entire water body.

For all methods, comments about the condition of the macrophytes, water clarity and weather are recorded as standard.

Results

Transect Survey

The results of the 2006 survey indicated that over 50% of the lakes visited had degraded water plant populations, characterised by low plant abundance. Data from three broads: Hickling, Barton and Cockshoot, show different trends up until 2006.

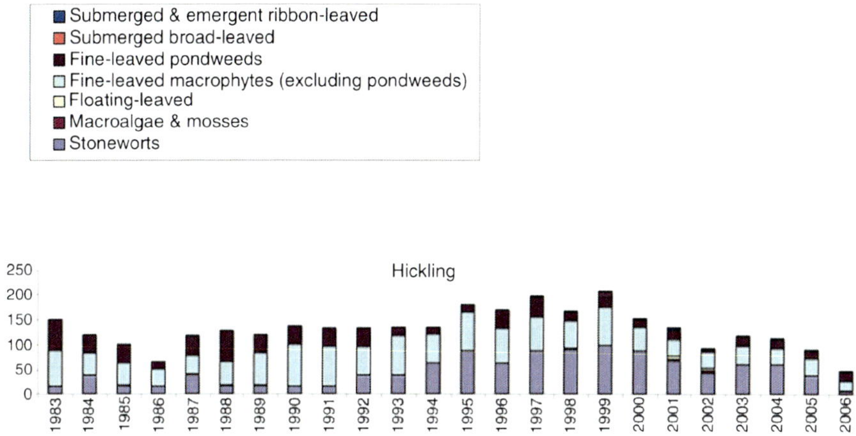

Fig. 24.3 Aquatic macrophytes at Hickling Broad 1983–2006

Hickling Broad – Upper Thurne Valley

Upper Thurne broads are home to the richest population of stoneworts in the UK, with 16 species recorded. Stoneworts are recorded in broads outside of the Thurne, but only the more common species and in much lower abundance. The Red Data Book species recorded include, three 'vulnerable' species: *Chara baltica*, *C. connivens*, *Nitellopsis obtusa,* one 'insufficiently known': *C. curta*, and one 'Rare': *C. intermedia* (Stewart and Church 1992). Thurne broads also provide a stronghold for the rare holly-leaved naiad *(Najas marina)* as well as more common vascular plants such as spiked water milfoil (*Myriophyllum spicatum*) and mare's-tail (*Hippurus vulgaris*).

Within the Broads, this association of plants is unique to the Upper Thurne. Figure 24.3 shows the trend revealed by the transect survey for aquatic macrophytes at Hickling Broad.

Barton Broad – Ant Valley

Barton has taken over 20 years to respond to external nutrient control and until recently had a very low abundance (and occasional complete absence) of recorded aquatic macrophytes (Fig. 24.4). Up to the mid-1990s, only the occasional yellow water lily (*Nuphar lutea*), rigid hornwort (*Ceratophyllum demersum*) and macro-algae were recorded. Since then *C. demersum* and curly-leaved pondweed (*Potamogeton crispus*), have been noted more regularly. Since 2003, over 10 species of aquatic macrophyte have been recorded, at increasing levels of cover. This increase may be a result of improved water quality as well as from biomanipulation creating clear water and aquatic plant recovery.

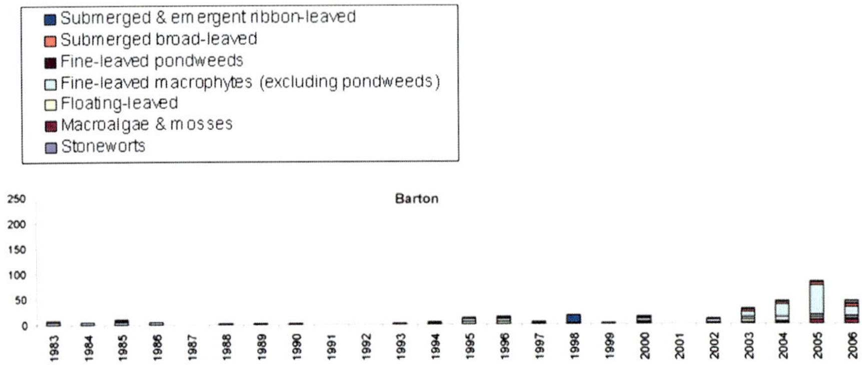

Fig. 24.4 Aquatic macrophytes at Barton Broad 1983–2006

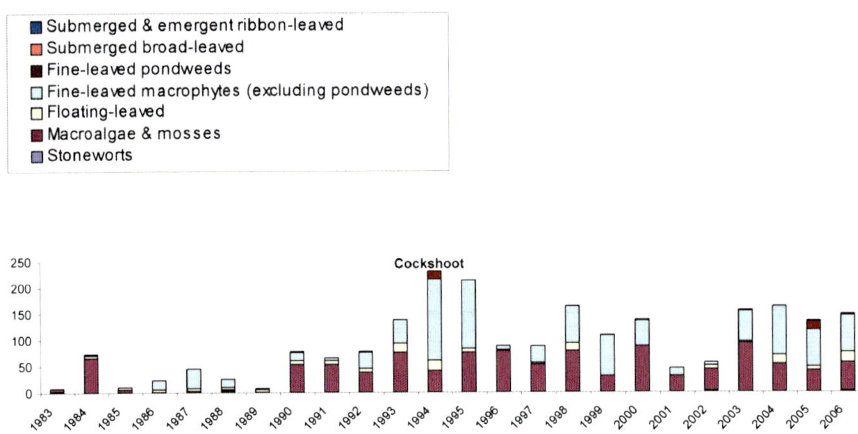

Fig. 24.5 Aquatic macrophytes at Cockshoot Broad 1983–2006

Cockshoot Broad – Bure Valley

Despite mud pumping and isolation from the River Bure in 1982, aquatic plants did not immediately respond, with low abundance of rigid hornwort (*Ceratophyllum demersum*) and water lilies recorded. Biomanipulation resulted in clear water conditions for most of the 1990s (Kelly 2008). The clear water reverted to turbid water when fish entered the broad in 2001/02, entering via channels around the dams that isolate Cockshoot from the River Bure.

These dams were repaired and both the plant growth and the number of species have improved from 2003 (Fig. 24.5), with abundance of fine-leaved pondweeds

(*Potamogeton* spp) and other fine-leaved submerged plants now found in the broad. Although macro-algae, which have the potential to smother plant beds or prevent the germination of seedlings, remain part of the community in parts of Cockshoot, the rare *Najas marina* has formed extensive beds since 2006.

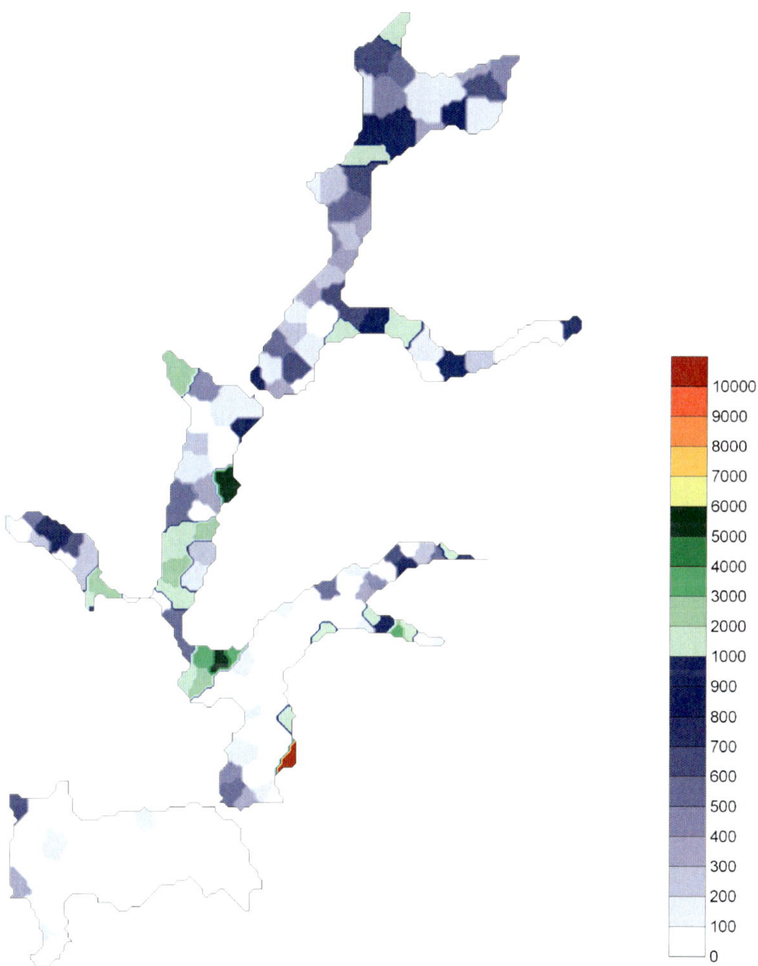

Fig. 24.6 Abundance of aquatic plants in Ormesby Broad measured by weight collected on standard rake hauls.

Point Survey

The point survey data has been mapped for each of the Trinity Broads (using Surfer software). Figure 24.6 shows an example of the plant abundance in Ormesby Broad within the Trinity Broads in 2008. Recent increase in plant abundance in all broads has resulted in increased survey time, thus a statistical analysis of the data will determine if sampling effort can be reduced without significantly affecting the overall results. In addition a statistical analysis of the plant, fish, zooplankton, algae and water quality data is ongoing.

SCUBA Survey

Changes in the aquatic plant communities in Hickling Broad were monitored over a 13-year period using this method. During this period, the abundance of intermediate stonewort *Chara intermedia* increased markedly. We produced maps of the devel-

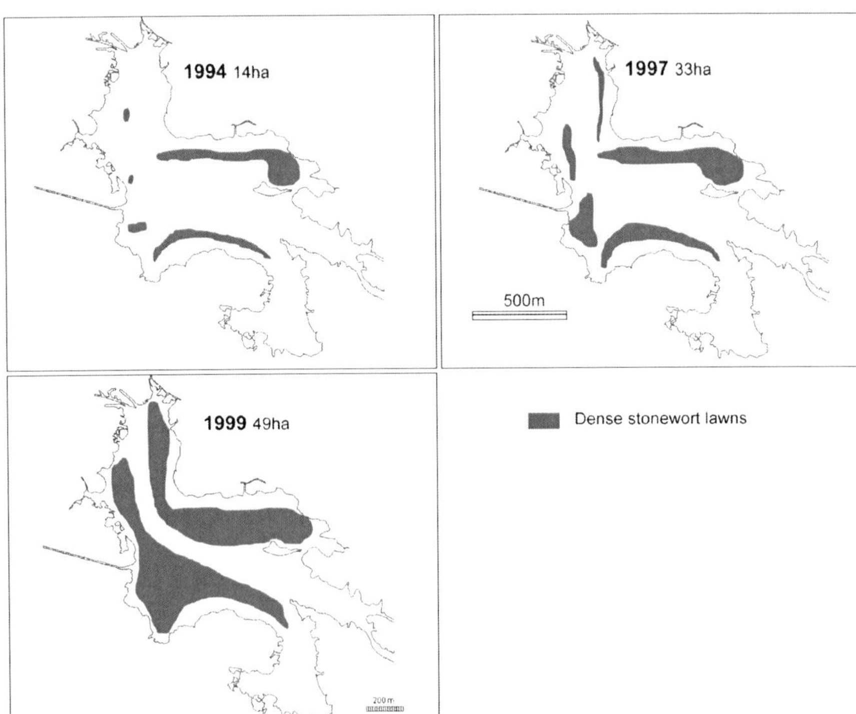

Fig. 24.7 Development of dense *Chara intermedia* lawns in Hickling Broad.

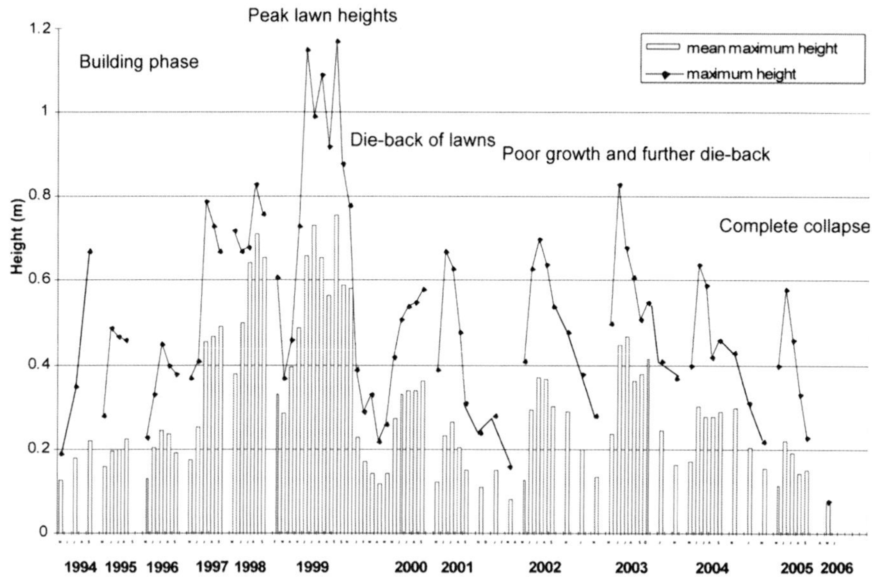

Fig. 24.8 Changing height of *Chara intermedia* lawns in Hickling Broad

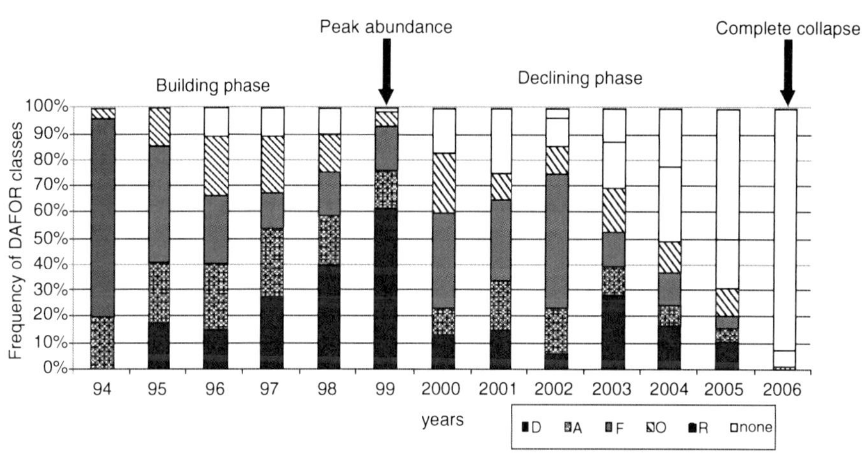

Fig. 24.9 Changing abundance of *Chara intermedia* in Hickling Broad

oping dense stonewort lawns by plotting abundance (plant cover) at the 112 sample stations (Fig. 24.7).

The increasing height of the stonewort lawns was also clearly demonstrated from the monitoring data. Additional winter monitoring showed that the lawns partly collapsed in the winter due to senescence of plant material at the bottom, and that spring growth took place from the top of the overwintering lawn. The amount of overwintering material increased annually, so that spring growth started at a higher point in the water column each year, thence contributing to greater summer heights of the stonewort lawns (Fig. 24.8).

Monitoring has recorded the 'boom and bust' growth pattern of *Chara intermedia* in Hickling Broad over a 13-year period. This is shown by plotting the frequency of abundance classes for the whole broad (Fig. 24.9).

Discussion

Transect Survey

This survey clearly shows a long-term trend of gradual ecological recovery of the river-connected shallow lakes in the Broads, such as Barton Broad, as phytoplankton responds to lower total phosphorus concentrations. In addition, these survey results have allowed us to quantify recovery from restoration programmes and to detect declines elsewhere (as seen at Hickling Broad).

This long-term dataset has become internationally valuable, following its use in characterising shallow, lowland alkaline lakes for the EU Water Framework Directive process. At a national and local level, the data has provided a base of evidence to underpin the SSSI condition assessment, and to inform both the government Public Service Agreement targets and the Broads Authority Lake Restoration Strategy. The transect survey provides a rapid survey methodology for a large number of lakes using a simple, replicable methodology, that is easy for surveyors to use and is relatively inexpensive to operate.

The relative simplicity of the transect survey facilitates participation across Broads Authority staff and between partner organisations as well as engaging with volunteers. These types of liaison promote communication and a broad understanding of conservation aims.

The results also provide important information for other functions. As the Broads Authority is also the Navigation Authority, it gains a clear picture of any potential conflict between the water plant beds and the navigation channels, allowing it to mitigate any negative impact on the plants.

The only real limitation of the transect survey is that at high plant biomass, the amount of plant material on the rake quickly achieves the maximum score. Although the score is divided by the distance travelled, the high biomass that can be achieved in these shallow productive lakes is therefore under-represented by this survey. This was one of the reasons for developing the more detailed point, SCUBA and hydroacoustic surveys that weigh, determine the PVI and/or map the height of the plant beds accordingly.

With extended growing periods predicted as a consequence of climate change, there may be justification for extending the survey period to accommodate the earlier emergence and longer growth periods of the species.

Point Survey

This long-term survey is an effective way of monitoring an ongoing lake restoration programme and designated site targets, with the spatial approach allowing for mapping by inferring between points. It provides greater detail than the transect survey and is better at quantifying biomass measures by recording the weight of plant material.

SCUBA Survey

SCUBA surveys require qualified divers with plant identification and surveys skills working in accordance with the Diving and Work Regulations, and are consequently expensive. They do, however, produce powerful evidence largely due to the *in situ* viewing and measuring of aquatic plants. Careful consideration is needed before initiating this method due to the ongoing financial commitment, but there are instances where *in situ* survey is the only means of monitoring specific important parameters such as plant height.

Underwater viewing also has the advantage of being able to assess plant vigour, growth form, impacts of grazing by herbivorous birds, vegetative and sexual reproduction, sediment deposition on plants and changes in plant cover and species at fixed points.

Hydroacoustic Survey

A new method, hydroacoustic monitoring, is being used and developed by the Broads Authority to provide figures on PVI and to assess aquatic plant height. The transect surveys have limitations when estimating biovolume, especially at high plant biomass, and the hydroacoustic survey also compares favourably to the dive survey particularly in terms of expense and availability of appropriately qualified dive surveyors. Using Sonar5-Pro equipment, this technique, when operating alongside existing methodologies in key sites, has proved particularly important for monitoring targets associated with the cutting of aquatic plants for navigation and for resolving issues with water users, such as sailing clubs and motorboat owners.

References

George M (1992) The land use, ecology and conservation of Broadland. Packard Publishing, Chichester

Jeppesen E, Ma S, Mo S, Christoffersen K (1998) The structuring role of submerged macrophytes in lakes. Springer, New York

Kelly A (2008) Lake restoration strategy for the broads. Broads Authority Report

Kennison GCB, Dunsford DS, Schutten J (1998) Stable or changing lakes? A classification of aquatic macrophyte assemblages from a eutrophic shallow lake system in the United Kingdom. Aquat Conserv Marine Freshwater Ecosyst 8:669–684

Phillips GL, Kelly A, Pitt J-A, Sanderson R, Taylor ED (2005) The recovery of a very shallow eutrophic lake, 20 years after the control of effluent derived phosphorus. Freshwater Biol 50(10): 1628–1638

Pollard P, Huxham M (1998) The European water framework directive: A new era in the management of aquatic ecosystem health? Aquat Conserv Marine Freshwater Ecosyst 8:773–792

Stewart NF, Church JM (1992) Red data books of Britain and Ireland. The Joint Nature Conservation Committee, Peterborough

Willby N, Pitt J, Phillips G (2006) Summary of approach used in LEAFPACS for defining ecological quality of rivers and lakes using macrophyte composition. Draft Report to the Lakes Task Team UK

Chapter 25
Monitoring Stoneworts *Chara* spp. at Bosherston Lakes

Bob Haycock and George Hinton

Introduction

Bosherston Lakes, an outstanding shallow marl lake system, are situated on the coast of south Pembrokeshire in south-west Wales (Fig. 25.1). The lakes were created at intervals in the late eighteenth and mid-nineteenth century by damming and drowning three valleys in the Carboniferous Limestone. The National Trust (NT) owns the lakes and they are within Stackpole National Nature Reserve (NNR) managed in partnership with the Countryside Council for Wales (CCW).

The lake system is fed by three small streams flowing in deeply incised valleys, sandwiched between a Carboniferous Limestone plateau and an Old Red Sandstone ridge. Parts of the water body are fed by Calcium-rich ground-water sources. The lakes are isolated from the sea, but only by a small sand dune ridge at Broad Haven.

The lakes support a strong population of rooted submerged and floating macrophytes, their distribution largely reflecting the differing degrees of nutrient enrichment (eutrophication) within the lake system and its surrounding catchment (Leach and Dyke 1978; Holman *et al.* 2009; Countryside Council for Wales, Bosherston Lakes water chemistry data 1977–2008).

The largest arm (Eastern Arm) running north from Central Lake is characterised by variable dense stands of curled pondweed *Potamogeton crispus*, fennel pondweed *P. pectinatus*, spiked water-milfoil *Myriophyllum spicatum* and Canadian waterweed *Elodea canadensis*.

In contrast, the two smaller arms (Western and Central Arms) linked to the Central Lake, are mainly spring-fed. Levels of phosphorus (P) (one of the main nutrients) in this southerly section of the water body are generally below 25 μg/l. With a high

B. Haycock (✉)
Countryside Council for Wales, The Old Home Farmyard, Stackpole, Pembroke, SA71 5QD, UK
e-mail: b.haycock@ccw.gov.uk

G. Hinton
Natural England, Northminster House, Northminster Road, Peterborough, PE1 1UA, UK

C. Hurford et al. (eds.), *Conservation Monitoring in Freshwater Habitats: A Practical Guide and Case Studies*, DOI 10.1007/978-1-4020-9278-7_25,

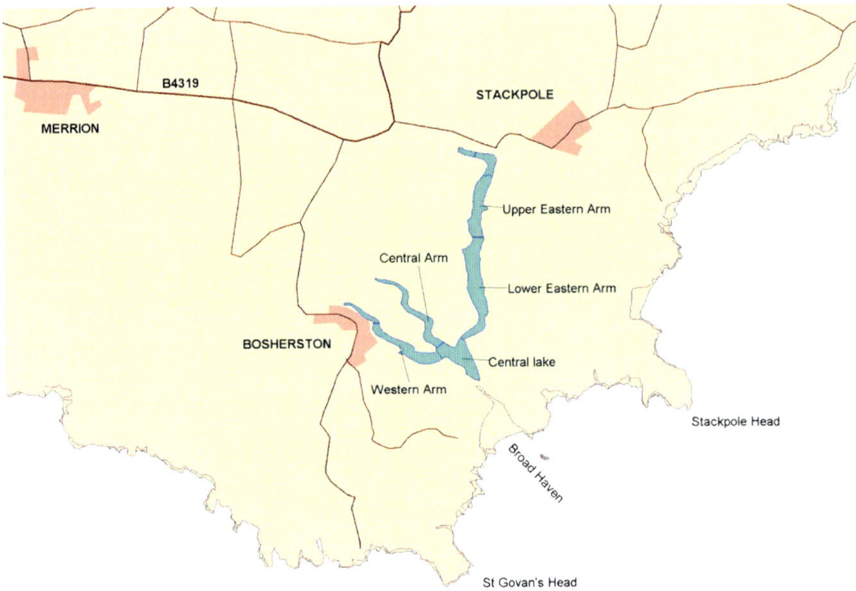

Fig. 25.1 The Bosherston Lakes system and its topographic setting. © Crown Copyright and/or database right. All rights reserved. Licence number 100043571

pH (around 8) and low nutrient levels, the Western and Central Arms support characteristic submerged macrophyte vegetation dominated by stoneworts (*Chara*) – including probably the largest population of *Chara hispida* (bristly stonewort) in Wales.

Chara hispida is a species that is generally uncommon in the Wales (Duigan and Haycock 1995) (Fig. 25.2). The C. *hispida* beds form part of a clear-water, oligo-mesotrophic marl-lake vegetation community of European importance.

Quite extensive beds of white water lily *Nymphaea alba* fringe the *C. hispida* beds. There are also smaller and more variable quantities of *C. globularis, C. virgata* and *C. vulgaris* though these species are not as common as *C. hispida. Chara* species are also found in a few small man-made pools in an adjacent drier valley behind Broad Haven dunes where an additional charophyte species *Tolypella glomerata* has also been recorded.

Much of the shoreline is wooded, and by the inflow streams and lake-shoreline there is emergent vegetation of common reed *Phragmites australis*, bulrush *Typha latifolia*, common spike-rush *Eleocharis palustris*, branched bur-reed *Sparganium erectum* and greater pond-sedge *Carex riparia* as swamp and fen communities develop.

Otters *Lutra lutra* are resident within and around the lake margins and have at least one breeding holt: Bosherston Lakes are a stronghold for this species. The lake system and its extensive linear shoreline are also attractive to bats, notably horse-shoe bats *Rhinolophus* spp. which utilise sheltered wooded flight-lines for feeding purposes.

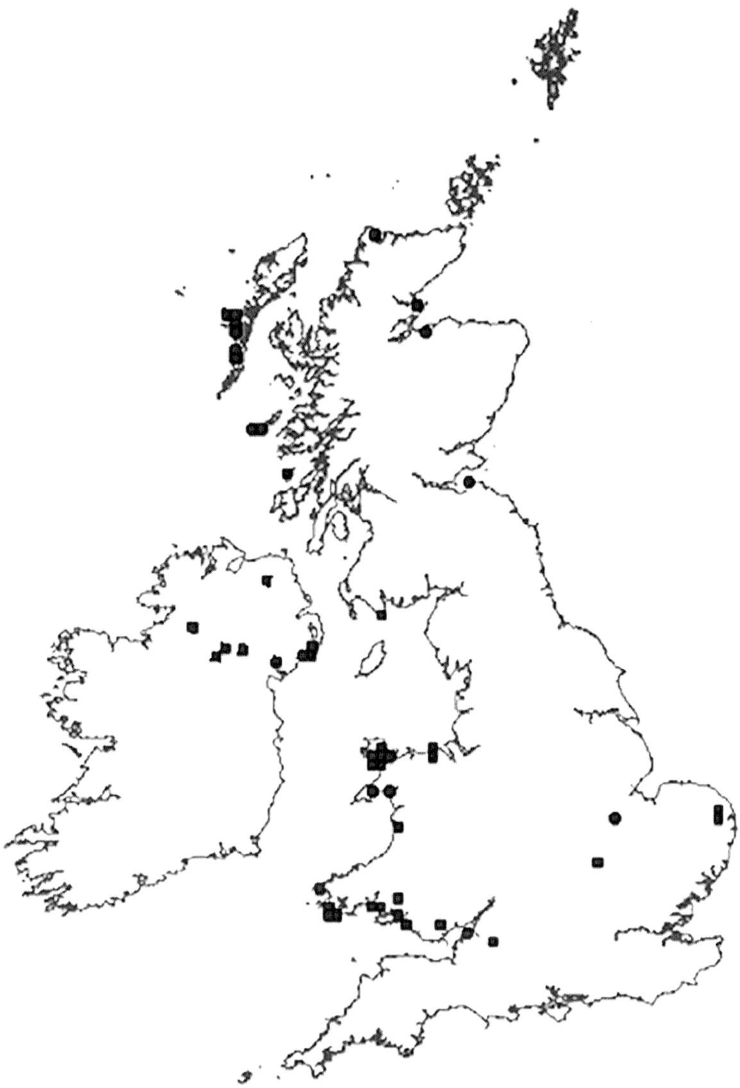

Fig. 25.2 10km squares with records for *Chara hispida* (Bristly Stonewort) in Great Britain and Ireland. (Source: National Biodiversity Network).

The Distribution of Chara hispida at Bosherston Lakes

Chara hispida forms dense beds up to 1m high (Fig. 25.3), with individual plants up to 3.5m long within the Central Arm, throughout much of the Western Arm and more intermittently within the Central Lake area (Fig. 25.4). Other *charophytes*

Fig. 25.3 A *Chara hispida* bed in the Central Arm of Bosherston Lakes. *Photo by Bob Haycock*

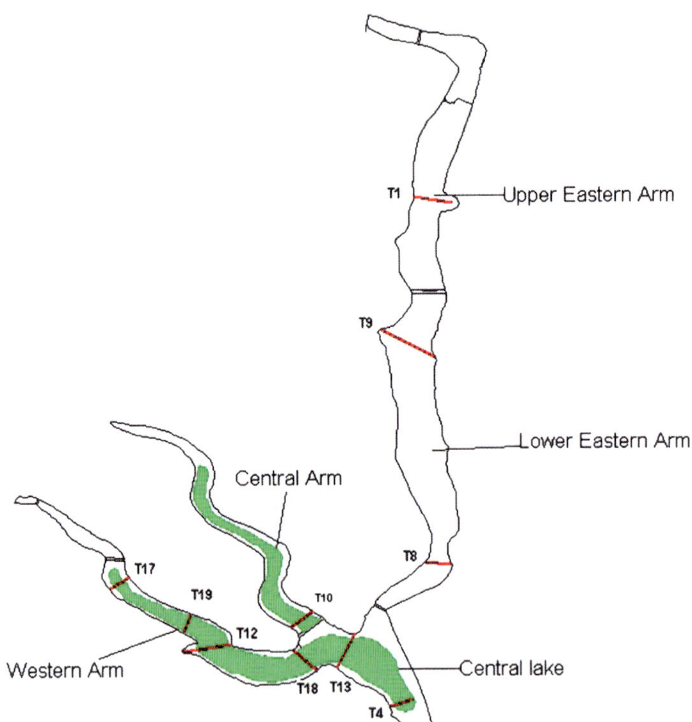

Fig. 25.4 The distribution of *Chara hispida* (*green*) and submerged transect locations within the Bosherston Lakes system. © Crown Copyright and/or database right. All rights reserved. Licence number 100043571

(notably *C. globularis*) occur in quite low quantities in these same areas and also in the Eastern Arm, where *C. hispida* is currently presently absent or rare. Most of the *Nymphaea alba* beds are in the Western and Central Arms.

Survey History – the Development of the Monitoring Methods

Charophytes have been recorded from Bosherston Lakes since at least the 1930s (Wade 1937; Stewart 1997). But it was not until the late 1970s, during the establishment phase of Stackpole National Nature Reserve, that any systematic recording of submerged macrophytes (including *Chara* species) was achieved.

The first detailed surveys of submerged and emergent vegetation within the Lake system were undertaken in the summer of 1977 (Dyke and Leach 1978; Dyke *et al.* 1978) as part of a series of base-line surveys to record and interpret the biological and physical features of Stackpole NNR. These surveys were undertaken by two graduates employed by the NT (funded by the Manpower Services Commission), and jointly supervised by the Nature Conservancy Council (predecessor of CCW) and the Field Studies Council. Surveys of macrophytes were repeated in 1978 (Leach 1978).

The following method summary is based on an abstract from Dyke and Leach (1978). The aim of this survey was to describe the submerged macrophyte communities, get some idea of their distribution and abundance, and establish permanent surveillance transects.

Transects

Transect-lines were selected through all the submerged macrophyte communities identified during preliminary survey. Eleven transects were identified for recording macrophytes out of twenty recorded for emergent as well as submerged communities.

The survey line was marked by strong nylon cord, tied on opposite banks of the lake to permanent stakes. The cord was marked at 2 m intervals with bitumen. Submerged macrophytes were sampled along this line using a small four-pronged grapnel opened to a grab diameter of approximately 45 mm. They recorded all species retrieved with three drops of the grapnel. The sampling interval was usually 2–4 m, with all transects surveyed from a canoe.

The results were given as % frequency for each species at each transect - imagining each transect to be a large "points frame". The results provided a quantitative measure of the distribution and abundance of each species recorded.

Underwater Survey

Initially, they feared that the "grapnel" technique may have been too insensitive to pick up changes in the vegetation. To test this, two SCUBA divers surveyed one of the selected transects (Transect 9) using subdivided ½ × ½ m quadrat frames to estimate % rooted frequency of each species. This method proved to be very time-consuming; eight quadrats were recorded this way, in addition to providing a running commentary on the character of the vegetation along the whole transect.

It became clear that while the grapnel method may "underestimate" the abundance of species with low cover values, it was surprisingly sensitive at picking up the more important changes in community structure. It also had the overriding advantage of being relatively simple, inexpensive and easily repeatable.

Of the transects sampled in 1977, those in the Central Lake; in the Central Arm; and in the Western Arm crossed through *Chara hispida* beds, an important component of the submerged macrophyte community.

Method Refinements

Further studies were carried out on Bosherston Lakes between 1980 and 1986, including surveys of submerged macrophytes. Henshilwood (1982) re-surveyed the 1977 transect lines in 1982, then Hinton (1987, 1989) re-surveyed most of them at regular intervals between 1984 and 1987. Three extra transects were also added in the Western Arm (T17, T18 and T19) where it was determined that more locations were needed to record the presence and frequency of *Chara hispida* beds, which had markedly declined since 1977.

As in 1977, a line was tied to permanent stakes on each bank, but the cord was marked at 5 m intervals, instead of 2 m intervals. Submerged plants were sampled at each point using a similar grapnel to that used in previous surveys. Hinton recorded those species collected from 10 grapnel drops at each point (compared with three drops in the previous surveys). Thus the frequency of each macrophyte was recorded as a number between 0 and 10 (equivalent to 0–100%) at each point on the transect line. Cumulative percentage frequencies were then compiled for each transect surveyed; for example a 100 m long transect would be represented by 200 grapnel drops (20 × 10).

Further testing of the linear transect grapnel method showed that it is consistent in determining species dominance, producing results in agreement with quantitative estimates made using two biomass sampling methods (a modified grab operated from the surface of the lake and a hand-held box quadrat deployed using SCUBA divers). Major changes in species dominance, whether seasonal or longer term, could be detected using the linear transect method, however variability (inherent in the linear transect method) meant that subtle long-term changes could be overlooked.

The main transects considered to be representative of the series appropriate for future surveillance/monitoring were: T1, T4, T8, T9, T10, T12, and T13 of the Leach and Dyke series; plus T17, T18 and T19, added by Hinton (Fig. 25.4).

Hinton (1987) recommended that these transects should be surveyed at least once every 5 years, and that more frequent effort is required for the *Chara hispida* beds - notably T4, T10, T12, T13, T17, T18 & T19,..... ideally annually, or at least every two years. This has method has been used to the present day (Perry 1994; Countryside Council for Wales, Bosherston Lakes submerged macrophyte transect records 1977–2008).

For several years this was the adopted method, utilising a large reel of nylon cord clamped to the stern of a rowing boat to make deployment easier between regular lake-side anchor points (identified by photographs and permanent shoreline concrete markers) on opposite banks. Not less than two personnel were required, one to hold the line/fasten the boat into position and record the species, etc.; the other to drop the grapnel and identify species captured. We also collected records of depth and water clarity (using a Secchi disk).

Whilst bank-side markers and photographs made it fairly easy to re-find the transect lines, the downside was the long time needed to deploy the nylon cord and to do the survey especially in a strong breeze!

During the last 10 years portable GPS units have become readily available and relatively inexpensive, and it is now possible to store hundreds of 10 figure British Ordnance Survey (O.S.) grid reference (co-ordinate) locations as waypoints, making such devices excellent survey tools. For the last five years, fixed nylon cord has been abandoned in favour of GPS to locate and record positions along the transect routes. Although there is some inherent "wobble" in the new recording lines, this is more than made up for by the time saved and the ability to record and store more points, including sample points between the regular transect routes. Furthermore, there is the added value of being able to easily transfer, store and display the sampling locations into computer-generated maps in Geographical Information Systems (GIS).

SCUBA divers were employed again in 1989 and in 1993, to record changes in *Chara hispi*da biomass since Hinton's earlier survey. Known volumes of *Chara* were collected from box quadrats placed on the bed of the lake in the Western and Central Arms and Central Lake. The retained samples were bagged, dried in a low temperature oven and then weighed to assess their biomass as a comparison with results reported by Hinton (Moore 1989, 1993). This method, as well as requiring additional health and safety measures and trained divers, was relatively expensive.

Monitoring Targets

Because of the initial survey work and regular surveillance of submerged macrophytes along standardised transects since 1977, we know quite a lot about the distribution of *Chara* within the lake system. The lake water chemistry and hydrology have also been recorded regularly over the last 30 years and we also know how the health and distribution of the population has fluctuated over this time due to various factors – including changes in water quality (increased eutrophication); changes in water levels and climate; and responses to the management of these factors.

This enabled us to establish a conservation objective and monitoring targets for the *Chara* beds, a component of the Pembrokeshire Bat Sites and Bosherston Lakes Special Area of Conservation (SAC). The key Natura 2000 habitat associated with the lakes is '*Hard oligo-mesotrophic waters with benthic vegetation of Chara* spp./ *Calcium-rich nutrient-poor lakes, lochs and pools*' (EU Habitat Code: 3140).

A "Vision" for Chara hispida Within Bosherston Lakes

Chara hispida will be considered as having favourable conservation status, when all of the following general conditions are satisfied:

'Standing open freshwater habitats and communities, supporting stoneworts are being maintained within the Bosherston Lake system. Water remains clear in the ground-water fed Western and Central Arms of the lake revealing quite striking beds of submerged *Chara hispida* vegetation, fringed by *Nymphaea alba* beds. *Chara* is also present within the Central Lake.'

We don't expect *Chara* beds to be present within the stream-fed parts of the lake system though these areas will be generally well-vegetated with other submerged and marginal plants. In the Eastern Arm, *Potamogeton crispus* and *Myriophyllum spicatum* are both quite common. Other submerged plants, include *P. pectinatus* and small pondweed *P. berchtoldii*. Smaller quantities of *Chara* species (not necessarily *C. hispida*) are also found along surveillance transects.

Water depth varies from about 3.5 m (winter maximum) to about 0.5 m or less in places in summer. Because the Western and Central Arms are spring-fed, nutrient levels here are low. One of the main nutrients (total phosphorus) (TP) reaches no more than 25 μg/l at regular sampling points.

We know that nutrient limits above this limit can tip the balance in favour of more nutrient tolerant species and filamentous algae that can shade out *Chara*. However, we expect that in the stream-fed Eastern Arm, levels of TP will be higher, though if actions are being maintained to remove point and diffuse sources, levels of TP are no more than 50 μg/l.

There is evidence that nitrogen (N) is also an important nutrient affecting vegetation growth; current estimates suggest N should be less than 1 mg/l at all sampling points to maintain "healthy" *Chara* beds. Open water is a key requirement so the lakes should not be infilling from silt eroding from adjacent farmland.

Rationale Underpinning the Condition Indicators

The extent of *Chara* beds as mapped in 1995 was used to inform the lower limit for habitat extent, with *Chara hispida* used both as a marker species for the habitat and as an indicator of habitat condition. The lower limits for the frequency of *Chara hispida* were informed by the results from a long-standing series of surveillance transects on the site dating back to 1977. Similarly, we know that the negative indicator species are likely to be present in the more nutrient-enriched parts of the lake system and that their presence at high frequency in the 'clear-water' lakes could indicate a worsening of water quality and loss of *Chara* (Table 25.1). NB. We have also set performance indicator targets for water quality, hydrology and sedimentation

Table 25.1 The condition indicators for the 'Hard oligo-mesotrophic waters with benthic vegetation of *Chara* spp./calcium-rich nutrient-poor lakes, lochs and pools' habitat at Bosherston Lakes

Condition indicator table	The 'Hard oligo-mesotrophic waters with benthic vegetation of Chara spp./calcium-rich nutrient-poor lakes, lochs and pools' habitat at Bosherston Lakes will be in favourable condition when:	
Extent	Lower limit	>30% of the lake bottom in the Western and Central Arms should be covered by [1]healthy *Chara hispida*
Vegetation composition	Lower limit	[2]Clear water plant communities should be present along seven standard transects sampled in the Western and Central Arms and Central Lake, of which: *Chara hispida* should have a frequency of at least: 50% in 3 out of 4 transects in the Western Arm (T12, 17, 18 & 19); 70% in the Central Arm transect (T10); 20% in the Central Lake (T4 & T13). And *Myriophyllum spicatum*, *Potamogeton pectinatus*, *Elodea canadensis*, *Lemna spp.* should have individual frequencies of: <30% along transects 4, 10, 12, 13, 17, 18 and 19; And Filamentous algae should have a frequency of <20% along transects 4, 10, 12, 13, 17, 18 and 19; And Invasive alien species, e.g. *Azolla filiculoides*, should be absent from all transects (For the purposes of this attribute, *Elodea canadensis* is considered an 'honorary native')
		Site-specific definitions
[1]Healthy *Chara hispida*		Bright green spiny shoots (not dull blue-green shoots that are coated with algae, e.g. blue-green algae, and apparently rotting).
[2]Clear water plant communities		Vegetation characterised by the presence of *Chara hispida*

(physical factors that may influence the *Chara* community) but these are not presented here.

Results

Over the last ten years, trends in repeat transect measurements of macrophytes and *Chara* species indicate that *Chara hispida* maintains a very healthy population within the Central Arm of the lake system (its core area). It is also maintaining its target condition in the Western Arm (Fig. 25.5), but is regularly failing to meet the target condition in the central lake (Fig. 25.6).

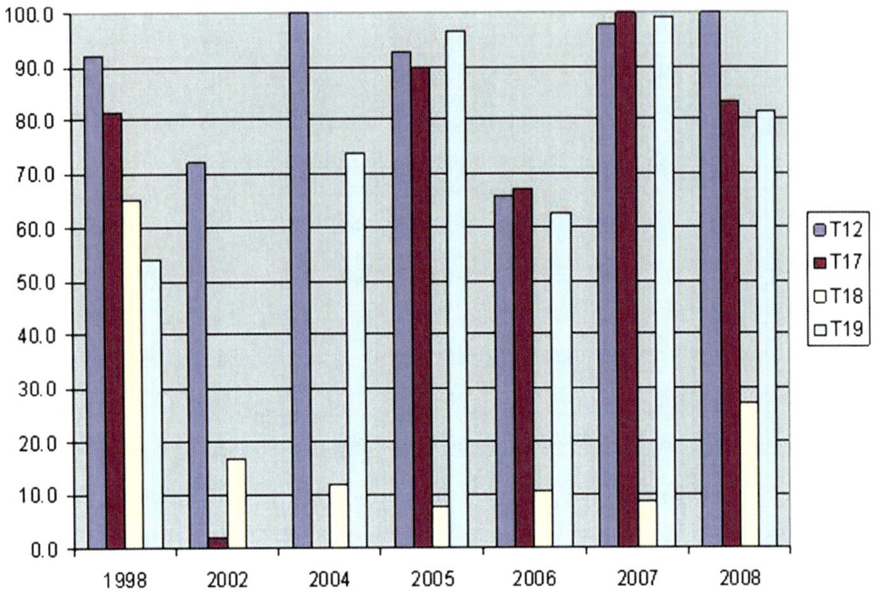

Fig. 25.5 Mean percentage frequency of *Chara hispida* at four submerged macrophyte surveillance transects in the Western Arm 1998 to 2008

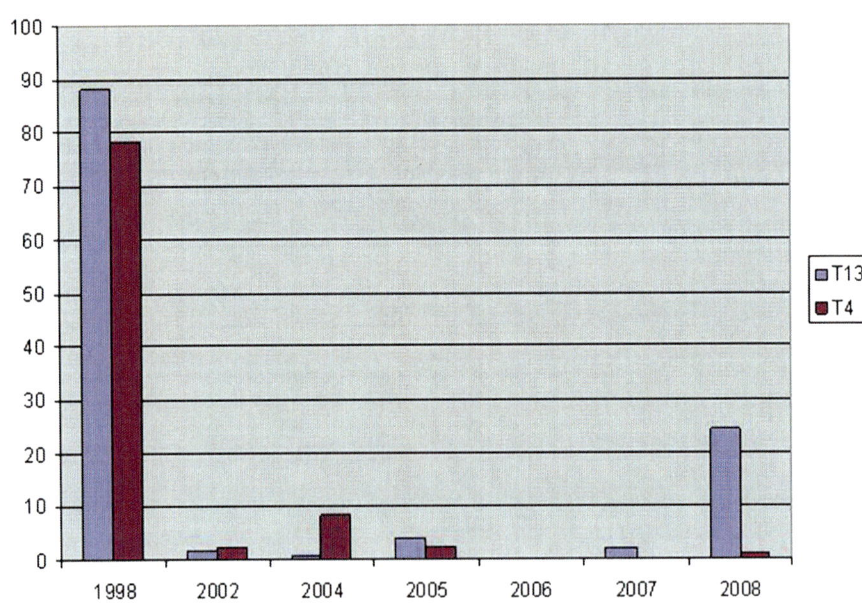

Fig. 25.6 Mean percentage frequency of *Chara hispida* at two submerged macrophyte surveillance transects in the Central Lake 1998 to 2008

Variable amounts of *Myriophyllum spicatum*, *Potamogeton crispus* and *Elodea canadensis* now regularly dominate much of the Central Lake. *Chara hispida* is relatively scarce, but in 2008 *Chara globularis* was more evident along most transects than it is in some years (Fig. 25.7). Other negative indicator species, such as blanket weed algae *Cladophora* sp. (often combined with ivy-leaved duckweed *Lemna trisulca*) have also become more abundant over the last 10 years. However, in 2008 neither of these was quite as frequent – possibly due to a cooler, wetter summer. Both seem to grow well in summers when water levels are lower.

Water quality targets for TP of <30 μg/l are generally being met in the Central and Western Arms, and P has declined during the last c. 30 years – helped by improvements to stream water quality at a sewage works upstream of the lake system. However, the Central Lake levels of TP are still occasionally higher than the desired state and so fail to meet overall target condition. Lake water clarity and pH target levels have been maintained in all sections.

Lake levels fluctuate considerably, usually reaching low conditions within the Central Lake and the Western Arm during late summer. Water leaks away due mainly to natural processes, but wet summers in recent years have maintained slightly higher

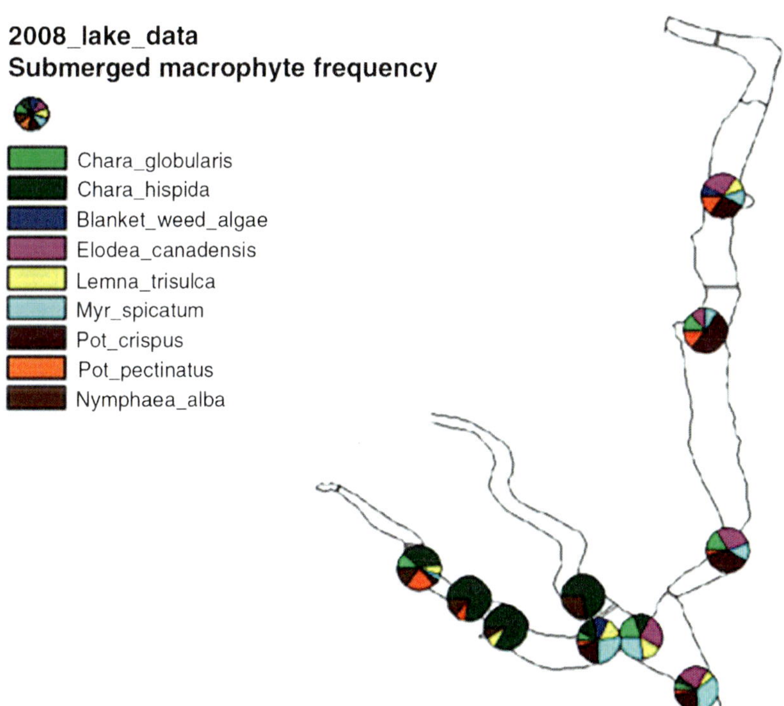

Fig. 25.7 Mean percentage frequency of macrophytes along transects within Bosherston Lakes in 2008.

levels than usual. Low summer water levels, combined with algal and macrophyte die-back, have increased the amount of sedimentation at the head of the Western Arm over the last 30 years.

Overall therefore, the condition of the *Chara hispida* "habitat" in 2008 was unfavourable.

Discussion

Long-term surveillance (over 30 years) of regular transects has enabled us to collect a lot of comparable data, and whilst it used to take more than a week to survey the whole lake, with the advent of GPS it is now possible to survey all the transects and collect additional records form other points within 3–3.5 days.

Furthermore, the availability of GIS means that it is now possible to map and display the results for each species. This makes comparison between years relatively straightforward. Digital photography also enables us to obtain regular images of the *Chara* beds, although there are limited view-points where we can employ fixed-point photography.

Due to the density of *Chara* beds in the Central and Western Arms and the falling summer water levels, it is important to record these areas no later than the end of June, or else it could be impossible to row and manoeuvre a boat over the *Chara* and lily beds in these areas.

The *Chara* beds are still reasonably healthy over much of the spring-fed parts of the lake, though currently not in the Central Lake (which is at the interface between low nutrient ground-water and higher nutrient stream-water influences): the periods when it is abundant here are very few.

Significant steps have been made to reduce nutrient enrichment and levels of P entering the lakes. A number of point-sources for enrichment have been identified and managed – including diversion of treated sewage which, prior to 1984, discharged into a stream entering the lakes Eastern Arm. A pipeline was installed on the bed of Central Lake to by-pass stream-fed water from the Eastern Arm from the spring-fed lower lakes (Haycock and Duigan 1994; Moss *et al.*, 1996). For a time (in the early-mid 1990s) this appeared to be having a significant positive affect on *Chara hispida*, which re-colonised the Central Lake to levels not seen since 1977/78. Steps have also been taken to intercept nutrient-enriched silt eroding from the intensively farmed catchment, using several silt-trap pools. Within-lake sediment has also been excavated at the heads of the Eastern and Western Arms (Haycock and Ellis 2002; Haycock and Bennett 2008).

Over the last 10 years, *Chara* has struggled to maintain a toe-hold in the Central Lake. Here, filamentous algal (*Cladophora*) growths together with *Lemna trisulca* (this species was unknown in the lakes until the late 1980s) can now become abundant in late summer, as water levels fall and water temperatures rise. Also, there are still issues affecting Central Lake water quality. TP is still not as low as it could be in the Central Lake and one of the other possible nutrients (N) could also now be influencing plant growth generally within the lake system. However, early in 2009, the Bosherston Lakes catchment was designated a Nitrate Vulnerable Zone, so hopefully it should be possible to regulate levels of N more closely in the future.

We still don't completely understand the hydrology of the lakes, but we are aware that water leaks out of the lake system – possibly through very many small natural fissures in the lake bed (Thomas 2007; Husband and Cassidy 2008). Warmer summers, rises in water temperature (combined with regularly low summer water levels, and gradually increasing sedimentation within the lake) as well as predicted sea level rise over the next 50–100 years, means that the Central Lake could become more regularly saline in future. It is also possible that the *Chara* beds will retreat naturally, and if this occurs we will need to review the conservation objectives and monitoring targets accordingly.

References

Countryside Council for Wales. Bosherston lakes water chemistry data 1977–2008 (unpublished data)

Countryside Council for Wales. Bosherston Lakes submerged macrophyte transect surveillance records 1977–2008 (unpublished data)

Duigan CA, Haycock RJ (1995) Reserve focus – The Bosherston lakes. British Wildlife 6(4): 231–234

Dyke J, Leach S (1978) Bosherston Lakes Report 7, Survey of submerged macrophytes. Report resulting from joint project by the Nature Conservancy Council, Field Studies Council and the National Trust

Dyke J, Leach S, Holden P (1978) Bosherston Lakes Report 6, Survey of emergent vegetation. Report resulting from joint project by the Nature Conservancy Council, Field Studies Council and the National Trust

Haycock B, Bennett I (2008) In at the deep end – restoring open water at Bosherston Lakes. Natur Cymru 27:36–41

Haycock B, Duigan CA (1994) Saving stoneworts at Bosherston. Enact Manag Land Wildlife 2:21–23

Haycock B, Ellis R (2002) Bosherston lakes – An introduction to their ecology and management. The National Trust and the Countryside Council for Wales

Henshilwood DA (1982) Bosherston lakes – Interim report on study of macrophytes in open water. Nature Conservancy Council Unpublished Report

Hinton GCF (1987) Report on resurvey of Chara beds in Bosherston lakes, Dyfed on 6/7 August 1987. A report to the nature conservancy council

Hinton GCF (1989) A eutrophication study – Bosherston Lakes, Dyfed, Wales. Chapter 5. Biota. Report no. 936. A report to the nature conservancy council

Holman IP, Davidson T, Burgess A, Kelly A, Eaton J, Hatton-Ellis TW (2009) Understanding the effects of coming environmental change on Bosherston Lakes as a basis for a sustainable conservation management strategy. CCW Contract Science Report No. 858

Husband C, Cassidy N (2008) Geophysical surveys of Bosherston lakes (Lily Ponds) Stackpole National Nature Reserve October 2007. Final Report to Countryside Council for Wales

Leach S (1978) Bosherston Lakes – Report 9. Changes in status of submerged macrophytes 1977–1978. Report resulting from joint project by the Nature Conservancy Council, Field Studies Council and the National Trust

Leach S, Dyke J (1978) Bosherston lakes – Reports 2 and 3. Lake Chemistry. Reports resulting from joint project by the Nature Conservancy Council, Field Studies Council and the National Trust

Moore J (1989) Report on a re-survey of charophyte beds in Bosherston Lakes, Stackpole NNR, on September 1st. 1989. Nature Conservancy Council Unpublished Report

Moore J (1993) Report on a re-survey of charophyte beds in Bosherston Lakes, Stackpole NNR, on September 8th 1993. CCW Contract Science Report No. 83

Moss B, Madgwick J, Phillips G (1996) A guide to the restoration of nutrient-enriched shallow lakes - Chapter 10 The Collective Experience – a Compendium of European Case Studies. Environment Agency and Broads Authority 146–148

Perry K (1994) Monitoring the recovery of the charophyte beds in Bosherston lakes, Stackpole NNR, Pembs. 1994. Report, CCW Spécies and Monitoring Branch

Stewart NF (1997) A review of Welsh stoneworts. CCW Contract Science Report March, 1997

Thomas E (2007) Bosherston lily ponds, Stackpole Pembrokeshire. Geophysical survey. Report No. 530.1/0/0607. A Report to the Countryside Council for Wales

Wade AE (1937) The charophyta in Wales. Report of the Botanical Exchange Club 1937:596–603

Chapter 26
Monitoring Wetland Mammals: An Ecological Case Study

Penny Neyland, Dan Guest, Charles Hipkin, and Dan Forman

Introduction

The water vole (*Arvicola terrestris*) is a medium sized rodent that displays remarkable ecological plasticity throughout its European range. In stark contrast to its common name, water voles adopt a fossorial (underground dwelling) lifestyle in many regions of mainland Europe in which the occurrence of water is not a defining factor in their distribution (Strachan and Jefferies 1993; Strachan 1998). There, water voles inhabit mountainous terrain and grassland habitats and are regarded as a serious pest species of vegetable root crops in some European regions (Giraudoux *et al.* 1997; Morilhat *et al.* 2008). In contrast, the distributions of water vole populations in Britain (the species is absent from Ireland), the Netherlands and parts of Spain and France are closely associated with wetland habitats providing suitable opportunity for burrowing and abundant structured riparian vegetation (Carter and Bright 2003; Moorhouse and Macdonald 2008). Alarmingly, over the last 100 years, the British water vole population has undergone a dramatic and widespread decline (Jefferies *et al.* 1989; Strachan and Jefferies 1993) and the species is currently a priority Local Biodiversity Action Plan (LBAP) listed species of significant national conservation concern in Wales, England and Scotland (Strachan 1998; Strachan and Moorhouse 2006).

Research undertaken by the Vincent Wildlife Trust and the Environment Agency has clearly shown that the number of sites historically occupied by water voles is reducing significantly in all regions of Britain (Jefferies *et al.* 1989; Strachan and Jefferies 1993). The factors that have driven this alarming decline, in what was previously a relatively common mammal in Britain, are complex and not completely

P. Neyland, C. Hipkin and D. Forman (✉)
Conservation Ecology Research Team (CERTS), Institute of Environmental Sustainability, Swansea University, Singleton Park, Swansea, SA2 8PP, UK
e-mail: D.W.Forman@swansea.ac.uk

D. Guest
Countryside Council for Wales, Llys Tawe, King's Road, Swansea, SA1 8PG, UK

C. Hurford et al. (eds.), *Conservation Monitoring in Freshwater Habitats: A Practical Guide and Case Studies*, DOI 10.1007/978-1-4020-9278-7_26,

understood. However, it is generally agreed that modification and loss of wetland habitats, coupled with the active predation of water voles by feral populations of American mink (*Mustela vision*), are significant determining factors in the widespread decline of water voles (Woodroffe *et al.* 1990; Macdonald and Strachan 1999). Since the plight of the water vole in Britain was recognised and highlighted, significant effort has been invested in the identification and monitoring of local populations by a wide range of statutory and non-statutory organisations. In addition, a considerable amount of autecological research has been conducted on water voles, predominately on populations inhabiting linear wetland habitats such as rivers, ditches and canals. This research has revealed important insights into how water vole populations behave at both the local and broader landscape scale (e.g. Bonesi *et al.* 2002; Telfer *et al.* 2003) and into the relationships between water vole distribution, population density, different vegetation community types and mink (e.g. Woodroffe *et al.* 1990; Lawton and Woodroffe 1991; Rushton *et al.* 2000; Moorhouse and Macdonald 2005). This and other ongoing work is beginning to provide the information and tools necessary to successfully monitor, appropriately manage and safeguard the remaining British populations of water voles with the ultimate aim of restoring this important species to as much of its former range as possible.

Current Techniques Used to Monitor Water Vole Populations

The most widely used technique to monitor the activity of water voles is the standardised transect survey in which the presence/absence of distinctive field signs (including food caches, burrows, foot-prints, and, perhaps more reliably, faeces and latrines) are recorded over a defined distance of wetland edge (see Strachan 1998; Strachan and Moorhouse 2006). Live capture and release programmes (under controlled and licensed conditions) can also provide an extremely valuable and detailed source of population-level data. These techniques have recently been used by Oxford University to study several water vole reintroductions (Moorhouse *et al.* 2009). Over time this approach can provide information on the number of animals present, their respective range sizes and movement patterns in different water vole populations.

Study Site

This case study describes a recent water vole project undertaken at the National Wetlands Centre Wales (grid reference SS 532 984), a National Key Site for water voles in Wales, United Kingdom. This site provides an ideal location to study the water vole as it supports an established meta-population within a number of diverse, interconnected habitats including ponds, ditches and reed beds of varying size and complexity. Long-term data collection across multiple connected areas of the wetlands has provided a valuable insight into the dynamics of wild populations. This study focuses on two interconnected ponds, one relatively large and complex (Pond

A – 360 m circumference – Fig. 26.1) and one smaller and circular (Pond B – 80 m circumference). The two ponds are adjacent to one another and connected by a broad expanse of soft rush *Juncus effusus* pasture.

Water voles captured within the study site used the two ponds concomitantly and thus we regarded the ponds both as two separate entities and as a combined system when interpreting the results. This study highlights the need to consider the proximity of other potential habitats when monitoring the activities of water voles as they frequently move between suitable patches, provided they have the necessary vegetation corridors.

Aim

To develop a vegetation-based sampling approach for monitoring water vole activity within a complex pond system.

Methods

This study builds upon previous live capture and release approaches used to monitor water vole populations. Here however, we describe a new method of studying the movement patterns of water voles in relation to discrete stands of vegetation. In order to do this, it was necessary to identify a set of site-specific vegetation types, defined by the dominant plant species within each stand. These 'dominant vegetation types'

Fig. 26.1 Pond A – Optimal habitat with dense bankside vegetation. *Photo by Penny Neyland*

(DVTs) were mapped in the field onto recent ortho-rectified aerial photographs at 1:1250 scale and the boundaries subsequently digitised using Geography Information System (GIS) software. The respective movements of individually micro-chipped voles tracked over 16 months were superimposed onto this DVT matrix to establish the use by individual water voles of each DVT. Little information is available describing the population size, movement patterns and other ecological aspects of water vole populations occupying non-linear habitats such as ponds. We anticipated that our study would provide a detailed assessment of the local movements of a water vole population in distinct plant patches that may assist in the construction of empirically derived models that could potentially be used to predict water vole population densities under different ecological conditions. This case study and the approach it describes should augment the knowledge required for the effective monitoring and management of pond systems containing water voles and provide useful comparative data on the species' ecology.

Vegetation Surveys

Since water voles depend on the vegetation surrounding the water body as both a source of food and cover from predation, we mapped all of the bankside and emergent vegetation of the ponds together with the surrounding land use (tracks, trees/scrub). This provided a vegetation map onto which other activities such as trapping events and field signs could be mapped. In effect, this is a bottom-up description of habitat utilisation. Homogenous stands of vegetation were identified in the field and drawn onto the aerial photographs. Each stand was labelled according to the dominant vegetation type (DVT). Plant community associates were also noted as these often feature in the water vole diet, but were not used to define the map. The resulting field map was then digitised using Mapinfo Version 8.5 (MapInfo GIS is a product of the MapInfo Inc.) which creates colour-coded polygons corresponding to individual plant species and provides a way of visualising the vegetation patterns.

The DVTs serve as the experimental blocks or sampling units. Intensive field surveys together with capture data can be overlaid onto these sampling units to provide a multidimensional map of water vole habitat selection and utility. This multilayered approach offers a holistic interpretation of water vole ecology and can serve as a baseline from which to develop further studies of wetland mammals.

Live Trapping

A total of 22 numbered single entry rat cage traps were positioned at 20 m intervals in the dense vegetation close to the edge of the ponds. The number of traps positioned in each study pond was restricted by the size of the pond, accordingly 18 traps were placed around pond A and 4 traps around pond B. All traps were provided with abundant dry hay and circa 150 g of apple. Traps were set for continuous periods of at least 5 days every month over a 16-month period. During each trapping

Fig. 26.2 Water vole in vegetation at National Wetland Centre Wales. *Photo by Penny Neyland*

period, every trap was regularly checked and fresh hay / apple provided as required (field voles, *Microtus agrestis* frequently consume apple bait without triggering the trap mechanism). Any water voles caught were examined whilst in the trap for injuries, parasites and other notable features, before being transferred gently to a netting bag. Each vole was then sexed visually and individually tagged using a single Passive Interrogation Tag (PIT) injected between the scapular. All re-captured voles were scanned using a hand held PIT reader to determine the identity of marked animals. In order to minimise the potential exposure risks to voles caught in traps, trapping was not conducted in either very hot or cold weather, or during periods of heavy and prolonged rainfall. The total amount of trapping effort expended during this project was 1,760 trap nights.

Bulrush *Typha latifolia* and soft rush *Juncus effusus* are the most common dominant vegetation types (providing both food and cover) with yellow flag *Iris pseudacorus* and water-pepper *Persicaria hydropiper* as examples of community associates (and seasonal components of the water vole diet).

Analysis

Population densities are presented as the mean number of individuals per 100 m of habitat and calculated separately for each pond. The minimum number of animals alive (MNA) provides the most conservative population estimates (i.e. the least number of water voles on a pond during a given trapping session) and includes adults and juveniles. MNA was used as the population estimate for the site thus:

population density = (population estimate for site/length of trapped habitat) × 100 (Moorhouse and Macdonald 2008). If an animal was not trapped for a particular month the MNA was calculated from recaptures thereafter. The location and movements of individuals were plotted on the vegetation map (Fig. 26.3).

Current knowledge suggests that adult females exclude same sex individuals from their range during the breeding season (February–October) but overlap with males (Strachan and Jefferies 1993; Strachan and Moorhouse 2006). Females have been seen to overlap with other females when establishing territories at the onset of the breeding season or after territoriality breaks down during the onset of winter (PN personal observation). Live trapping throughout the year allows us to observe these overlaps that may be overlooked by other studies that focus on trapping during the breeding season only.

Results

Populations

The traps were occupied on 87 occasions, and 35 water voles caught over the duration of the study: these comprised 16 females, 14 males and 5 juveniles unable to be sexed (Fig. 26.4). Fourteen water voles were recorded only once. Seven voles moved between this site and adjacent ponds not included in this study.

Three voles were previously marked elsewhere within the wetland complex and immigrated into the study area. Eight voles (five males and three females) were caught repeatedly in the same trap within a given trapping session. Breeding was confirmed on both ponds by the presence of sexually active males and lactating females, and by the capture of young water voles.

The mean density of water voles (on both ponds as a combined system) decreased from approximately 3.5 voles per 100 m to less than one animal per 100 m during the breeding season of 2007 and subsequently dropped to 0.5 voles per 100 m, where it remained until early in the summer of 2008 (Fig. 26.4). The MNA data indicate that water vole numbers decreased on pond A from between two to four animals after November 2007 and then dropped to only one animal throughout the winter, until February. Two animals were then recorded on pond A from March onwards (in comparison to the previous March when 12 animals were caught), with a total of six recorded there in June 2008. There were fewer water voles on pond B, with a maximum of three animals recorded. No animals were caught on pond B between July 2007 and June 2008. Several water voles (both males and females) occupied ranges that were restricted to only one or two trap locations, most noticeably after the summer of 2007. This restricted movement of animals and overall low density of voles recorded throughout the latter part of the monitoring period may have been the result of an increase in rat *Rattus norvegicus* presence at the ponds (see Discussion) or simply one of the natural oscillations in population cycles common to most species of Microtine rodents (Lambin *et al.* 1998; Oksanen *et al.* 1999).

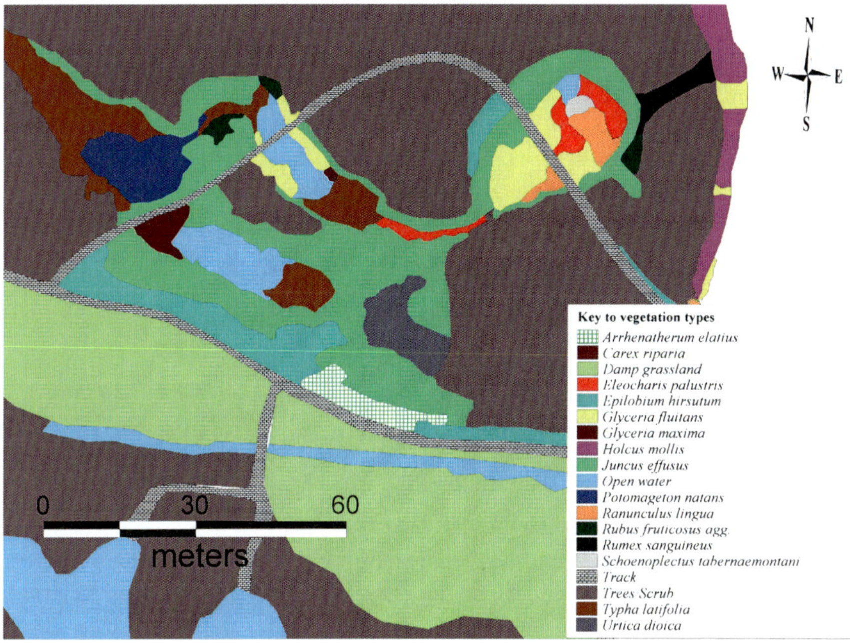

Fig. 26.3 The Dominant Vegetation Type (DVT) map for the water vole study area covered in this chapter.

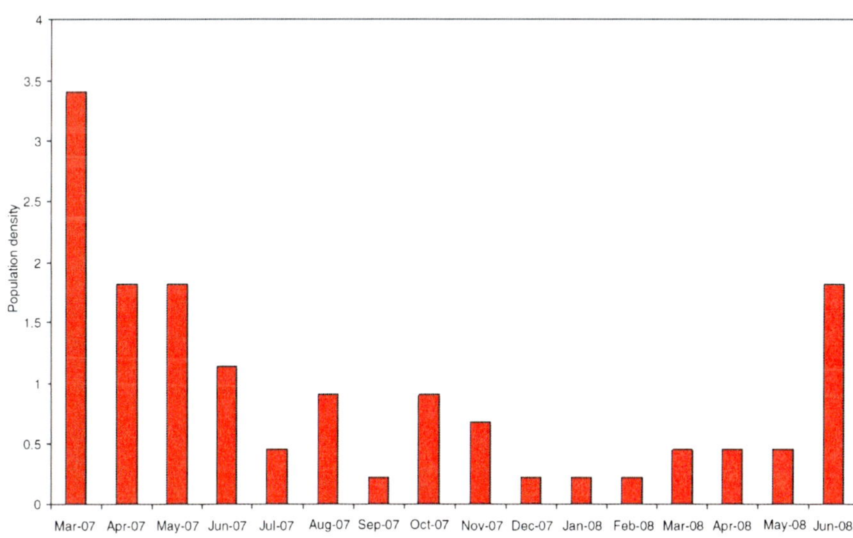

Fig. 26.4 Population density of water voles trapped on ponds A and B as a combined system (mean number of individuals per 100 m of habitat)

Dominant Vegetation Types

A greater number of water vole captures took place in soft rush (*Juncus effusus*) than any other DVT, reflecting to some degree that more traps were located in *Juncus effusus* than any other DVT. However, the number of captures was adjusted for unequal sample distribution in each patch to provide a true reflection of patch preference (Fig. 26.5).

Figure 26.5 clearly illustrates that water voles show a preference for specific DVTs. *Juncus effusus* supports the highest relative number of animals, particularly females (Fig. 26.6). *Typha latifolia* is also an important component of the habitat, particularly for males (Fig. 26.7) and juveniles. Females were associated with four DVTs whilst males and juveniles were restricted to two.

Discussion

During the course of the study the numbers of brown rats (*Rattus norvegicus*) caught (and associated field signs) increased significantly. During October 2007 a routine clearance of *Typha latifolia* was undertaken on pond A. During the months after the clearance a substantial increase in rat activity was seen around the ponds. Furthermore, water voles trapped during this period occasionally exhibited injuries atypical of intra-specific fighting injuries (Forman and Brain 2006), and on one occasion the remains of a dead water vole were found in a rat food cache, although it is not clear whether rats predated or scavenged this vole. However, as it is highly likely that rats influenced the distribution and behaviour of the water voles on both ponds during the

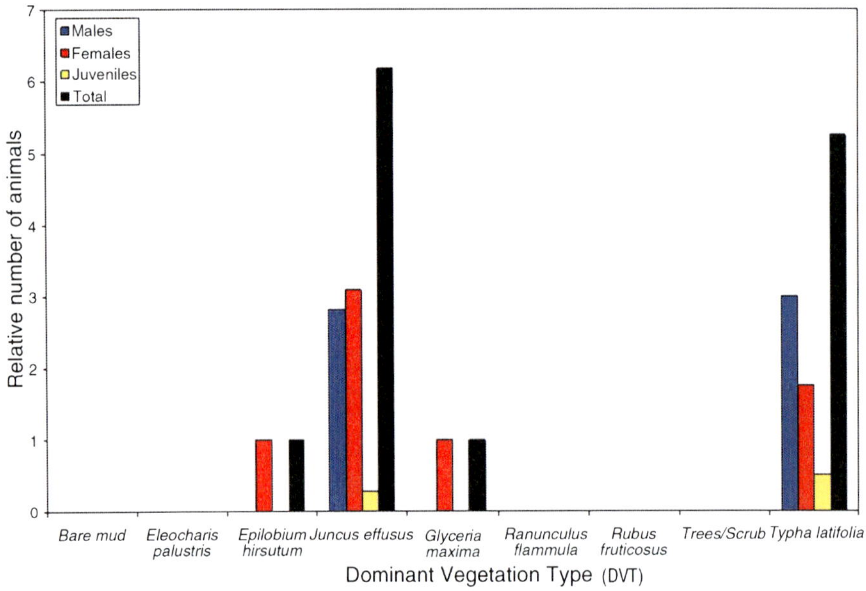

Fig. 26.5 Relative number of water voles caught per DVT (adjusted for unequal sample distribution in each patch)

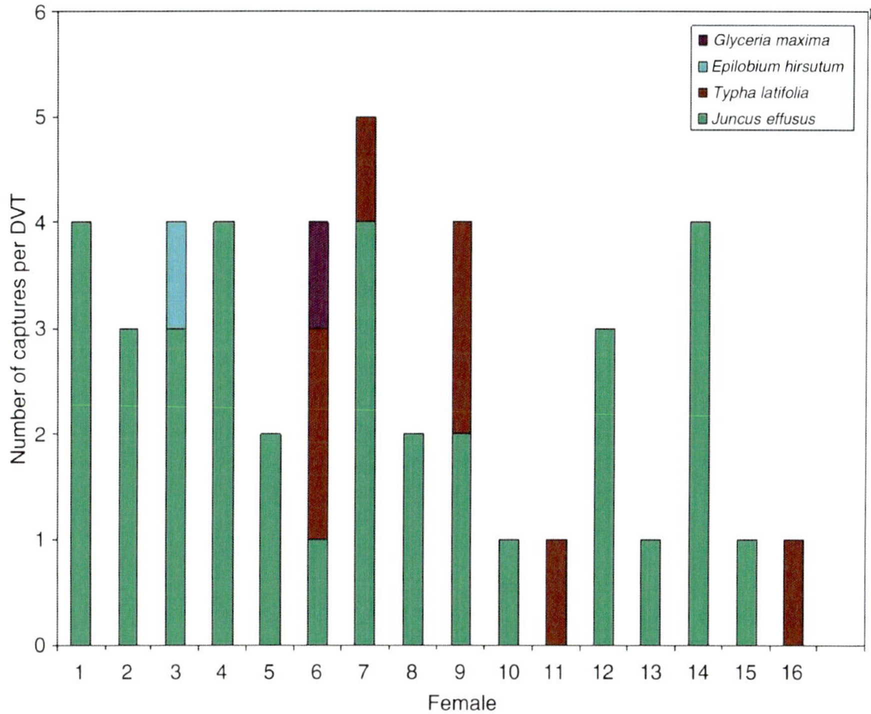

Fig. 26.6 The number of captures in each Dominant Vegetation Type for each female water vole

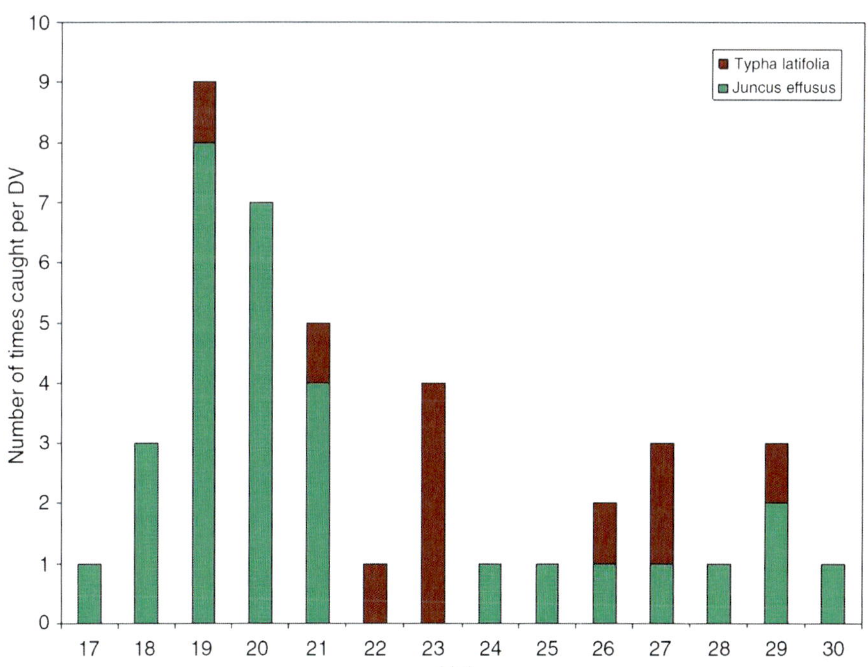

Fig. 26.7 The number of captures in each Dominant Vegetation Type for each male water vole

study period, density estimates are likely to have been influenced by this. Accordingly the following discussion will focus on the DVT approach rather than comparatively interpreting our data in light of similar studies describing water vole populations.

Water Vole Activity in DVTs

Water voles were trapped most frequently in two DVTs, *Juncus effusus* (primarily) and *Typha latifolia*. Both of these vegetation types provide consistent cover and food throughout the year and their importance to water voles has been recognised in other studies (Carter and Bright 2003; Strachan and Moorhouse 2006).

Females used four different DVTs namely *Juncus effusus, Typha latifolia, Epilobium hirsutum* and *Glyceria maxima*. Most use was made of *Juncus effusus*, which provides year-round food and cover, with the soft pith used to line the nests during breeding season (DWF & PN, personal observation): perhaps unsurprisingly, all sexually active females were associated with this DVT.

Epilobium hirsutum was occasionally used by females; this plant is rich in nitrogen (PN, unpublished data) and is therefore of particular nutritional benefit to water voles (particularly breeding females) that require this element for the production of proteins and nucleic acids.

Monthly field surveys revealed that young leaves of *Epilobium hirsutum* are frequently found in food piles from February onwards (when fresh growth occurs) and the species is foraged upon throughout the breeding season (PN, personal observation). Males were only caught in two DVTs, *Juncus effusus* and *Typha latifolia*, whereas newly weaned juveniles were captured only once in these DVTs and were presumably operating in the vicinity of mothers. The importance of *Typha latifolia* to water voles should not be underestimated, as recent observations indicate that this plant is exploited as an important source of food throughout the winter months (DWF & PN unpublished data). Furthermore, as we only studied above ground habitat utilisation, the field survey results should be treated with caution as many food caches (and sources of food) are located underground. Frustratingly, our inability to study them underground limits our understanding of their behaviour.

At the pond scale (in this study c. 440 m in total perimeter), water voles displayed different ranging behaviours – with some permanently on the ponds, some regularly moving between the study ponds and other ponds nearby (c. 200 m away), and other more transient animals that were caught only once or very infrequently. Water voles have been seen to use both underground tunnel systems and overland routes to travel between ponds and ditches separated by concrete paths and rank grassland habitats at the study site. As such, to ensure that water voles are free to move between ponds, it is important to protect terrestrial areas that link wetland patches together, both above and below ground.

Analysis of patch-based data gathered over a longer time period and larger spatial scale at the study site will provide further clarification of the extent and nature of localised water vole movement patterns within and between different ponds and other habitat patches. However, based on the results from our study, the following section outlines our recommendations for management and monitoring water voles at NWCW.

Recommendations for Conservation Management and Monitoring

Water vole distribution and activity is strongly related to a variety of habitat parameters, particularly vegetation structure and composition. It is important that vegetation is mapped in order to monitor successional processes of change as well as water vole population density and distribution in given areas to be managed. The approach described in this case study provides a useful and easily standardised method of monitoring activity of water voles at the habitat level. It also provides a common ground to facilitate comparisons between locations and habitat types with varying plant species composition and physiognomy and water vole population density. Live trapping provides the most robust and accurate estimate of population density and combined with field surveys can give a comprehensive and synergetic description of habitat utilisation, and given adequate resources is the recommended method of monitoring water vole populations.

However, live trapping is time consuming and must be undertaken by skilled (and licensed) individuals who may not be available or affordable to site managers. Where live trapping is impractical field surveys of vole activity can be undertaken, although it should be noted that these only give an indication of habitat utilisation and not occupation, unless drum-marked latrines are present (indicative of breeding females). It is important to note that the absence of water vole field signs does not automatically imply the absence of water voles as much water vole activity is confined to underground burrow systems. The Condition Indicator Table (Table 26.1) and recommendations are based on the best available data and techniques at the current time and are specific to the NWCW. Ongoing research involving a number of habitat types and water vole population densities over a 3-year period is currently being conducted and analysed. This additional research will facilitate further refinements to the DVT approach and will ultimately provide a more robust and generic monitoring tool for this and other wetland species.

Rationale Underpinning the Condition Indicator Table

The population targets are set to take account of known annual fluctuations in rodent populations and set at a level that we would expect to see exceeded in a healthy population at least once in every 3 years. These figures are based on research on the ponds described plus observations from six other ponds within the metapopulation complex at NWCW. Habitat targets set to take account of distribution and extent of various vegetation types. Stands of vegetation (i.e. DVTs) should be spread across each pond.

A pond in isolation may be declared unfavourable if all recommended DVTs are not present and there are no habitat corridors linking to more favourable ponds.

Table 26.1 The Condition Indicator Table for the water vole population at the National Wetland Centre for Wales

Condition indicator table	The water vole *Arvicola terrestris* population at the National Wetland Centre for Wales will be in favourable condition when:	
Population size	Lower limit	In any 3-year cycle Minimum of 20 individuals (MNA) including at least one juvenile are known to have been present in Areas A and B (collectively) (see map) And/or Signs of water vole activity are present in at least 20% of the bankside; with a minimum threshold latrine density typically 6.4/100 m (25% of these drum-marked)
Habitat quality	Lower limit	In each of Areas A and B >1% of the total pond area is open water (with at least 1 continuous patch of open water over 1.5 m^2 in extent) At least 95% of the bankside is vegetated (no more than 5% of the bank should be covered by bare ground/mud, with no areas of bare ground >1 m^2) >50% of the bankside vegetation should be dominated by *Juncus effusus* and >5% by *Typha latifolia* – with at least one DVT present of *Epilobium hirsutum* (>3.5% of the pond/ditch edge) and one DVT of *Carex riparia* (>1% of the pond/ditch edge) <10% of the bankside vegetation should be dominated by trees/scrub Habitat corridors should be present between areas A and B
	Site-specific definitions	
Signs of water vole activity	Food-piles (aggregations of cut vegetation of identified plant species circa 10 cm in length) And/or Latrines (water vole faeces of varying age in aggregations) that may/may not be drum marked (See field survey recommendations)	
Bankside	Area of habitat within 5 m of open water with/without submerged macrophytes (e.g. *Potomageton natans*) or with emergent vegetation (e.g. *Typha latifolia/Eleocharis palustris/Ranunculus lingua/Glyceria fluitans*)	
Open water	Open water includes water with / without submerged macrophytes (e.g. *Potomageton natans, Glyceria fluitans, Lemna* sp., etc.)	
Habitat corridors	Continuous terrestrial vegetation / necessary substrate in which to burrow linking different ponds, providing protection from predation whilst travelling above/below ground	
Juvenile	Water vole weighing less than 140 g	

Water Vole Field Survey Recommendations

In the first instance a DVT map of the site that requires management or monitoring should be created. This provides a baseline onto which movements of individuals and the results of field surveys can be plotted. In addition to the vegetation map, plant community associates should be noted for each DVT as these will assist in species identification of plants in food-piles during field surveys.

Note that after surveying each DVT ensure that vegetation remains undisturbed; return any swards of vegetation back to their original arrangement ensuring that the ground below is not exposed otherwise voles returning to their food-piles (Fig. 26.8) and latrines (Fig. 26.9) are more visible to aerial predators. If field surveys suggest

Fig. 26.8 A food-pile of *Iris pseudacorus* in a stand of *Juncus effusus*. *Photo by Penny Neyland*

Fig. 26.9 Water vole latrine with fresh and drum-marked pellets. *Photo by Penny Neyland*

that the habitat is occupied by a breeding population (i.e. in a favourable condition) then a trapping survey can be undertaken. Trapping should be conducted for at least one breeding season however at least three breeding seasons are required to reveal true population processes. Recommendations for water vole food-piles and latrine field surveys are outlined in Tables 26.2 and 26.3 respectively.

Table 26.2 General information on water vole food-piles

Food-piles - mark location of each on DVT map	
Plant species	Identify to species level. Unskilled surveyors can use a key to identify any unknown species. There are a number of plants that may be confused; using a key, hand lens and internal structure of the stem will aid in classification. Once a food-pile has been located there is usually evidence of feeding on nearby vegetation (since water voles are predominantly patch based foragers). This can also aid in identification of the species in the food-pile.
Length	c. 10 cm long N.B. Take care not to confuse water vole food-piles with those of field voles, *Microtus agrestis* (which are much shorter and often covered in small faecal pellets).
Angle of cut	Rodents cut vegetation at a 45° angle across the stem.
Distance from edge of water body	Mostly within 50 cm, but up to 200 cm on this site.
Notes	Care should be taken when comparing species that are superficially similar, particularly if the flowering spike is not present, e.g. *Juncus effusus* and *Eleocharis palustris* (the two can be distinguished by the arrangement of the pith, which is cylindrical and fills the stem in *Juncus*, but cross-hatched and paler in *Eleocharis*). *Typha latifolia* and *Iris pseudacorus* may also be confused in food piles, (the former has a fleshier and less glaucous blade).

Table 26.3 General information on water vole latrines

Latrines – mark location of each on DVT map	
Faecal pellet size	About 10 mm long, cylindrical, rounded at tips N.B. Take care not to confuse with rat pellets which are of a similar size but have a pungent smell, or field vole pellets which are a similar shape but only a few mm long
Drum-marked	Look for drum-marked pellets that have been marked with scent and crushed into the ground, these indicate the presence of breeding animals
Distance from edge of water body	Mostly within 50 cm, but up to 200 cm on this site
Notes	Latrines themselves are not necessarily indicative of breeding but are deposited at range boundaries. Small aggregations of faeces may deposited by animals that are moving through more transient areas, rather than actually occupying them

Acknowledgements We would like to thank Clive Hurford for his assistance with vegetation surveys, together with the numerous assistants that helped out with live trapping in the field. We would also like to thank NWCW for allowing us access to the site and their continuous support of this research, and CCW for use of their GIS software.

References

Bonesi L, Rushton S, Macdonald D (2002) The combined effect of environmental factors and neighbouring populations on the distribution and abundance of *Arvicola terrestris*. An approach using rule-based models. Oikos 99(2):220–230

Carter SP, Bright PW (2003) Reedbeds as refuges for water voles (*Arvicola terrestris*) from predation by introduced mink (*Mustela vison*). Biol Conserv 111(3):371–376

Forman DW, Brain PF (2006) Agonistic behaviour and bite wound patterns in wild water voles (*Arvicola terrestris* L.). Aggr Behav 32(6):599–603

Giraudoux P, Delattre P, Habert M, Quéré S, Deblay R, Defaut R, Duhamel MF, Moissenet D, Salvi D, Truchetet D (1997) Population dynamics of fossorial water vole (*Arvicola terrestris scherman*): A land use and landscape perspective. Agric Ecosyst Manage 66:47–60

Jefferies DJ, Morris PA, Mulleneux JE (1989) An inquiry into the changing status of the water vole *Arvicola terrestris* in Britain. Mammal Rev 19:111–131

Lambin X, Elston DA, Petty SJ, MacKinnon JL (1998) Spatial asynchrony and periodic travelling waves in cyclic populations of field voles. Proc Roy Soc B Biol Sci 265(1405):1491–1496

Lawton JH, Woodroffe GL (1991) Habitat and the distribution of water voles: Why are there gaps in a species range? J Anim Ecol 60:79–91

Macdonald D, Strachan R (1999) The mink and the water vole: Analyses for conservation. Oxford University, Environment Agency and Wildlife Conservation Research Unit

Moorhouse TP, Macdonald DW (2005) Temporal patterns of range use in water voles: Do females' territories drift? J Mammal 86(4):655–661

Moorhouse TP, Macdonald DW (2008) What limits male range sizes at different population densities? Evidence from three populations of water voles. J Zool 274(4):395–402

Moorhouse TP, Gelling M, Macdonald DW (2009) Effects of habitat quality upon reintroduction success in water voles: Evidence from a replicated experiment. Biol Conserv 142(1):53–60

Morilhat C, Bernard N, Foltete JC, Giraudoux P (2008) Neighbourhood landscape effect on population kinetics of the fossorial water vole (*Arvicola terrestris scherman*). Landscape Ecol 23(5):569–579

Oksanen T, Schneider M, Rammul Ü, Hambäck P, Aunapuu M (1999) Population fluctuations of voles in North Fennoscandian tundra: Contrasting dynamics in adjacent areas with different habitat composition. Oikos 86:463–478

Rushton SP, Barreto GW, Cormack RM, Macdonald DW, Fuller R (2000) Modelling the effects of mink and habitat fragmentation on the water vole. J Appl Ecol 37:475–490

Strachan R (1998) Water vole conservation handbook. English Nature Environment Agency, Wildlife Conservation Research Unit, Oxford

Strachan R, Jefferies DJ (1993) The water vole arvicola terrestris in Britain 1989–1990: Its distribution and changing status. Vincent Wildlife Trust, London

Strachan R, Moorhouse T (2006) Water vole conservation handbook, 2nd edn. The Wildlife Conservation Research Unit, Oxford University, Oxford

Telfer S, Piertney SB, Dallas JF, Stewart WA, Marshall F, Gow JL, Lambin X (2003) Parentage assignment detects frequent and large-scale dispersal in water voles. Mol Ecol 12(7):1939–1949

Woodroffe GL, Lawton JH, Davidson WL (1990) The impact of feral mink on water voles *Arvicola terrestris* in the North Yorkshire Moors National Park (England, UK). Biol Conserv 51:4–62

Chapter 27
Ringed Seals in the Gulf of Bothnia

Michael Schneider

Introduction

Seals in the Baltic Sea are of great concern. There are three reasons for this: firstly, we want to conserve seal populations for their own sake. Populations are small, red-listed and protected according to national and international regulations. Secondly, we use seals as environmental indicators (cf. Chapter 4). Seal populations should be studied thoroughly, as the status of their health can give an early warning of environmental problems. Thirdly, we should manage seal populations to avoid damage to fisheries. Seals eat fish and damage catches as well as fishing gear. When seal populations increase, the damage to fisheries also increases, unless appropriate actions are taken to counteract the problems.

Seals in Sweden

Three species of seal occur in Swedish waters; the grey seal (*Halicoerus grypus*) in most of the Baltic Sea, the harbour seal (*Phoca vitulina*) in the southern parts, and the ringed seal (*Phoca hispida*) (Fig. 27.1) in the northern parts of the Baltic Sea. During the first decades of the twentieth century, the numbers and distribution of all three species had been greatly reduced due to the effects of over-hunting and environmental contamination, but seal populations have started to increase again, as the intensive hunting has stopped and as levels of pollutants have decreased. With increasing seal populations and increasing damages inflicted by seals upon coastal fisheries, there is an increasing demand for a management of the species.

M. Schneider (✉)
County Administration, Västerbotten County, SE-901 86, Umeå, Sweden
e-mail: Michael.Schneider@lansstyrelsen.se

C. Hurford et al. (eds.), *Conservation Monitoring in Freshwater Habitats: A Practical Guide and Case Studies*, DOI 10.1007/978-1-4020-9278-7_27,

Fig. 27.1 A ringed seal sun-bathing on a rock in the Bothnian Bay. The species is named after its coat pattern. *Photograph by Jörgen Wiklund*

The Ringed Seal

The ringed seal is the smallest of all seal species. It has a circumpolar distribution and is resident in North Pacific, North Atlantic and Arctic waters. There are five subspecies, of which three are adapted to living in brackish and freshwater environments. *Phoca hispida botnica* occurs in the Baltic Sea including the Gulf of Bothnia, the Gulf of Finland and the Gulf of Riga (Figs. 27.2 and 27.3). Two other subspecies live in the freshwater habitats of Lake Saimaa in Finland (*P. h. saimensis*) and Lake Ladoga in Russia (*P. h. ladogensis*). These subspecies have been isolated since post-glacial land uplift reduced the extent of the Baltic Sea about 8,000 years ago. The world population of ringed seals is estimated to be 6–7 million, while the population in the Baltic Sea, which we focus on here, consists of about 10,000 animals (Schneider *et al.* 2009 and references therein).

The Baltic subspecies is larger than other subspecies. Adult animals are up to 150 cm in length and weigh up to ca. 140 kg. Females are slightly smaller than males. Pups are born in February and March, and mating takes place shortly afterwards. Ringed seals give birth in snow lairs over holes carved in the ice far out on the sea. Lairs are predominantly found in dense drift ice with ridges, where snow drifts are formed by wind action (cf. Fig. 27.4). These lairs protect pups from cold temperatures and predators (Reeves 1998).

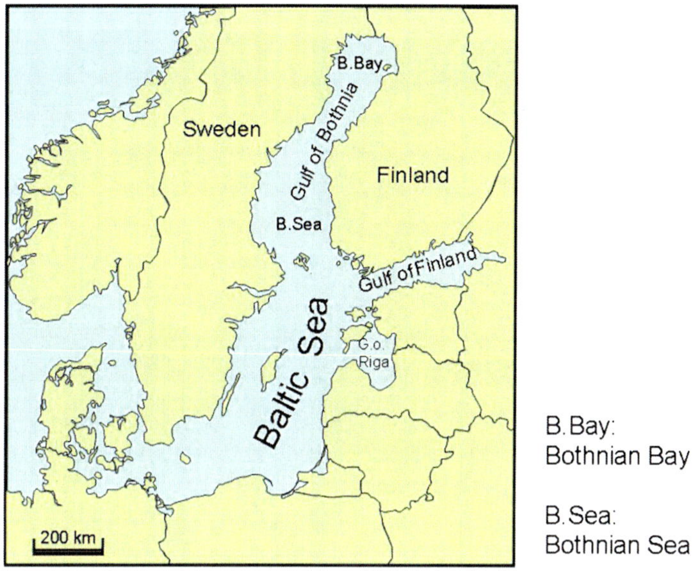

Fig. 27.2 Geography of the Baltic Sea

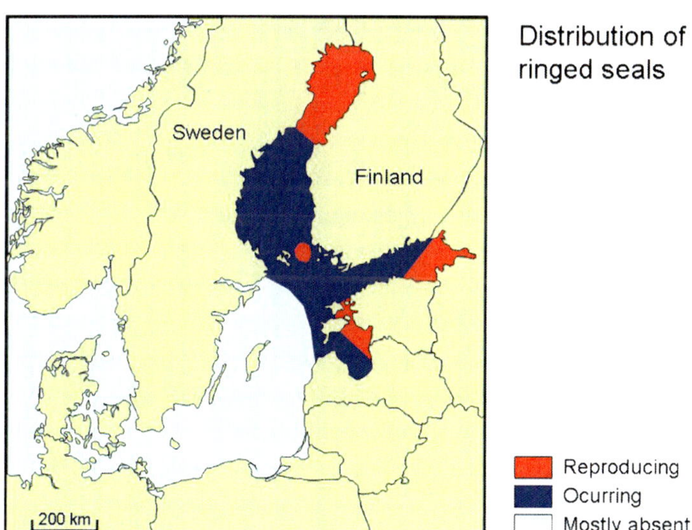

Fig. 27.3 The ringed seal inhabits the northern and eastern parts of the Baltic Sea. Reproduction occurs mostly in four restricted areas. The Bothnian Bay is the most important refuge for the species in the Baltic Sea

Fig. 27.4 Drift ice in the Bothnian Bay in April 2009. Far out amongst the ridges and not to be seen on the photograph, ringed seals are basking in the sun. *Photograph by Michael Schneider*

The Gulf of Bothnia

The Gulf of Bothnia is the northernmost part of the Baltic Sea, bordered by Sweden in the west and Finland in the east (Fig. 27.2). The Gulf is subdivided into the Bothnian Bay in the north and the Bothnian Sea in the south. The waters in the Bothnian Bay are relatively shallow and brackish (Salomonson *et al.* 2006). Large parts of the Gulf of Bothnia are covered by ice during winter (Fig. 27.5), from December until April/May (SMHI 2008). The ice cover reaches its largest extent in February/March.

Simulations of sea ice development in the northern Baltic predict a reduction of 83% on average of ice-cover by the end of the current century. As a consequence, most of the current breeding areas of the ringed seal will become ice-free (Schmölcke 2008 and references therein). Only in the northernmost part of the Bothnian Bay will ice persist long enough for ringed seals to breed. Thus, the breeding population of the species in the Baltic Sea is likely to decline and to shift northwards, with extinction possible in the more southerly parts of the Baltic Sea (HELCOM 2007). It seems likely that only the Gulf of Bothnia will persist as a fairly good winter sea ice habitat for the species.

Changing Focus

Until recently, the ringed seal was mostly viewed from a conservation perspective. Surveys focused on population development and its underlying mechanisms, especially environmental contamination and health problems. Today, we acknowledge that we have to actively manage the species. For practical management to work at both a national and regional scale, we even have to look at fisheries and how these are affected by the seals. We have learned lessons from the management of large

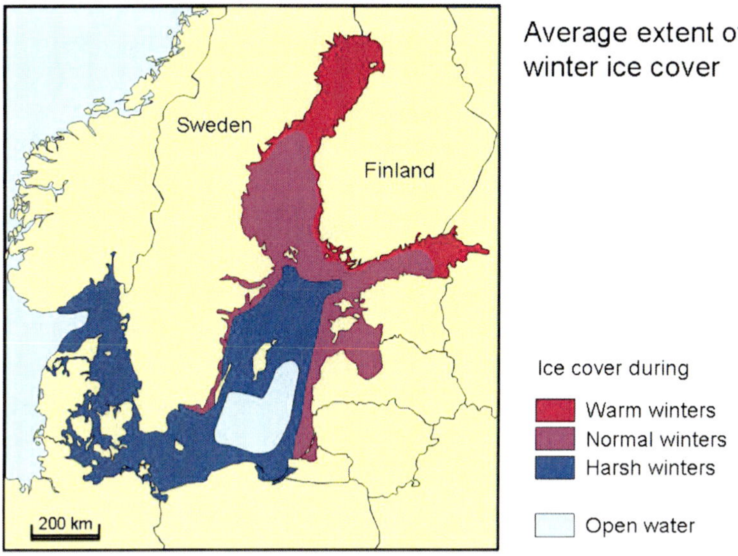

Fig. 27.5 The extent of the ice cover in the Baltic Sea is variable, depending on winter conditions. The Bothnian Bay and the inner parts of the Gulf of Finland are covered by ice even during warm winters. Adapted from a map compiled by the Swedish Meteorological and Hydrological Institute (SMHI 2008)

terrestrial carnivores in northern Sweden (see Schneider 2006) regarding the involvement of stakeholders in management planning. Consequently, we have started to discuss the possibilities of hunting ringed seals and to develop a platform for international co-operation between Finland and Sweden regarding the practical management of the species in the Gulf of Bothnia (Schneider *et al.* 2009). This change of focus poses new questions and new challenges for monitoring activities within ringed seal management.

Key Factors for the Management of Ringed Seals

There are four key factors that influence management decisions for the ringed seal. These include the conservation status of the species, an overall lack of knowledge, damage to fisheries, and the fact that Finland and Sweden share a common population.

Conservation Status

The Baltic ringed seal is listed as 'vulnerable' by the International Union for the Conservation of Nature (IUCN), and as 'near threatened' in the latest version of the Swedish red list of threatened species (Gärdenfors 2005). It is also listed in the EU

Habitats Directive and the Bern Convention, and is covered by the Helsinki Commission (HELCOM) recommendation on the conservation of seals in the Baltic Sea.

Lack of Knowledge

The existing knowledge about the biology and ecology of the Baltic ringed seal is limited: questions about diet, summer habitat, behaviour, and the exchange of individuals between subpopulations have still to be answered (Schneider *et al.* 2009).

Damage to Fisheries

Seals can cause reduced catches by taking fish from nets or traps, and they can cause expensive damage to fishing equipment. This is known for grey seals, but it is unclear the degree to which ringed seals are engaged in these activities (Hemmingsson and Lunneryd 2006).

International Population

Sweden and Finland share a common population of ringed seals in the Gulf of Bothnia. Management decisions made in one country may also affect the status of ringed seals in the other country. Therefore, a co-management of ringed seals is necessary, which has been acknowledged by authorities in both countries (Jord- och skogsbruksministeriet 2007; Schneider *et al.* 2009).

Objectives for Our Work

Objectives for the work with ringed seals are different when we look at an international, a national or a regional scale.

International Obligations

HELCOM, the governing body of the Helsinki Convention, works to protect the marine environment of the Baltic Sea from all sources of pollution through inter-governmental co-operation between Denmark, Estonia, the European Community, Finland, Germany, Latvia, Lithuania, Poland, Russia and Sweden.

HELCOM adopts recommendations for the protection of the marine environment, which the governments of the contracting parties must act on in their respective national programmes and legislation (www.helcom.fi).

In 2006, HELCOM issued recommendations regarding the management of seals in the Baltic region. These recommendations suggest specific reference levels for population size, distribution and health status and are meant to provide guidance for current and future management actions. For the Baltic ringed seal, two management units were defined: the sub-population in the Gulf of Bothnia, and the subpopulation in the more southerly parts of the Baltic Sea (HELCOM 2006).

National Targets

To successfully manage the population of ringed seals in the Gulf of Bothnia, we must learn much more about the species' biology and ecology. To maintain a positive population trend, we have to achieve a good health status among ringed seals. We also have to keep track of a possible switch by the seals from ice to land for reproduction and we have to try to understand what this means for the species in terms of pup survival, productivity and population growth rate. Ringed seals and grey seals occur together in the Gulf of Bothnia, with the latter apparently the more problematic species. However, as it often is hard to distinguish between problems caused by different seal species, the co-management of ringed and grey seals is a necessity. Also, ringed seals in Sweden and Finland should be managed in a uniform way, as they belong to the same population.

Regional Necessities

At a regional scale, the number of seals drowning in fishing gear should be low. At the same time, seal damages inflicted on fisheries should be kept at a low level. Local fishermen as well as regional groups, organisations and authorities with an interest in seals should be involved in the management of the species. Hunters should have the possibility to market seal products (skin, meat, etc.). Illegal hunting should be rare.

Ongoing Activities

The ringed seal is a species of concern from both a conservation and a management point of view. Therefore, we must make the effort to learn more about the species in the Gulf of Bothnia.

Population Surveillance

Three different methods have been used to estimate the development of the population of ringed seals in the Bothnian Bay: hunting statistics, modelling and aerial surveys. For the first half of the twentieth century, hunting statistics reflect the decline of the species (Harding and Härkönen 1999). A modelling approach has then been used to look at the population during the period 1950–1990 (Schneider *et al.* 2009 and references therein). Since the mid 1970s, ringed seals in the Bothnian Bay have been censused using aerial surveys (Helle 1980). Since 1988, this survey has been conducted in a more structured way (Härkönen and Heide-Jørgensen 1990; Härkönen and Lunneryd 1992) (Fig. 27.6). During April–May each spring, ringed seals are counted by researchers using fixed-wing aircrafts patrolling above the last remnants of the winter ice, where the seals are emerged to change coat. These counts are weather-dependent, and only parts of the sea ice (at least 10%) are surveyed each year. Even under optimal conditions for surveillance, only parts of the population will be observed. This results in estimations of the population size, not in exact figures (Härkönen *et al.* 1998).

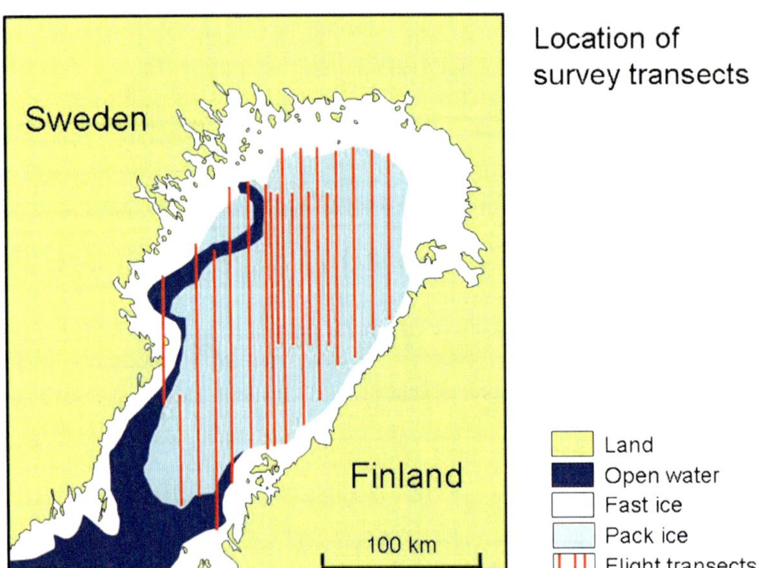

Fig. 27.6 Flight transects for the aerial survey of ringed seals are oriented in a north-southward direction in the Bothnian Bay. Surveys are conducted annually. At least 10% of the ice sheet has to be covered during each survey. The figure depicts the transects used in 1996. Adapted from Härkönen *et al.* (1998)

Survey of Health Status

The Swedish Museum of Natural History in Stockholm is responsible for collecting and examining dead seals in Sweden. During the last 30 years, the museum has conducted necropsies on 50–200 seals from the Baltic Sea annually, comprising those that had drowned in fishing gear (by-catches), that had been found dead or that had been shot during the grey seal hunting season. Most of the seals received by the museum were grey seals. During necropsy, different measurements are taken and the seals are weighed. The cause of death is determined and pathological changes are looked for in the bodies. Samples are then taken for long-term storage and for different research projects (Karlsson *et al.* 2007).

By-Catches

In 2002, the Swedish Board of Fisheries carried out telephone interviews to investigate the number of by-catches of seals, porpoises and birds in the Swedish fishing industry. In total, 220 randomly selected commercial fishermen were interviewed. By comparing the fishing efforts of the respondents in the survey with the total national fishing effort, it was possible to calculate the total number of seals that drowned in Swedish commercial fisheries in 2001 (Lunneryd *et al.* 2004).

Damage to Fisheries

Fishermen blame seals for causing reduced catches by taking fish from nets or traps, as well as for causing expensive damage to fishing equipment. This has been scientifically verified for grey seals, but not for ringed seals (Hemmingsson and Lunneryd 2006). In the Bothnian Bay, seals are believed to affect the coastal salmon and whitefish fisheries. Current research is aiming at determining the extent to which damages are caused by ringed seals, and to develop fishing tackle that can withstand the attention of seals (Schneider *et al.* 2009).

Attitudes

A survey of attitudes towards seals (grey seal and ringed seal) has been conducted in Finland (Storm *et al.* 2007). This survey targeted people that are affected by seals in their daily life, as well as organisations and authorities that deal with nature conservation, that exploit natural resources, or that are responsible for natural resource management. A questionnaire was sent out to groups, organisations, institutions and authorities at a regional and national scale. About 250 responded.

Furthermore, ca. 10 meetings were organised along the Finnish coast to collect information on local attitudes. In total, 440 people participated and 640 comments were made in these meetings.

Available Information

Population Size

As estimated from hunting statistics, the population of ringed seals consisted of about 200,000 individuals in the beginning of the twentieth century. Due to intensive hunting, the population decreased to a few tens of thousands of animals in the 1950s. Environmental contamination caused a further decline of the population and a low population size between 1960 and 1980. By modelling, the size of the population has been estimated at 5,000 animals in the Baltic Sea for that period. Since the 1980s, annual surveys have been conducted and the population is now estimated from numbers of animals observed on the ice (Fig. 27.7). The current population in the Bothnian Bay is increasing at a rate of about 5% per year, numbering some 8,000 animals (Fig. 27.8). Populations in the Gulf of Finland (ca. 300 animals) and the Gulf of Riga (ca. 1,000 animals) are much smaller and appear to be stable or decreasing.

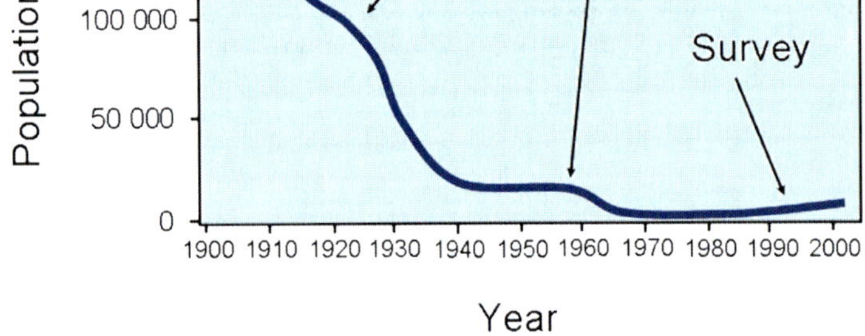

Fig. 27.7 Three different methods have been used to estimate the size of the population of ringed seals in the Baltic Sea: hunting statistics, modelling and aerial surveys. From an estimated 200,000 animals, the population crashed to about 5,000 individuals, but it has started to increase again during recent decades

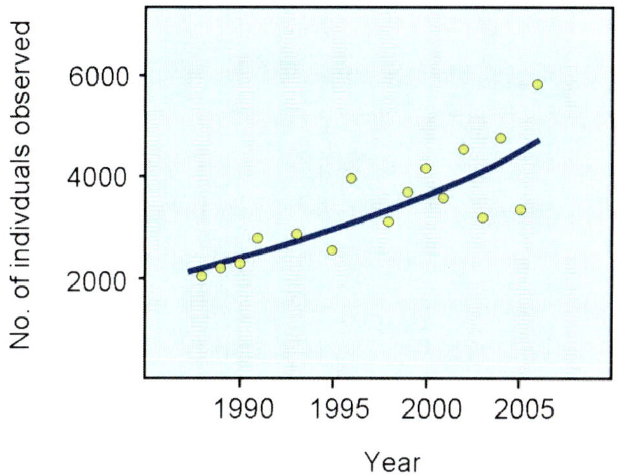

Fig. 27.8 The numbers of ringed seals observed during aerial surveys in the Bothnian Bay. Survey results (*yellow dots*) vary between years. Since 1988, the population has increased with about 4.5% per year (*blue trendline*). Adapted from Karlsson *et al.* (2007)

Distribution of Breeding Habitat

Ice is crucial for the reproduction and the moult of Baltic ringed seals. In the Bothnian Bay, the winter distribution of ringed seals varies only slightly between years, and during the last 10 years, the highest concentrations of seals have been observed in an area south-east of the island of Haparanda Sandskär. Figure 27.9 shows the winter distribution of ringed seals in the Bothnian Bay in 1996. Surveys conducted before and after 1996 show the same results, as do data from the 1930s. The highest densities of ringed seals have been found on the pack ice in the central parts of the Bothnian Bay (Härkönen *et al.* 1998).

Health Status

In total, about 400 ringed seals were delivered to the Swedish Museum of Natural History for necropsy during the last 30 years. Increased levels of PCB have been made responsible for a disease complex (pathological kidney alterations, uterine occlusions) found in ringed seals between 1965 and 1985 (Bergman *et al.* 2001). Since then, levels of contaminants have been decreasing, but still 30–40% of adult females are sterile (Mattson and Helle 1995). This infertility in females is believed to be responsible for a reduced rate of population growth. The population increase has been estimated to be 4–5% per year, as compared to a possible 10% for the species (Karlsson *et al.* 2007, Härkönen 2006).

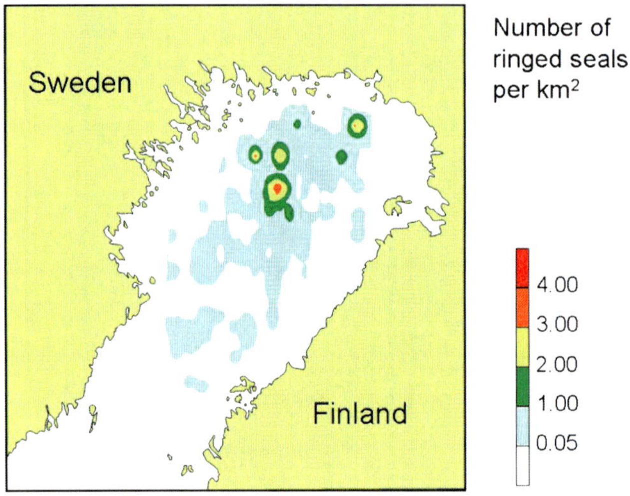

Fig. 27.9 The winter distribution of ringed seals in the Bothnian Bay in 1996. Adapted from Härkönen *et al.* (1998)

By-Catches

Every year, several ringed seals are caught in different types of fishing equipment in the Gulf of Bothnia. In the survey carried out by the Swedish Board of Fisheries in 2002, there was a large variation in the number of animals caught in the by-catch between different fishermen and different areas. A large majority of fishing vessels (73%) did not report any seals in their by-catch, while a few (3.5%) caught more than five seals each. In total, a by-catch of 10 ringed seals was reported for 2001, and these reports came from the northern Baltic Sea only. The majority of by-catches occurred in fixed traps for salmon. By comparing the fishing efforts of the respondents in the survey with the total national fishing effort, the Swedish Board of Fisheries calculated that 52 (34–70) ringed seals drowned in Swedish commercial fisheries in 2001. These figures are supposed to be an under-estimation of the total loss, as they do not include part-time and non-commercial fishing, and as they do not cover adequately by-catches in fishing methods that are known to have low by-catches (Lunneryd *et al.* 2004).

Damage to Fisheries

Seals hunting around fishing equipment can chase away fish that could otherwise have been caught in the traps or nets. Seals can also remove fish that have already been trapped, or just take bites off the fisherman's prey. Seals can also damage fishing equipment when they try to gain access to captured fish. It is well known that grey seals can inflict great damage on fisheries. According to fishermen operating in the Bothnian Bay, even ringed seals can be problematic, and some hold that ringed seals pose a much bigger problem than grey seals. The academic knowledge about the level of damage caused by ringed seals is restricted, but there are video recordings of ringed seals inspecting fish traps and nets (Hemmingsson and Lunneryd 2006).

Attitudes

A survey of attitudes of people, organisations and authorities towards seals and seal management has been conducted in Finland, but not in Sweden (Storm *et al.* 2007). Many different opinions emerged during the survey, but fishermen and hunters were over-represented among respondents. Generally, opinions were rather negative and seals were looked upon as being problematic. Many respondents stressed the damage inflicted by growing seal populations on fisheries and how these damages jeopardize the livelihood of the fishermen. Criticism was directed not only at the seals, but also at the managing authorities, researchers, conservationists, the European Union, and especially the shortcomings of the compensation system for seal damage. Seal hunting was put forward as the main solution to the problems (Storm *et al.* 2007).

Discussion

Unclear Objectives

The seal expert group within HELCOM has suggested three population reference levels for the ringed seal in the Baltic Sea, which are intended to aid the management of the species. The Target Reference Level is set at the population size where the population growth rate starts to level off to an asymptotic population level, which in ecological terms is called "carrying capacity". The Limit Reference Level is the population size when the long-term persistence of the population is ensured. The Precautionary Approach Level is the population size where the population is at maximum productivity (HELCOM 2006).

However, with our current and limited knowledge of ringed seal ecology and population dynamics, it is not very clear what these levels mean in terms of actual seal numbers. What is the current carrying capacity of the environment? Which factors determine the carrying capacity today, and how and why may this upper limit for population size change over time? Thus, the suggested reference levels are not very easily applied in practice.

Today, we lack concrete and measurable objectives for ringed seals in the Gulf of Bothnia regarding:

- Population increase, population size, and distribution
- The frequency and extent of damages that are tolerable to coastal fisheries
- The maximum number of by-catches
- Possible future hunting quota

All of these objectives have to be developed in the near future.

Unsophisticated Political Ideas

In a Swedish–Finnish project on the management of grey seals, participants focused on the sustainable harvest of seals and the marketing of seal products (Kvarkenrådet 2008). However, as a reaction to unethical hunting methods outside the European Union, the EU-commission presented recently a proposal to prohibit the trade with seal products (Carlberg and Lundquist 2008). If seal products cannot be sold, there is little reason for Swedish hunters to shoot seals. If seals are not hunted to regulate their population, damage to fisheries may increase, local discussions may get more infected, fewer seal corpses will be available for health status examination and screening for environmental contaminants, and the regional management of the species will get more difficult.

Current and Future Activities

Looking at population size and health status of ringed seals only is not enough. For practical management to work, we even have to look at fisheries and how these are affected by seals. Schneider *et al.* (2009) identified a list of activities in order to start up a functioning management system for ringed seals in Sweden. To acquire the information needed for informed management, we have to continue with what we already do, we have to develop certain activities, and we have to start up new ones.

Activities to Be Continued

- Annual aerial surveys of population size
- Surveys of health status

Activities to Be Developed

- The process for reporting survey results to managing authorities and the public has to be streamlined to guarantee a fast and unhampered flow of information.

- Regular surveys of the frequency and extent of damage to fisheries inflicted by ringed seals have to be conducted.
- Structured surveys of causes and rates of mortality in ringed seal populations have to be carried out.
- Research on ringed seal ecology has to be intensified.

Activities to Be Started

- The factors that determine carrying capacity of ringed seal habitats have to be identified.
- Ecological research and population viability analyses have to determine current and future carrying capacity.
- Attitude surveys have to be conducted also on the Swedish side of the Gulf of Bothnia. Attitudes reflect the success of the management system and may give some indication on the extent of illegal hunting.
- The willingness of hunters to shoot seals with and without the possibility to market seal products has to be analysed.
- The suitability of regional seal councils to survey traditional knowledge and opinions and attitudes of stakeholders has to be analysed. Regional seal councils are meeting places for fishermen, hunters, researchers, conservationists and managers to discuss seal management.

The implementation of the existing Finnish management plan for ringed seals (Jord- och skogsbruksministeriet 2007) and the presumable adoption of the Swedish action plan for the species (Schneider *et al.* 2009) in 2009 hopefully will make things happen.

Acknowledgements I am grateful to Johnny Berglund and Clive Hurford for valuable comments on the manuscript.

References

Bergman A, Bergstrand A, Bignert A (2001) Renal lesions in Baltic grey seals (*Halichoerus grypus*) and ringed seals (*Phoca hispida botnica*). Ambio 30:397–409

Carlberg T, Lundquist T (2008) EU-förslag kan krångla till svensk skyddsjakt på säl. Fauna & Flora 103/3:16–18 (in Swedish)

Gärdenfors U (ed) (2005) Rödlistade arter i Sverige 2005. The 2005 Red list of Swedish species. ArtDatabanken, SLU, Uppsala

Harding KC, Härkönen TJ (1999) Development in the Baltic grey seal (*Halichoerus grypus*) and ringed seal (*Phoca hispida*) populations during the 20th century. Ambio 28:619–627

Härkönen T (2006) *Phoca hispida botnica*, vikare. ArtDatabanken, SLU (in Swedish), Faktablad, Uppsala

Härkönen T, Heide-Jørgensen M-P (1990) Density and distribution of the ringed seal in the Bothnian Bay. Holarctic Ecol 13:122–129

Härkönen T, Lunneryd SG (1992) Estimating abundance of ringed seals in the Bothnian Bay. Ambio 21:497–510

Härkönen T, Stenman O, Jüssi M, Jüssi I, Sagitov R, Verevkin M (1998) Population size and distribution of the Baltic ringed seal (*Phoca hispida botnica*). In: Heide-Jørgensen MP, Lydersen C (eds) Ringed seals (*Phoca hispida*) in the North Atlantic, vol 1. NAMMCO Scientific Publications, Tromsø pp 167–180

HELCOM (2006) Recommendations on the conservation of seals in the Baltic Sea area. Available at: www-helcom.fi/Recommendations/en_GB/rec27-28_2/ (accessed on 1 Dec 2008)

HELCOM (2007) Climate change in the Baltic Sea area – HELCOM thematic assessment in 2007. Baltic Sea Environment Proceedings No. 111

Helle E (1980) Aerial census of ringed seals *Pusa hispida* basking on the ice of the Bothnian Bay, Baltic. Holarctic Ecol 3:183–189

Hemmingsson M, Lunneryd S-G (2006) Inventering av sälskadesituationen i Västerbottens och Norrbottens län. Report from the Swedish Board of Fisheries, 2006-01-04 (in Swedish)

Jord- och skogsbruksministeriet (2007) Förvaltningsplan för Östersjöns sälstammar. Jord- och skogsbruksministeriet, rapport 4a/2007 (in Swedish)

Karlsson O, Härkönen T, Bäcklin B-M (2007) Sälar på uppgång. Havet 2007:84–89 (in Swedish)

Kvarkenrådet (2008) Sälen vår gemensamma resurs. Slutrapport. Report from Kvarkenrådet (in Swedish)

Lunneryd S-G, Königson S, Sjöberg NB (2004) Bifångst av säl, tumlare och fåglar i det svenska yrkesfisket. Finfo 2004:8, Fiskeriverket, Göteborg (in Swedish with English summary)

Mattson M, Helle E (1995) Reproductive recovery and PCBs in Baltic seal populations. Eleventh biennial conference on the biology of marine mammals. 14–18 Dec 1995, Orlando USA, p 74

Reeves RR (1998) Distribution, abundance, and biology of ringed seal (*Phoca hispida*): An overview. In: Heide-Jørgensen MP, Lydersen C (eds) Ringed seals (*Phoca hispida*) in the North Atlantic, vol 1. NAMMCO Scientific Publications, Tromsø pp 9–45

Salomonson A, Katajista J, Sedin A (2006) Coastal lagoons in Sweden and Finland. In: Hurford C, Schneider M (eds) Monitoring nature conservation in cultural habitats. Springer, Dordrecht, pp 361–376

Schmölcke U (2008) Holocene environmental changes and the seal (Phocidae) fauna of the Baltic Sea. Coming, going and staying. Mammal Rev 38:231–246

Schneider M (2006) Monitoring the Brown Bear *Ursus arctos* in Västerbotten County. In: Hurford C, Schneider M (eds) Monitoring nature conservation in cultural habitats. Springer, Dordrecht, pp 195–214

Schneider M, Härkönen T, Hemmingsson M, Karlsson M, (2009) Åtgärdsprogram för förvaltning av vikare *Phoca hispida botnica*. Unpublished manuscript (in Swedish with English summary)

SMHI (2008) Genomsnittlig isutbredning. Available at: www.smhi.se/oceanografi/istjanst/isutbredning.pdf (accessed on 30 November 2008) (in Swedish)

Storm A, Routti H, Nyman M, Kunnasranta M (2007) *På tal om säl – regionala och nationella synpunkter och förväntningar på förvaltningen av sälstammen*. Kala- ja riistaraportteja 396B (in Swedish)

Section 6
Integrated Surveillance, Monitoring and Management at Doñana Natural Space

Chapter 28
An Integrated Monitoring Programme for Doñana Natural Space: The Set-Up and Implementation

Ricardo Díaz-Delgado

Donana Natural Space – A Brief Overview

Protected since 1968, Doñana National Park (537 km^2) is a UNESCO Biosphere Reserve, a Ramsar Site, a Natural World Heritage Site and is integrated in the Natura 2000 network. It contains the largest wetland in Western Europe (García Novo and Marín Cabrera 2005), an intricate matrix of marshlands (270 km^2), phreatic lagoons, and a 25 km-long dune ecosystem with its respective shoreline and representative Mediterranean terrestrial plant communities (Fig. 28.1). The conservation objectives include the preservation of (a) critically endangered species (Iberian lynx *Lynx pardina*, Spanish imperial eagle *Aquila adalberti*, marbled teal *Marmaronetta angustirostris*), (b) the abundance of waterfowl, and (c) the Mediterranean wetlands and terrestrial ecosystems. Furthermore, Doñana is both a critical stopover site for Palearctic birds migrating to Africa and an important over-wintering site for waterfowl.

Doñana marshlands have a typical Mediterranean climate: the hydrological cycle starts in September and usually reaches maximum inundation levels during February, mainly driven by the rainfall regime. In late spring, evaporation becomes the most important factor in the water balance, and the marshes dry up slowly until they are completely dry by the end of July. At this time, the aquifer plays a central role in maintaining water levels and permanent lagoons (Grimalt *et al.* 1999). As is the case for most continental wetlands, interannual variability is driven by meteorological patterns.

R. Díaz-Delgado (✉)
Head of Landscape & Remote Sensing Monitoring, Doñana Biological Station-CSIC, c/Américo Vespucio s/n. Isla de la Cartuja, Sevilla, 41092, Spain
e-mail: rdiaz@ebd.csic.es

C. Hurford et al. (eds.), *Conservation Monitoring in Freshwater Habitats: A Practical Guide and Case Studies*, DOI 10.1007/978-1-4020-9278-7_28,

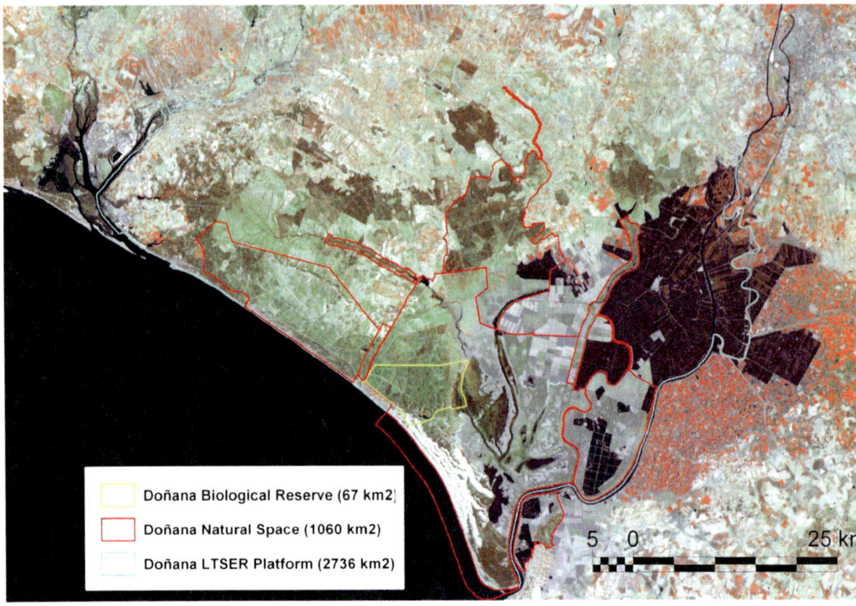

Fig. 28.1 Limits of Doñana Natural Space, Doñana LTSER platform and Doñana Biological Reserve

Historical Monitoring Initiatives and Gaps in Knowledge

The protection of Doñana was originally promoted by José Antonio Valverde (Fig. 28.2), who faced many challenges and threats before the administration finally declared Doñana a National Park (DNP) in 1969, making it the largest protected area in Spain. Just 6 years previously, J.A. Valverde had already preserved Doñana Biological Reserve (DBR), a 6794 ha plot inside the current National Park (see Fig. 28.1), with the essential help of the World Wildlife Fund. At the same time, he created the Doñana Biological Station (DBS), a specific research centre belonging to the Spanish Research Council (CSIC). Thus, since the very beginning, Valverde, as the first DNP manager and DBS director, firmly linked conservation to research.

Initially, DBS scientific research focused on conservation biology, and especially on threatened species and management benefits. In subsequent years, DBS became a leading research centre on biological conservation topics. From the outset, Valverde passionately promoted the systematic collection of data on relevant conservation variables, such as Iberian lynx population size and distribution, bird banding and band readings or cork oak inventory and marking (Solís 1996). As a consequence, local staff working at DBR began to focus systematically on understanding the most relevant species present in DNP, with staff and visitors keeping field diaries to record both occasional events and daily field work either inside

Doñana or even overseas (Valverde led many exploratory campaigns in different continents). Also, during that period, monthly aerial waterfowl censuses were introduced to record habitat selection and migratory trends of the most abundant aquatic birds in Doñana marshlands. These, in fact, were the first attempts to establish periodical monitoring activities as the basis for research and management of Doñana's fauna and flora.

In the period from 1970 to 1990, many management decisions for DNP were based on information from field diaries, aerial census reports and assessments. Systematically collected data derived from these research activities have since contributed to publications in international scientific journals (Almaraz and Amat 2004a, b; Rendon *et al.* 2008).

Despite the obvious interest in, and effort invested in collecting, long-term datasets, these mostly focused on the abundance and occurrence of birds.

It was only in the 1980s that Doñana National Park began focusing more on ecological, limnological, hydrological or ethological issues (García Novo and Marín Cabrera 2005). After the eminent Spanish ecologist Ramón Margalef had described the planktonic community of Doñana lagoons (Margalef 1976), other taxa and natural processes came to be accepted as relevant conservation issues for Doñana.

However, even at the end of the last century, there were still gaps in our knowledge, both from a research and monitoring perspective, particularly relating to topics such as plant physiology, ecophysiology, plant ecology, landscape ecology, land

Fig. 28.2 José Antonio Valverde, the first director of DBS and first manager of DNP

use cover changes, soil sciences, plant–soil interactions and land cover energy balance. Many of these have since been addressed through the development of an integrated monitoring programme and associated research projects.

The Framework for an Integrated Monitoring Programme: Policy and Opportunity

Vaughan *et al.* (2001) defined integrated monitoring as monitoring that uses detailed sets of ecological information, unlike simple monitoring, survey monitoring and proxy monitoring. Although many of the periodical surveys carried out at Doñana acquired relevant data on population condition and breeding success, by the end of the 1990s, the need for an integrated program of long-term ecological monitoring was evident. Therefore, in 2001, a meeting held among DBS and DNP heads led to the proposal for a joint project to set up the integrated ecological long-term monitoring of DNP. Representatives from both institutions (CSIC-DBS and Spanish National Parks Network) agreed on the main features to be monitored according to conservation priorities. The main goal was to achieve long-term knowledge on the dynamics of Doñana natural processes and the conservation management effects on its biodiversity. The conceptual approach focused on monitoring the following features: species, habitats and ecological processes. The species level monitoring focused on the endangered and threatened species, while the foci habitats were mostly the representative and under-represented ones. Finally, the ecological processes did not only include the natural processes interlinking ecosystem functioning and structure but also the human driven impacts (resulting from management decisions – including the conservation measures). Thus, the joint project, entitled "Design and refinement of the integrated programme for monitoring natural processes and resources in Doñana National Park", was initiated in 2002 and devoted 3 years of research to the validation of monitoring protocols and the conceptual model of long-term ecological monitoring. The specific goals of the project were:

1. To achieve an exhaustive bibliographical review of available, up-to-date and standard protocols for the selected features to be monitored.
2. To designate scientific supervisors with recognised expertise in the ecological monitoring targets.
3. To propose feasible monitoring protocols and test their validity and adequacy for monitoring the proposed targets.
4. Final adoption/rejection of the tested methodological protocols and the features proposed for monitoring.

Originally, both institutions agreed on a number of proposed features for monitoring (Tables 28.1 and 28.2). The conceptual approach led to addressing the features under three main monitoring themes and 11 monitoring targets. In addition to flora, fauna, management and geophysical monitoring, a landscape scale monitoring

Table 28.1 The Biological Monitoring theme, identified targets and features of Doñana Integrated Long Term Ecological Monitoring Programme. Asterisks indicate the features approached at landscape scale

Monitoring theme	Target	Feature	Feature description
Biological Monitoring	Vegetation	NPP	Grass biomass
		Plant cover & structure	Pine, cork oak & juniper woodlands, shrubland, marshland & riparian vegetation (*)
		Flora	Aquatic plants distribution and richness
			Rare threatened plant ssp. distribution and abundance
			Singular trees inventory
			Alien spp. distribution and abundance (*)
	Fauna	Terrestrial invertebrates	Coprophagian Coleoptera abundance
			Argentine ant abundance and distribution
			Diurnal butterflies
			Demographic insect blooms
		Aquatic invertebrates	Aquatic invertebrates abundance and richness
			Red crayfish abundance and distribution
		Fish	Fish spp. communities & abundance
		Amphibians	Amphibians spp. communities & abundance
		Reptiles	Greek tortoise distribution
			Greek tortoise population structure
			Native turtle distribution
			Exotic turtle distribution
			Lizard spp. distribution & abundance
		Birds	Vulnerable and endangered birds
			Passerine communities
			Bird breeding
			Wintering birds
		Birds key spp.	Red-legged partridge and Eurasian coot abundance and distribution
		Mammals key spp.	Rabbit and hare abundance
		Mammals	Wild ungulates abundance (red deer and wild boar)
			Water vole abundance & distribution
			Carnivore relative abundance
			Otter abundance and distribution

Table 28.2 The Geophysical and Management Monitoring themes, identified targets and features of Doñana Integrated Long Term Ecological Monitoring Programme. Asterisks indicate the features approached at landscape scale

Monitoring theme	Target	Feature	Feature description
Geophysical Monitoring	Climate	Meteorology	Meteorological stations
	Atmosphere	Air Quality	Pollutant & Aerosol concentration
	Surface water	Flooding dynamics	Marshland flooding dynamics (*)
			Temporary pools
		Water quality	Water quality
	Underground water	Water table	Water table measurements
	Geomorphology	Erosion/Sedimentation	Sedimentation on marshland/Dunar system dynamics/ Shoreline dynamics (*)
Management Monitoring	Hydrology	Water management	Water management evaluation
			Managed artificial pools
			Doñana 2005 marshland restoration programme
	Vegetation management	Management effects	Plant mowing effects
			Shrubland clearing effects
			Silvicultural activities
			Reforestation with autochtonous plant spp.
			Alien spp. eradication
	Fauna management		Fauna response to silvicultural activities
			Iberian lynx population: breeding and mortality
			Imperial eagle population: breeding and mortality
	Land uses and services		Cattle management
			Pinenuts harvest management
			Pilgrimage activities and effects
			Traffic and public access
			Beekeeping management
			Wedge shell fishing activities
			Land use change around Doñana Natural Space (*)
			Hunt management

approach was proposed to monitor broad-scale processes in the DNP and in its surroundings (unprotected area, see Fig. 28.1). The rationale behind this relies on the need for understanding underlying ecological processes such as connectivity, fragmentation or habitat loss at the landscape scale.

Thus, the main purpose of the programme was to assess individual features using a set of proposed indicators, usually relevant informative variables such as relative and absolute abundance, species distribution, species richness and life condition.

Programme Implementation: Protocol Validation and Arising Issues

A thorough bibliographical review revealed a complete set of available methods for Doñana features (at least suitable after minor modifications). Most reference sources were supplied by the UK Environmental Change Network (http://www.ecn.ac.uk/), US National Park Service (http://www.nps.gov/), Environment Canada (http://www.ec.gc.ca/) and the Europarc handbook (Atauri *et al.* 2005), reference guidelines and books as well as scientific papers. During the period 2002–2005, the guidelines of the joint project were to test the validity of the programme. Every feature was then assigned to at least one methodological protocol consisting of:

- The rationale behind the need for feature monitoring (narrative)
- Specific objectives
- State indicators
- Location of sampling sites on the study area, minimum sampling unit and minimum required sampling frequency
- Material and staff needed (Standard Operation Procedures according to Oakley *et al.* 2003)

Validation mainly consisted of evaluating (through the preliminary monitoring results from the first phase (2002–2005) of the indicators' suitability) the protocols' feasibility and statistical reliability and the actual relevance of proposed features to be monitored.

During this period it became evident which features had been improperly approached, either by using an unsuitable spatial scale, by using unfeasible sampling frequencies or by using inappropriate sampling methods. Often, accessibility was the critical factor to overcome in order to ensure continuous monitoring: for example, to reach the most remote areas inside DNP, trail and track networks often cross deep sand banks, mud covered areas and deep water pools, implying long and harsh journeys to reach monitoring plots and the waste of a full day of labour.

The monitoring protocol for juniper (*Juniperus phoenicea* ssp. *turbinata*) woodlands provides an example of how we carried out the validation. The original proposal for this habitat, which is characteristic of top-stabilized coastal sand dunes, was to carry out monitoring every 5 years. However, data gathered during the first 3 years stressed the need for a higher sampling frequency to properly record the disturbance effects on the demographic dynamics of this xeric plant community. Many other features were also subjected to a change of sampling frequency and sample size. Other features originally planned as landscape scale projects, such as monitoring of erosion of the Guadalquivir river bank, had to be reassessed due to the inability of automatic remote sensing or digital aerial photograph analysis to easily delineate changes on the river shoreline. Finally, certain standard protocols, in particular those using trapping nets, had to be refined according to the behavioural traits of specific species, e.g. for monitoring the population and abundance of aquatic macroinvertebrates.

After the validation exercise, a further set of monitoring activities were proposed for features not originally considered, such as monitoring phenology, or plant physiology: this is an ongoing process. However, among the leading monitoring institutions there is an implicit awareness of constraining the relevant features to be monitored, based mainly on available funding and resources. From the outset, the responsibility for implementing the long-term ecological monitoring programme in Doñana was assigned to five managers – one for each work area, i.e. landscape scale (which included vegetation dynamics and geomorphological monitoring), amphibians and reptiles, limnology, birds and mammals. In developing the programme, each of these managers had to employ 12 staff members, most of whom were assigned for birds monitoring. During this process, we ran specific training courses for the Doñana monitoring team (ESPN) to facilitate participation, motivation and increase basic monitoring skills. One of the first courses introduced the practical use of GPS and Personal Digital Assistant (PDA) devices. Other courses addressed the identification of Doñana's flora (both aquatic and terrestrial), driving in harsh conditions and health & safety criteria. Two recent courses have provided the monitoring staff with an introduction to data representation standards (map visualisation of sampling units and variables) and on simple guidelines for reporting results (charts, graphs and tables). A short introduction was also provided on the use of Cybertracker software for data collection through PDAs in the field (see below).

Nowadays, the monitoring programme relies on 15 staff members, ten of whom occupy permanent positions funded by CSIC (Fig. 28.3). However, the programme aims to increase the number of permanent positions in order to achieve the long term monitoring objectives. DBR also runs a volunteer programme that often benefits the monitoring team. Skilled volunteers are periodically recruited to help staff members in baseline monitoring activities, though always after training and under strict staff supervision. This help is gratefully received.

Critically, after a favourable hearing from the Spanish Constitutional Court in 2005, full environmental competences were transferred from the central administration to autonomic governments and, in 2008, the responsibility for DNP management was transferred to the Andalusian regional government. Under this new regime, Doñana Natural Park, (a protected area of 55,300 ha managed by the Andalusian administration that forms a buffer around DNP) was merged for monitoring purposes with DNP. This new and expanded protected area, named Doñana Natural Space (DNS), is now the target area for the monitoring programme, with

Fig. 28.3 Group picture of most of the staff members devoted to Doñana Integrated Long-Term Ecological Monitoring Programme

major implications for most of the monitoring activities. As a consequence, we have had to enlarge sampling areas, and reduce both the sampling effort and the number of sampling locations inside DNP to achieve the integration. Today, this remains the most pressing challenge for the implementation phase of the DNS monitoring programme.

Automatic Monitoring and Data-Quality Assessment: Enhancing the Programme

As a consequence of the issues outlined above, we identified the need for automatic procedures to make the monitoring programme cost-effective. Concerns about data quality assessment and metadata management were already evident. Both issues were addressed by different opportune solutions.

The first issue was resolved by Cybertracker: free software (http://www.cybertracker.org/) that uses PDA devices to digitally record any type of observation in the field. It was originally conceived as a tool to improve environmental monitoring by increasing the efficiency of data collection and observer reliability (see Chapter 32 for more information). Since the Doñana monitoring programme began using it there have been two major efficiencies achieved by the combined use of PDAs and

Cybertracker software: firstly by avoiding the loss of required data, i.e. occasionally required records were just skipped or ignored on site; and secondly by reducing the time needed to digitally archive data by rapid synchronization to the database. However, efficient sequential procedures have to be designed prior to on-site application, and this requires advanced skills in Cybertracker database management.

The second issue was resolved by the recognition of DBR as a Scientific and Technological Singular Infrastructure (ICTS) by the Spanish Ministry of Science and Innovation. Under the Spanish Research Framework Programme, projects of this type qualify for the award of an initial budget to enhance services from such infrastructures to the scientific and technological communities. DBR started to improve the infrastructure by setting up a wireless communication network inside DBR, this was essential and allowed us to install a real-time network of probes, sensors and devices to cover the gaps that we had identified in the monitoring programme. Up to 1 M€ was allocated to monitor new features such as soil, below-ground water content and temperature, water quality monitoring, tropospheric O_3, CO_2 or enhancing sensor networks such as the meteorological network. ICTS is also providing support for integrating the wireless network with other sensors working in Doñana, sensors that were established a long time ago like the piezometer gauges or hydrological stations. Besides the integrative role played by ICTS for long term ecological monitoring in Doñana, the implementation programme also provides an opportunity for European scientists to apply for short-term prospective projects to be carried out in DNS. Such projects may be proposed under a specific call for projects open twice per year, covering travelling expenses and accommodation. Many of the current ongoing research projects are tightly linked to features already being monitored by enhancing scientific research on global change effects on the Doñana biota.

Interestingly, these solutions have combined to advantage the whole programme, notably through the development of a wireless protocol to easily transfer data gathered in the field (by means of PDA devices) to the central database at the moment of data collection (see Chapter 32). We expect innovative synergies like these to increase the cost-effectiveness of the monitoring programme.

Data Access and Publishing

Since the outset, one of the premises of the monitoring programme was to make the resulting data publicly available. The only restrictions would apply for conservation purposes, such as locating nesting sites of endangered birds for instance. This mandatory premise has been accomplished by producing reports in which results are presented in tables that are updated annually. Moreover, these reports provide concise interpretations according to the latest observed trends. DNP managers are therefore provided with an annual update on every feature being monitored. However, it is also essential to allow data access through a web page, and over the first 5 years, the ESPN developed a web portal for accessing data results, protocol documentation, annual reports and many other relevant documents at: http://www-rbd.ebd.csic.es/Seguimiento/seguimiento.htm.

This web page, which is still online at the time of writing, is being transferred progressively to the new ICTS website (http://icts.ebd.csic.es) where results and protocols from manually monitored features are merged in the same database with the results from the automatic procedures. The final web portal will soon allow any visitor to query the central database on results from both approaches (automatic and manual) providing quick insights on trends and the relationship between different indicators.

Success Stories on Data Use for Research and Management

The main overarching goal of the monitoring programme is to provide baseline information on long-term trends and changes of the monitored features. Accordingly, historical data might be used to set up robust criteria for effective conservation management.

Since 2005, the Doñana management board has considered monitoring results as valid criteria for informing management decisions. The use of baseline trends of ecological indicators in the decision-making process is considered a practical realisation of the monitoring programme's usefulness and success. Early warning monitoring for toxic algal blooms is one such success story. Fully implemented in 2006, the protocol activates a set of management decisions when the indicators reach a standard threshold. Outbreaks of toxic cyanobacteria blooms have occurred periodically in Doñana marshlands causing the poisoning and death of hundreds of different species. During the last 2 years, episodic events have been successfully managed, reducing dramatically the number of affected birds.

On the other hand, monitoring a cork oak (*Quercus suber*) population (named "La Pajarera") that is more than 200 years old, where up to 14,000 pairs of herons and storks nest every year causing tree decay, has provided evidence of the need for preventing the depletion of the acorn and seedlings bank by wild herbivores. With a mortality rate of 1.96 individuals per year in the last 43 years, the "Pajarera" site has only recruited 72 individuals by planting and none through natural regeneration. Intensive acorn predation occurs during fruiting season by fallow deer (*Dama dama*), wild boar (*Sus scrofa*) and red deer (*Cervus elaphus*). In 2005, an expert panel proposed a long term restoration programme for this site consisting of several actions, including an experimental design of exclosure fences for testing grazing effects and granting tree recruitment.

Donana LTSER Platform in the ILTER and LTER Europe Networks

In 2006, DBS joined the European Network of Excellence named ALTERNet (A Long-Term Biodiversity, Ecosystem and Awareness Research Network). ALTER Net aims to establish a lasting infrastructure for integrated ecosystem research.

This combines ecological and socio-economic approaches, with emphasis on communication with the relevant audiences focusing on the CBD (Convention on Biological Diversity) target of attaining a significant reduction of the current rate of biodiversity loss by 2010. Many of the partner research institutions also participated in the ILTER network (International Long-Term Ecological Research Network) and its regional section of LTER-Europe. During the project development, the LTER-Europe constitution in 2007 proposed the first nine LTSER (Long-Term Socio-Ecological Research) pilot platforms based on selected criteria. The Doñana LTSER platform emerged from this process as a relevant place to investigate socio-economic drivers and pressures on biodiversity following the DPSIR framework (Bugmann and Solomon 2000; Parr *et al.* 2003; Gobin *et al.* 2004; Nikolaou *et al.* 2004). The Doñana LTSER platform includes DNS and the immediate surrounding area, conforming to Doñana County (see Fig. 28.1), in order to assess the effects of human drivers and pressures on biodiversity. The Doñana LTSER platform has started to contribute by developing the socio-ecological long-term research and implementing the DPSIR framework (Haberl *et al.* 2006). Doñana, together with two other LTSER platforms (Pleine-Fougères in Brittany and Islands of Breila in Romania) have recently reported on an early assessment of EU regulations (Common Agricultural Policy) of unexpected effects on local biodiversity (unpublished ALTERNet Report). The Doñana LTSER platform has also successfully collaborated in the proposal of a conceptual socio-ecological model for LTSER platforms (Haberl *et al.* 2009) that will soon be implemented and tested. Both experiences are helping in spreading the LTSER concepts. Since then, DBS has played a major role in setting up the LTER process in Spain. Thus, LTER-Spain was formally accepted as an ILTER and LTER-Europe member in 22nd August 2008. It currently comprises 10 sites, including the Doñana LTSER platform. These sites represent the main ecosystems present in the Iberian Peninsula. LTER-Spain has agreed on the minimum common ecological parameters to be monitored by every LTER-site of the network. These parameters cover abiotic variables, primary producers, consumers and other relevant parameters. To date, meteorological variables are the most widely monitored at every LTER-Spain site, closely followed by forestry parameters, birds, invertebrates, soil characteristics, atmospheric deposition, phenology and water body characteristics.

Acknowledgements The author and the ESPN (DNS Monitoring Team) want to sincerely thank the Spanish National Parks Network and CSIC (Spanish Research Council) for the funding by which the joint monitoring project was originally initiated. Also, we want to express our gratitude for the ongoing support from Consejería de Medio Ambiente of Junta de Andalucía (Autonomous Andalusian Government) who are fully committed to the long term monitoring activities in Doñana. Crucial funding support was also provided by the Confederación Hidrográfica del Guadalquivir, the Guadalquivir Catchment Management Institution, through the "Doñana 2005" restoration project. The Ministry of Science and Innovation is also contributing to the long-term support of DBR-ICTS under the Ingenio 2010 initiative. We are also grateful to DNP management administration for its permanent assistance on field work and data sharing. And finally, the ESPN team wants to give special thanks to all the volunteers for their selfless aid in the daily monitoring tasks, sometimes under harsh conditions. The author wants also to thank Clive Hurford for reviewing the manuscript and for his suggestions for improving the readability of the chapter.

References

Almaraz P, Amat JA (2004a) Complex structural effects of two hemispheric climatic oscillators on the regional spatio-temporal expansion of a threatened bird. Ecol Lett 7:547–556

Almaraz P, Amat JA (2004b) Multi-annual spatial and numeric dynamics of the white-headed duck *Oxyura leucocephala* in southern Europe: Seasonality, density dependence and climatic variability. J Anim Ecol 73:1013–1023

Atauri JA, Múgica M, de Lucio JV, Castell C (2005) Diseño de planes de seguimiento en espacios naturales protegidos. Serie Manueales Europarc-España Nº 2. Fundación Fernando González Bernáldez, Barcelona, 175 pp (in Spanish)

Bugmann HKM, Solomon AM (2000) Explaining forest composition and biomass across multiple biogeographical regions. Ecol Appl 10:95–114

García Novo F, Marín Cabrera C (2005) Doñana: Agua y Biosfera. Ministerio de Medio Ambiente, Sevilla, 354 pp (in Spanish)

Gobin A, Jones R, Kirkby M, Campling P, Govers G, Kosmas C, Gentile AR (2004) Indicators for pan-European assessment and monitoring of soil erosion by water. Environ Sci Policy 7:25–38

Grimalt JO, Ferrer M, Macpherson E (1999) The mine tailing accident in Aznalcollar. Sci Total Environ 242:3–11

Haberl H, Winiwarter V, Andersson K, Ayres RU, Boone C, Castillo A, Cunfer G, Fischer-Kowalski M, Freudenburg WR, Furman E, Kaufmann R, Krausmann F, Langthaler E, Lotze-Campen H, Mirtl M, Redman CL, Reenberg A, Wardell A, Warr B, Zechmeister H (2006) From LTER to LTSER: Conceptualizing the socioeconomic dimension of long-term socioecological research. Ecol Soc 11:art. 13

Haberl, H., Gaube, V. Díaz-Delgado, R., Krauze, K. Neuner, A., Peterseil, J., Singh, S.J. and Vadineanu, A. 2009. Towards an integrated model of socioeconomic biodiversity drivers, pressures and impacts. A feasibility study based on three European long-term socio-ecological research platforms. Ecological Economics 68:1797–1812

Margalef R (1976) Algas de agua dulce de Doñana. Oecol Aquat 2:79–91 (in Spanish)

Nikolaou K, Basbas S, Taxiltaris C (2004) Assessment of air pollution indicators in an urban area using the DPSIR model. Fresen Environ Bull 13:820–830

Oakley KL, Thomas LP, Fancy SG (2003) Guidelines for long-term monitoring protocols. Wildlife Soc Bull 31(4):1000–1003

Parr TW, Sier ARJ, Battarbee RW, Mackay A, Burgess J (2003) Detecting environmental change: science and society – Perspectives on long-term research and monitoring in the 21st century. Sci Total Environ 310:1–8

Rendon MA, Green AJ, Aquilera E, Almaraz P (2008) Status, distribution and long-term changes in the waterbird community wintering in Donana, south-west Spain. Biol Conserv 141:1371–1388

Solís JC (1996) Plan de ordenación del alcornocal de Doñana. Doñana Biological Station, unpublished report, 80 pp (in Spanish)

Vaughan H, Brydges T, Fenech A, Lumb A (2001) Monitoring long-term ecological changes through the Ecological Monitoring and Assessment Network: Science-based and policy relevant. Environ Monitor Assess 67:3–28

Chapter 29
Monitoring Aquatic Ecosystems at Doñana Natural Space

Miguel Ángel Bravo-Utrera

Introduction

'Doñana Natural Space' (DNS), a protected area resulting from the merge of the former Doñana National and Natural Parks, covers a surface area in excess of 1,000 sq. km and includes a wide range of habitat types, from scrub to marshes and from pine forests to sand dunes. Within that heterogeneous landscape, aquatic ecosystems play such an important role that the whole of DNS cannot be properly understood without considering them, the biodiversity they harbour and the ecological processes depending on them. In fact, nearly half of the DNS surface is occupied by marshes linked to the Lower Guadalquivir River. Those marshes are vast wetlands in which estuarine and tidal ecosystems co-occur with freshwater systems located in the floodplain: mainly fed by the rain and the water entering the area through some small streams (e.g. Guadiamar, Rocina). The whole system was originally a coastal lagoon with a surface area of more than 150,000 ha. This lagoon progressively silted up over the last two millennia, producing a rather flat area where small topographic variations translate into important ecological differences (a descriptive synthesis can be found in García Novo and Marín 2005). Within this floodplain a large number of diverse water bodies can be found. This diversity is the outcome of differences in the origin and duration of the flooding period, in the type of soils, and in their capacity for exchanging salts with the water, thus creating an array of aquatic ecosystems ranging from ephemeral to permanent, and from subsaline to hypersaline (Bernués 1990). Furthermore, one half of the territory rests on silt-rich, non-permeable substrate, while the other half lies on porous, highly permeable detritic materials. This feature, in combination with a poor drainage network, results in the existence of an important aquifer system: a groundwater system that shows up dramatically in the 'El Abalario-Doñana' lake complex,

M. Á. Bravo-Utrera (✉)
Estación Biológica de Doñana (EBD-CSIC), CIECEM Parque Dunar, s/n 21760, Matalascañas, Spain
e-mail: mabravo@ebd.csic.es

C. Hurford et al. (eds.), *Conservation Monitoring in Freshwater Habitats: A Practical Guide and Case Studies*, DOI 10.1007/978-1-4020-9278-7_29,

where more than 600 small, low-salinity, temporary ponds occur. Their common origin might suggest that all these ponds are rather homogeneous; an assumption that is dispelled on examination of their ecological features. Within this lake complex, interdunal ponds that lie in the active front of the mobile dunes are particularly remarkable, as they overlap with already stabilised dune fields displaying a range of variability in both flood period and salinity (Serrano *et al.* 2006); of these, Santa Olalla lake stands out for its size (>30 ha in full flood), salinity and hypertrophic status (Bravo and Montes 1993).

In this context, and given that most fluxes of matter and energy in the area are heavily dependent on the functioning of these aquatic ecosystems, the implementation of a limnologically based monitoring plan is essential in order to understand and preserve the ecological integrity of DNS. Until relatively recently, however, the limnology was overlooked in the monitoring schedule and research agenda for the area. Geological and geomorphological processes were simply considered too stable and unchangeable to be worth monitoring. Instead, the main focus was on flood patterns, i.e. whether marshes and ponds filled on a seasonal or a permanent basis and how long they remained flooded, in the confidence that the biological diversity of the aquatic systems was tightly linked to this trait. Consequently, limnological monitoring protocols (*sensu lato*) were not implemented in the area until the 1970s, starting with the assessment of the aquifer phreatic levels – as this was considered an aspect of the water cycle highly sensitive to human activity. Flooding patterns on the marsh flood have been monitored, in an intermittent and uneven manner, since the 1980s, when water gauges were fitted in some ponds and in the main water bodies in the marshes. Water quality parameters were not intensively studied until the end of the 1980s and have only been monitored regularly during the last decade, following the mine spill in Alznalcóllar, when automatic water gauges and water quality sensors were installed in the marshes of the National Park. That event led to the implementation of a hydrological restoration project called 'Doñana 2005'. This project, with an initial budget of over 80 million Euros, included, together with other aims, monitoring the quality of the water entering the DNS (Saura *et al.* 2001). Since 2006, the efforts of the 'Doñana 2005' project have been complemented by the set up of the ICTS-Doñana Biological Reserve monitoring network that aimed to, among other goals, improve access to research data in the area and the establishment of facilities for the retrieval of environmental information for scientific use.

Background on Groups Selected for Biomonitoring

Hydrological data, like the quantity and quality of water flowing through the system and the flood patterns inferred via remote sensing techniques, are of great importance to any comprehensive limnological monitoring programme, and these are extensively covered elsewhere in this document. Here, therefore, we will focus exclusively on biological monitoring; i.e. the use of biological indicators (namely aquatic vegetation, aquatic invertebrate fauna and fish communities) to assess the condition or health of aquatic systems in the Doñana Natural Space.

The rationale behind using biological monitoring is that the integrity of the biota inhabiting aquatic ecosystems provides a direct, holistic and integrated measure of the integrity or health of the ecosystem as a whole. In addition to their practical value for biomonitoring purposes, these groups also provide relevant information on biodiversity, as they are rich in species, include many rare and endangered taxa, and are major contributors to the overall diversity of the area.

Aquatic Vegetation

Our focus for aquatic vegetation is on the so-called macrophytes, i.e. vascular plants (angiosperms), non-vascular plants (like liverworts and mosses) and macroalgae (Characeae). Beyond their role as primary producers, which transform solar energy into organic molecules and release oxygen, macrophytes have an important function as structural elements of the aquatic habitat by providing shelter to many organisms, increasing water transparency and enhancing the diversity of microhabitats (Carpenter and Lodge 1986). This is particularly so in the abundant shallow and ephemeral water bodies in Doñana, where macrophytes are of outstanding significance by comparison with phytoplankton.

Information on the vegetation of Doñana dates back to the 1940s (Vicioso 1946), and includes floristic and phytosociological studies (Castroviejo *et al.* 1980 and Rivas-Martínez *et al.* 1980, respectively). However, it was not until the work of González-Bernáldez and his colleagues (Allier *et al.* 1977) that the aquatic vegetation was considered to be a key functional element. Subsequently, many studies have contributed to our knowledge of the composition and distribution of macrophyte communities in the aquatic systems of Doñana (see references in Montes *et al.* 1998). Some of these studies applied an ecological approach, trying to relate environmental conditions, including the activity of vertebrate herbivores, to functional features in the aquatic vegetation, like space occupancy patterns and seed bank development (Duarte *et al.* 1990; Montes and Bernués 1991; Grillas *et al.* 1993; González 1999; Espinar *et al.* 2002).

Aquatic plants are known to be sensitive to changes in salinity, pH, trophic status and the duration of the flooding period, and have therefore been widely used for monitoring the aquatic systems in the area (Cirujano *et al.* 2005). With regard to their contribution to overall biodiversity, the DNS acts as a refuge area for many macrophyte species with a restricted distribution (e.g. *Hydrocharis morsus-ranae*, *Althenia filiformis*).

Benthic Macroinvertebrates

Aquatic invertebrates are, together with macrophytes, very good indicators of water quality and, as such, have been extensively used for monitoring lotic ecosystems (Rosenberg and Resh 1992; Usseglio-Polatera *et al.* 2000): this is less common, however, in shallow, temporary, lentic habitats (but see Boix *et al.* 2005; Gascón *et al.* 2008; Menetrey *et al.* 2008). Therefore, although some indicator indexes are being developed based on correlations between actual macroinvertebrate fauna in the DNS and environmental conditions, most effort has been devoted to assessing the diversity of the group, its distribution at a local scale and the dynamics of the most abundant populations. Background information is scattered and biased; some groups, such as gastropods and beetles, have been intensively studied, while others, such as mites and nematodes, are practically unknown. Jiménez (2000) lists a total of 120 species of insect in the orders Coleoptera (43 spp.), Odonata (29), Heteroptera (28), Diptera (14) and Ephemeroptera (6), and 14 species of gastropods. A recent study by Millán *et al.* (2005) provides accounts for 108 species of Coleoptera, giving a clear indication of the amount of information still lacking for this group alone.

Among the macroinvertebrates occurring in the DNS, some species exert a surmounting influence on its functional processes and are recognised as actual ecosystem engineers, capable of modifying the physical structure of the habitat where they live. This is particularly true of the red swamp crayfish *Procambarus clarkii*, an alien species that became to be a fundamental node linking primary resources (mainly organic matter stored in the sediment and aquatic vegetation) with larger sized secondary consumers, such as fish, birds and mammals (Gutierrez Yurrita and Montes 1997). The amount of energy currently transferred by crayfish to higher trophic levels suggests that changes in crayfish population density can trigger shifts between alternative ecological states of the aquatic systems; a topic that is currently under investigation (Crehuet *et al.* 2007).

Since this crayfish was first introduced in the area in 1974, its distribution range has grown to cover almost the entire marsh, and it can now be labelled as the best-known invertebrate species in Doñana. Like the Greek god Janus, the red swamp crayfish displays two heads. One is mild and gentle, due to its economic value as a fishery resource and that some endangered species in the area have benefited from it (like the otter *Lutra lutra* and many birds). The other one, however, is harmful and destructive; the species seems to be responsible for local extinctions and population declines in several native species, it has also contributed to the modification of some environmental features in the marsh, and its commercial exploitation put some undesired pressure on protected areas. Against this background, the red swamp crayfish is undoubtedly a key factor in the functioning of the aquatic ecosystems in Doñana, and the significant effort committed to this species in our monitoring programme reflects this.

Fish Communities

Inland fish species are among the most threatened groups of vertebrates in the world. On a global scale, fish populations in inland waters have declined due to overexploitation, flow modification, destruction of habitats, invasion by exotic species and pollution, all of which are interacting (Lévêque *et al.* 2008). Fish have also a significant functional role by producing large quantities of biomass that are supplied to higher trophic levels in the ecosystem, such as birds and mammals, especially in periods of high energy demand (such as during the breeding season). The study of their population fluctuations provides invaluable information on the functioning of the aquatic ecosystems.

Fish in Doñana are, by far, the least well-known group of vertebrates, with only a dozen studies devoted to them up to the 1980s (see references in Montes *et al.* 1998). Indeed, the first fish inventory of the National Park, including fish community composition and their conservation status, dates back to 1994 and remained as an internal report until its public release 6 years later (Fernández-Delgado *et al.* 2000). In total, 72 species of fish have been recorded in DNS; most of them marine or estuarine species that occasionally enter the marshes through tidal channels. Of the 16 species most frequently found in the marshes and ponds of DNS, 2 species are critically endangered (*Aphanius baeticus* and *Squalius pyrenaicus*), 1 is considered locally extinct (*Gasterosteus gymnurus*) and 6 are introduced exotic species (e.g. the ubiquitous mosquito fish, *Gambusia holbrooki*) (Fernández-Delgado 2005).

Monitoring an Unpredictable Environment

Covering the huge diversity of habitats in the area was the first challenge when planning the design of our sampling station network. The classification of ecotopes provided by Montes *et al.* (1998) was used as a basis for the design; and given the large number of different categories and in order to keep the network feasible, it was decided to restrict monitoring to those habitat types covering the largest possible surface areas of the DNS. Accordingly, monitoring stations were placed both in the marsh and in several ponds on sandy substrates: these are the most frequent and representative habitat types in the DNS. Within the marsh, a second habitat-type level was selected to accommodate the variability in salinity concentration and the hydroperiod (the duration of the wet period). The 'pond' category was also split in several types to incorporate different sorts according to their substrate, altitudinal location and degree of link to the aquifer (Borja Barrera and Díaz del Olmo 1987; Borja Barrera *et al.* 1999).

In order to guarantee continuity in the collection of data, other circumstances like all-year-around accessibility were also taken into account for deciding the exact location of the monitoring stations. That was not a minor issue given that sampling activities frequently require the use of vehicles, or other suitable means of transport, to carry large and heavy pieces of equipment. Because movements in the natural marsh are severely restricted during the flooding period, we made an effort to keep the number of sampling stations accessible only by foot or horse to a minimum. Finally, some additional stations were incorporated into the network as a consequence of the expansion of the study area when the DNS was established in 2007.

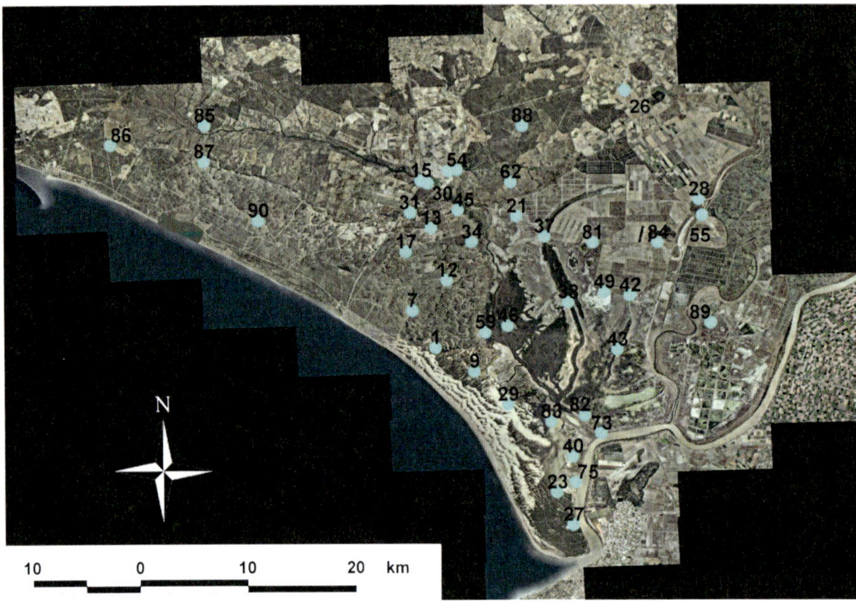

Fig. 29.1 Limnological network of level 1 sampling sites

The final outcome is an open design with monitoring stations allocated to one of the following categories:

- Level 1; this includes forty-one sites considered to be first-order sampling stations with regional importance (Fig. 29.1).
- Level 2; second-order sampling stations comprising all those sites that have been used at some time during previous studies and those for which there is historical information.
- Level 3; sites that have been sporadically used within the context of specific research projects.

Another concern during the establishment of the monitoring network was the frequency of sampling. It takes about 2 months to collect samples from all of the sites during the rainy season, so we decided to limit the sampling campaigns to only two per year at each station: one to be carried out in autumn–winter (actually between October and March), during the early wet period when water has been present for sufficient time to allow resistant life stages (i.e. spores, diapausing eggs) to hatch; and then a second sampling campaign in spring (from April until June), coinciding with the period of highest biological activity prior to the summer drought. An optional sampling campaign can be carried out in mid-summer to account for artificially flooded, permanent aquatic environments. Finally, and due to their special characteristics, tidal channels (three sampling stations) are each sampled on 30–45 days. These habitats are exceptionally important because fish use them extensively to move from the estuary to the marsh (and back).

Sampling Protocol

The sampling at each station takes no less than 2 days, a limit set by the minimum amount of time that traps for capturing crayfish must remain in the water (Bravo 1998). The sampling protocol starts with the assessment of some physico-chemical attributes (including water temperature, conductivity, pH, dissolved oxygen concentration, chlorophyll *a*, and nitrate and ammonium concentrations) and is followed by sampling the biological communities (macrophytes, benthic macroinvertebrates, crayfish and fish communities). In all cases, the sampling methods and the observed variables have been chosen on the basis of three criteria: easy use (i.e. not too time-consuming and not requiring a highly specialised team); fast results (most data are collected *in situ* and post-sampling processing is kept to a minimum); and relevance to the functioning of the system. Except for benthic macroinvertebrates, which are mostly taken and transported to the laboratory for identification, samples are directly identified in the field and returned to the water shortly after capture.

For aquatic vegetation, the following variables are recorded:

- Species (or genus) richness, identifying species to the lowest taxonomic level possible depending on the degree of development of the diagnostic characters
- Cover (valued from 0 to 5) of each taxon and different biological types; i.e. emergent (helophytes), submerged (hydrophytes) and free-floating macrophytes
- The phenological status of each species

Aquatic macroinvertebrate monitoring requires several methodological approaches. Hand-nets (0.5 mm mesh size) are used in shallow waters for qualitative sampling, gathering presence/absence data but not abundance. We use a core sampler (25 cm diameter, 3 replicates) to collect less mobile and smaller species from the upper sediment layer (first 5cm) and the water column, using a filtration net (also 0.5 mm mesh size) to sieve the sample. Finally, net traps (4 mm mesh size, 5–9 replicates), a modification of those regularly employed for the commercial fishing of eels and crayfish, are used for catching larger and more mobile specimens. Because these traps are also intended to capture crayfish, they remain in place for 24 h. Once removed, the invertebrates in the traps are identified *in situ*, counted and returned to the water. Wherever possible, the invertebrates are identified to species level, then either counted or weighed (as appropriate).

Due to its importance in the monitoring schedule, we have a more complex methodological protocol for the red swamp crayfish: this includes an analysis of the population structure by measuring the length (rostrum-telson), the identification of sex and the assessment of the degree of sexual maturity (Huner and Barr 1991).

Fish communities are monitored using the same large net traps used for invertebrates. Most species are identified to species level, then counted, and total biomass is recorded. We record additional data for particularly important species (e.g. eel, mugilids, barbel *Barbus sclateri* and carp *Cyprinus carpio*), such as total length, individual weight and reproductive state.

Preliminary Results: The Case for a Long-Term Monitoring Programme

The entire raw dataset is still to be fully analysed but it can be retrieved via the monitoring web page: http://icts.ebd.csic.es.

More than 50 macrophyte taxa have been recorded in the spring campaigns, although not all of them identified to species level. Helophytes (emergent macrophytes) and amphibious plants are the most frequent and widely distributed types. Although their floristic value is low, these species are responsible for most primary production in DNS aquatic ecosystems.

Given the high environmental diversity of the area and the uneven distribution of the sampling effort, we grouped the vegetation data into four habitat types: temporary ponds, brackish marsh, channels within the salt marsh, and transition areas where small streams run over the sandy substrate to enter the marshland (showing a mix of features between ponds and marsh). A fifth type, artificial ponds, has been added.

These artificial ponds, locally called 'zacallones', are holes dug to reach ground water so that livestock can drink. These can remain as small, permanent ponds for a long time, and during the very dry spring of 2005 were almost the only signs of surface water present in DNS.

Some results from the aquatic vegetation monitoring in DNS are presented in Table 29.1. The environments hosting the highest diversity are the temporary ponds, whereas artificial ponds and marsh channels are the least diverse. Figure 29.2 shows changes in average species-richness through time.

Table 29.1 Communities of aquatic vegetation recorded for some of the more extensive habitats. The dominant taxa in each habitat are listed first.

Marsh channels	Temporary ponds	Brackish and salt marsh	Streams
Scirpus maritimus	*Ranunculus peltatus*	*Scirpus maritimus*	*Scirpus maritimus*
Ranunculus peltatus	*Baldellia ranunculoides*	*Ruppia drepanensis*	*Lemna minor*
Chara sp.	*Myriophyllum alterniflorum*	*Chara* sp.	*Panicum repens*
Ruppia drepanensis	*Callitriche* sp.	*Callitriche* sp.	*Baldellia ranunculoides*
Scirpus lacustris	*Illecebrum verticillatum*	*Ranunculus peltatus*	*Callitriche* sp.
Callitriche sp.	*Eleocharis palustris*	*Azolla filiculoides*	*Iris pseudacorus*
Zanichellia obtusifolia	*Hydrocotyle vulgaris*	*Baldellia ranunculoides*	*Ranunculus peltatus*
Lemna minor	*Panicum repens*	*Scirpus littoralis*	*Agrostis stolonifera*

(continued)

Table 29.1 (continued)

Marsh channels	Temporary ponds	Brackish and salt marsh	Streams
Phragmites australis	*Scirpus maritimus*	*Zanichellia obtusifolia*	*Scirpus lacustris*
Scirpus littoralis	*Juncus heterophyllus*	*Iris pseudacorus*	*Apium nodiflorum*
Typha domingensis	*Chara* sp.	*Lemna minor*	*Eleocharis palustris*
Callitriche truncata	*Eleocharis multicaulis*	*Scirpus lacustris*	*Scirpus littoralis*
Azolla filiculoides	*Lemna minor*	*Myriophyllum alterniflorum*	*Typha domingensis*
Baldellia ranunculoides	*Typha domingensis*	*Apium inundatum*	*Myriophyllum alterniflorum*
Eleocharis sp.	*Agrostis stolonifera*	*Eleocharis multicaulis*	*Apium inundatum*
Ranunculus peltatus subsp. *saniculifolius*	*Hypericum elodes*	*Phragmites australis*	*Azolla filiculoides*
	Scirpus lacustris	*Ranunculus peltatus* subsp. *fucoides*	*Callitriche brutia*
Artificial ponds	*Apium inundatum*	*Agrostis stolonifera*	*Eleocharis multicaulis*
Chara sp.	*Eleocharis* sp.	*Callitriche brutia*	*Eleocharis* sp.
Lemna minor	*Zanichellia obtusifolia*	*Callitriche obtusangula*	*Lemna gibba*
Panicum repens	*Iris pseudacorus*	*Callitriche truncata*	*Phragmites australis*
Ranunculus peltatus	*Azolla filiculoides*	*Chara canescens*	*Ranunculus peltatus* subsp. *saniculifolius*
Azolla filiculoides	*Callitriche brutia*	*Hypericum elodes*	*Typha* sp.
Baldellia ranunculoides	*Lemna gibba*	*Panicum repens*	
Eleocharis palustris	*Phragmites australis*	*Ranunculus peltatus* subsp. *saniculifolius*	
Illecebrum verticillatum	*Ruppia drepanensis*	*Typha domingensis*	
Zanichellia obtusifolia	*Scirpus littoralis*	*Ruppia* sp.	
	Apium nodiflorum		

Changes in crayfish abundance in the main habitat types are shown in Fig. 29.3. The highest values in all cases correspond to the 2004 campaign: since then populations have suffered a severe decline, due to the drought in 2005. Their distribution in 2005 was restricted to sites where brooks entered the marshland and where groundwater reached the surface, i.e. the transition zone between the aeolian sand deposits and the silt-rich marsh.

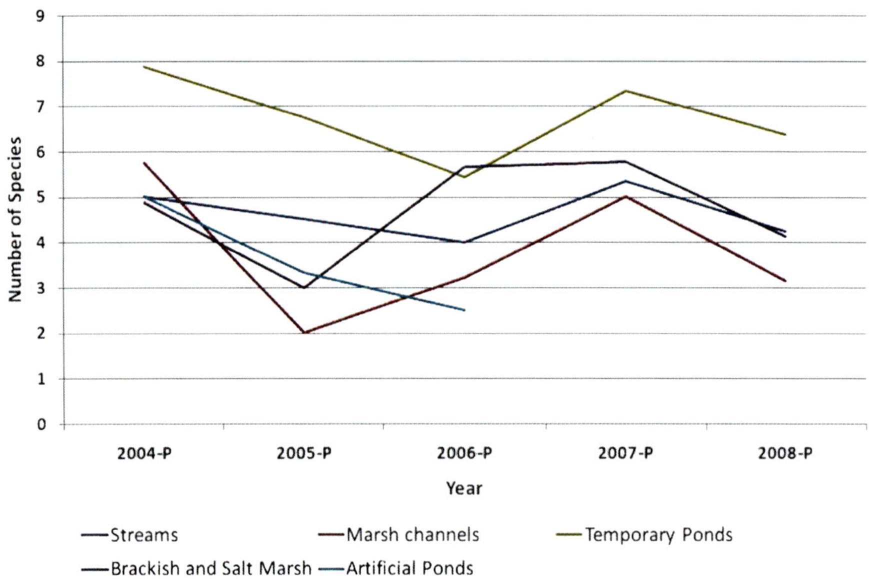

Fig. 29.2 Macrophyte species-richness in spring. The mean number of species recorded for each of the major habitat types defined

Populations of other invertebrate aquatic predators, such as beetles (families Dytiscidae and Hydrophilidae), follow similar patterns to crayfish (Fig. 29.4).

During the drought of 2005, most catches of these species were obtained in artificial environments, where the densities observed were an order of magnitude higher than in the natural systems – highlighting their function as refuges for many aquatic fauna. Figure 29.5 plots the catches of the shore crab *Carcinus maenas* in the marsh channels close to the mouth of the Guadalquivir River; high densities found in 2006 experienced a ten-fold decline to reach a level that remained stable until the most recent sampling (<10 Captures per Unit Effort).

Table 29.2 lists the fish communities found in some of the more extensive habitats. These species have been ranked according to their frequency of appearance in consecutive sampling events carried out since 2004. The most abundant species are the mosquito fish and the carp: both alien species and the only ones that are found in artificial environments. The marsh channels and deeper areas within the marsh host the most diverse fish community, with 23 species. This is mainly due to the presence of typically estuarine species in this ecotone. Among the native species, the Iberian loach (*Cobitis paludica*) and the eel (*Anguilla anguilla*), stand out for their wide distribution, and the catches of *Aphanius baeticus*, endemic to the South West of the Iberian Peninsula, are remarkable due to its importance for conservation. Table 29.3 lists the catches of the most interesting species in recent years and shows the frequency of occurrences, respectively.

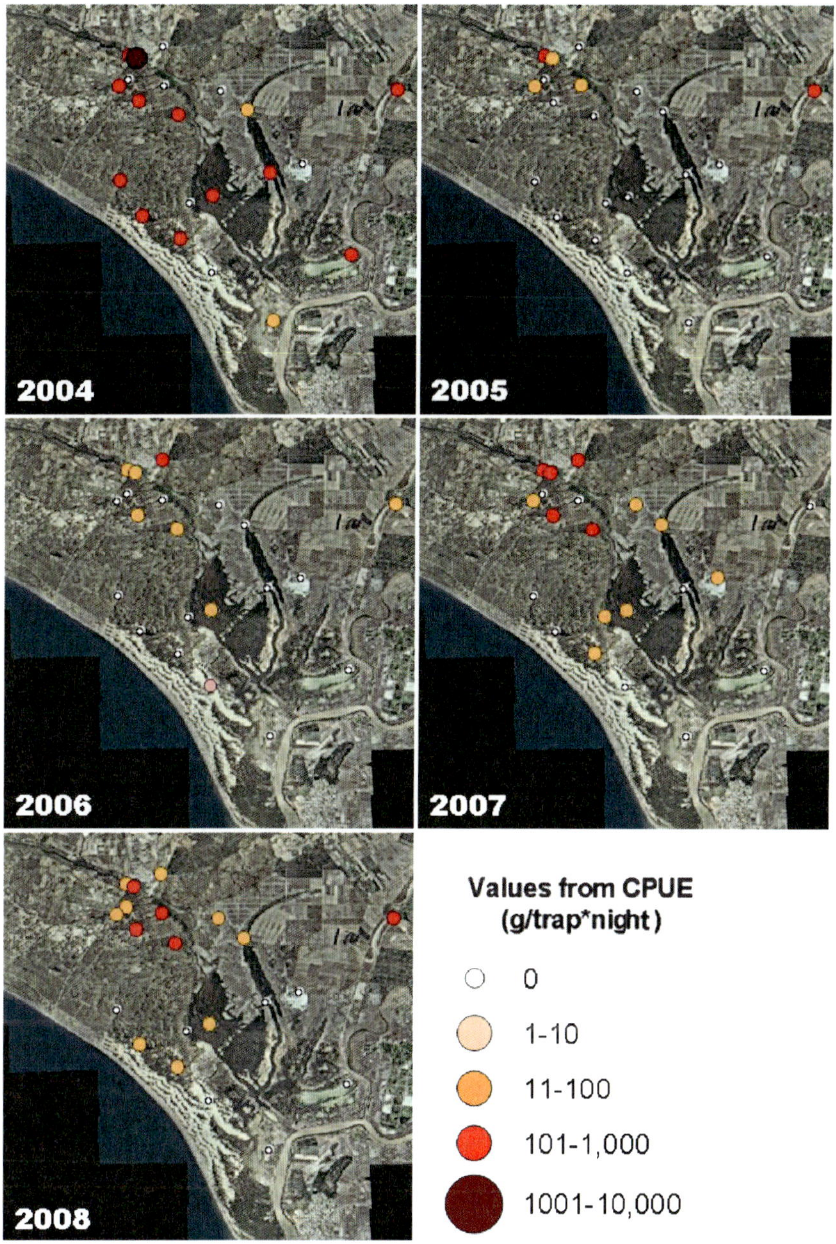

Fig. 29.3 Distribution and density on CPUE of red swamp crayfish populations by sampling sites in spring from 2004 to 2008. Shows a minimum in 2005 followed by a slow recovery that does not reach the values for 2004

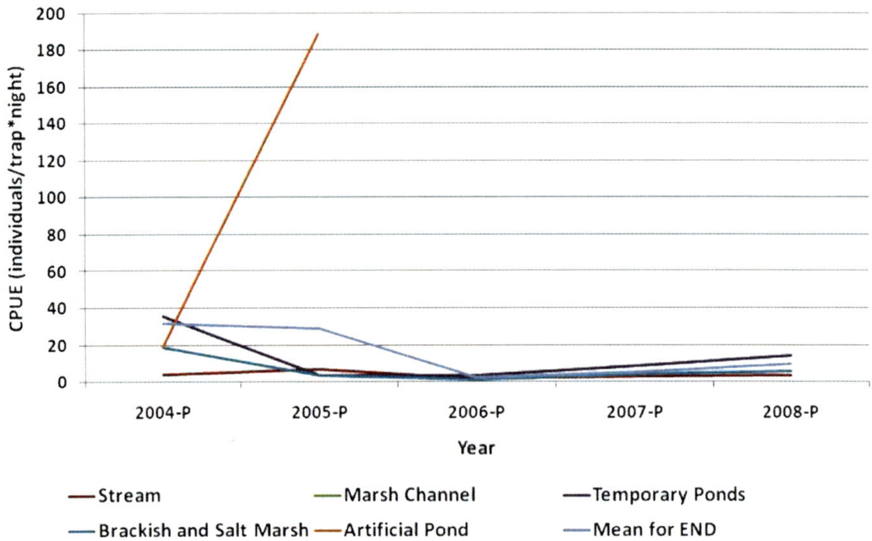

Fig. 29.4 Trends in catch per unit effort in the spring of Dytiscidae & Hydrophilidae in different habitats during the sampling period. This also shows the mean catches across all sampling sites

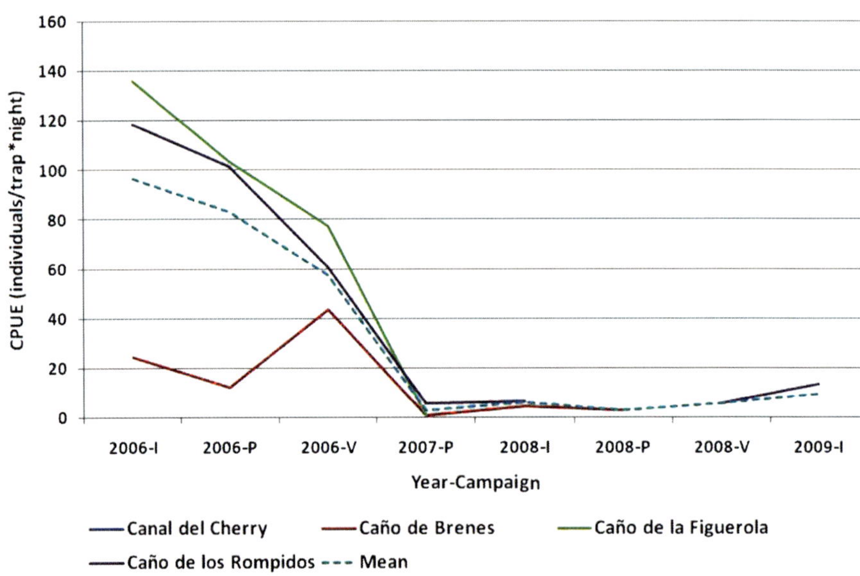

Fig. 29.5 Trends in catch per unit effort of *Carcinus maenas* in different habitats during the sampling period. This also shows the mean catches across all sampling sites

Table 29.2 The fish communities recorded from some of the major habitats. The species were ordered according to their frequency of occurrence in catches. The high diversity in the marsh channels is due to the exchange of species with the estuary. In artificial systems, only introduced species have been found

Marsh channels	Brackish marsh and saltmarsh	Stream
Gambusia holbrooki	*Cyprinus carpio*	*Gambusia holbrooki*
Cyprinus carpio	*Gambusia holbrooki*	*Cyprinus carpio*
Fundulus heteroclitus	*Fundulus heteroclitus*	*Cobitis paludica*
Anguilla anguilla	*Atherina boyeri*	*Anguilla anguilla*
Atherina boyeri	*Anguilla anguilla*	*Micropterus salmoides*
Mugil cephalus	*Liza ramada*	*Aphanius baeticus*
Liza ramada	*Micropterus salmoides*	*Barbus sclateri*
Pomatoschistus sp.	*Aphanius baeticus*	
Micropterus salmoides	*Cobitis paludica*	Artificial ponds
Carassius auratus	*Carassius auratus*	*Gambusia holbrooki*
Sparus aurata	*Chelon labrosus*	*Cyprinus carpio*
Liza aurata		
Dicentrarchus punctatus	Temporary ponds	
Barbus sclateri	*Gambusia holbrooki*	
Chelon labrosus	*Cyprinus carpio*	
Solea senegalensis	*Cobitis paludica*	
Dicentrarchus labrax	*Anguilla anguilla*	
Solea vulgaris	*Aphanius baeticus*	
Cobitis paludica	*Fundulus heteroclitus*	
Liza saliens		
Barbus bocagei		
Gobio gobio		
Lepomis gibbosus		

After six years of limnological monitoring, one of the preliminary conclusions that can be drawn from the data analysis is that climatic factors have a weight that significantly exceeds that of other stress factors.

Given the nature of the Mediterranean climate, i.e. highly unpredictable and with large inter-annual variations, and after having analysed the existing time series, it is not possible to unambiguously relate modifications in population densities to changes in management practices or to trends in average weather conditions. Indeed, most of the variability observed in species richness and other biological features with indicator values can be more easily related to local, short-term, climate driven variability than to consistent, stable, long-term drift. The only observed biological response to non-climate related phenomenon was due to the increase of total suspended solids (TSS) in the Guadalquivir River that occurred in 2007, and that correlates with a fall in the catches of *Carcinus maenas* in the tidal channels (Fig. 29.5).

Compared with the regularly occurring, highly predictable, summer drought, winter drought is a true disturbance event with consequences that show up immediately and last for some time. Its main effect concerns the absence of surface water that is readily reflected by indicators applied in the monitoring programme.

Table 29.3 Occurrence of each of the more important fish species in the 2004 – 2009 sampling campaigns. I, Fall-Winter; P, Spring; V, Summer

	04	05		06			07		08			09
Species Habitat	P	P	V	I	P	V	I	P	I	P	V	I
Anguilla anguilla	16	9	4	12	23	25	15	31	31	22	41	7
Stream			1					2		3	1	
Marsh channel	2	8		12	21	25	9	14	30	16	32	7
Temporary ponds	2	1	3								8	
Brackish and salt marsh	12				2		6	15	1	3		
Cobitis paludica	11	1		12	4		6	6	9	9		3
Stream	3	1		5	1		4	3	1	3		
Marsh channel									3	4		
Temporary ponds	5			7			2	3	1	2		3
Brackish and salt marsh	3				3				4			
Cyprinus carpio	159	12	2	7	71	7	43	120		30	21	1
Stream	12				9						3	
Marsh channel	64	7		7	46	7	26	56		30	6	1
Temporary ponds	16							5			12	
Brackish and salt marsh	67	5			16		17	59				
Fundulus heteroclitus	5			16	12	20	29	86	53	16	17	22
Marsh channel				16	12	20	13	36	29	15	16	22
Temporary ponds										1	1	
Brackish and salt marsh	5						16	50	24			
Gambusia holbrooki	121	22	17	44	39	4	135	101	98	107	11	36
Stream	15	10	5	15	11		25	9	21	24	1	
Marsh channel	33	2		20	7	4	69	38	41	42	9	26
Temporary ponds	43	7	9	1	6		15	20	26	28	1	10
Brackish and salt marsh	30	3		8	15		26	34	10	13		

Most interesting, nevertheless, is that recovery rates differ significantly between groups of organisms and between functional features. The vegetation recovers faster and reaches maximum cover levels as soon as flood patterns are re-established. The recovery of the fish population is slower and, unfortunately, with alien species performing better than natives.

The results analysed to date highlight the importance of the permanently flooded artificial ponds in the context of large seasonal variations. Although of minor significance with regards to overall plant diversity, these ponds are extremely valuable for several rare, and locally distributed, plant species. Invertebrate species, and particularly some large predatory beetles, typically use these artificial ponds as refuges, colonising the highly productive temporary waters as they become available. Similarly, some alien fish species retreat to these permanent habitats when most of the temporary ponds have dried up, only to re-invade them when they flood again.

Unsurprisingly, we have not unequivocally established any causal relationship for any of the observed patterns. Indeed, the large-scale heterogeneity (both spatial and temporal) which is characteristic of the ecological structure and functioning of the Mediterranean systems prevents us from naively assuming that DNS can be adequately understood from the results of a short-term series of observations. Obviously, long-term monitoring projects are needed to increase our levels of understanding and inform our management decisions. This, however, does not necessarily mean that nothing can be done until a long series of data is available. The whole approach must be interactive, using data gathered in the short term as the starting point in the search for plausible causal factors and for establishing cross analysis with observations derived from other components of the monitoring programme, such as those tracking water quality using monitoring probes and remote sensing.

Acknowledgements Obviously, a work like this is not the outcome of a one-man effort. Isidro Román, Edwin Riaño, Diego López and Antonio Alcaide have been deeply involved in the fieldwork and, under their supervision, a volunteer workforce too large to list here. In addition to other laboratory work, Arantxa Arechederra and Hugues Lefranc have overseen the sorting and identification of the macroinvertebrates. Hugues also risked his reputation to update the whole database and make it available, and succeeded.

References

Allier C, Bresset V, Lacoste A, Gonzalez Bernaldez F (1977) La Marisma de la Réserve Biologique de Doñana. Application de l'analyse factorielle á deux méthodes de prélèvements des données floristiques. In: Doñana: Prospección e inventario de ecosistemas. Ministerio de Agricultura, Madrid: ICONA, pp 111–129 (in French)

Bernués M (1990) Limnología del Parque Nacional de Doñana. PhD Thesis. Madrid: Departamento de Ecología Universidad Autónoma de Madrid (in Spanish)

Boix D, Gascón S, Sala J, Martinoy M, Gifre J, Quintana XD (2005) A new index of water quality assessment in Mediterranean wetlands based on crustacea and insect assemblages: The case of Catalunya (NE Iberian peninsula). Aquat Conserv Marine Freshwater Ecosyst 15:635–651

Borja Barrera F, Díaz del Olmo F (1987) Complejos Húmedos del Abalario (Entorno de Doñana, Huelva). Oxyura 4(1–1):27–44 (in Spanish)

Borja Barrera F, Zazo C, Díaz del Olmo F, Goy JL, Dabrio C, Lario J (1999) Holocene Aeolian phases and human settlements along the Atlantic coast of southern Spain. The Holocene 9(3):333–339

Bravo MA (1998) El cangrejo rojo de las marismas: biología, ecología aspectos socioeconómicos de su explotación comercial y análisis de compatibilidad con la conservación de los espacios naturales en el Bajo Guadalquivir. Tesis de Maestría. I Maestría en Conservación y Gestión del Medio Natural. Universidad Internacional de Andalucía. Sede Iberoamericana. La Rábida. Huelva. España (in Spanish)

Bravo MA, Montes C (1993) Inventario de las formaciones palustres del manto eólico del Parque Nacional de Doñana (SW España) In: Actas del VI Congreso Español de Limnología. Mayo 1993. Granada, pp 31–43 (in Spanish)

Carpenter SM, Lodge DM (1986) Effects of submerged macrophytes on ecosystems processes. Aquat Bot 26(3–4):341–370

Castroviejo S, Valdés Bermejo E, Rivas-Martínez S, Costa M (1980) Novedades florísticas de Doñana. Anales del Jardín Botánico de Madrid 36:203–244 (in Spanish)

Cirujano S, Cambra J, Gutiérrez C (2005) Protocolos de muestreo y análisis para macrófitos. Metodología para el establecimiento del Estado Ecológico según la Directiva Marco del Agua. Zaragoza Confederación Hidrográfica del Ebro, 43 pp (in Spanish)
Crehuet M, Alcorlo P, Bravo-Utrera MA, Baltanás A, Montes C (2007) Assessing the trophic ecology of crayfish: A case study of the invasive *Procambarus clarki*. In: Gherardi F (ed) Biological invaders in inland waters: Profiles, distribution, and threats. Springer, Dordrecht, pp 559–576
Duarte CM, Montes C, Agustí S, Martino P, Bernués M, Kalff J (1990) Biomasa de macrófitos acuáticos en la marisma del Parque Nacional de Doñana (SW España): Importancia y factores que controlan su distribución. Limnetica 6:1–12 (in Spanish)
Espinar JL, García LV, García-Murillo P, Toja J (2002) Submerged macrophyte zonation in a Mediterranean salt marsh: A facilitation effect from established helophytes? J Vegetat Sci 13:831–840
Fernández-Delgado C (2005) Los peces de Doñana. In: García Novo F, Marín Cabrera C (eds) Doñana: Agua y Biosfera. Ministerio de Medio Ambiente, Madrid pp 225–230 (in Spanish)
Fernández-Delgado C, Drake P, Arias AM, García D (2000) Peces de Doñana y su entorno. Organismo Autónomo de Parques Nacionales. Ministerio de Medio Ambiente, Madrid, 272 pp (in Spanish)
García Novo F, Marín C (eds) (2005) Doñana Agua y Biosfera. Ministerio de Medio Ambiente, Madrid, 353 pp (in Spanish)
Gascón S, Boix D, Sala J, Quintana X (2008) Relation between macroinvertebrate life strategies and habitat traits in Mediterranean salt marsh ponds (Empordà wetlands, NE Iberian Peninsula). Hydrobiologia 597(1):71–83
González MD (1999) Ecología del banco de semillas de hidrófitos de la marisma del Parque nacional de Doñana. PhD thesis, Madrid: Departamento de Ecología Universidad Autónoma de Madrid, 194 pp (in Spanish)
Grillas P, García-Murillo P, Geerzt-Hansen O, Marbá N, Montes C, Duarte CM, Tan Ham L, Grossmann A (1993) Submerged macrophyte seed bank in a Mediterranean temporary marsh: Abundance and relationship with established vegetation. Oecologia 94:1–6
Gutierrez Yurrita PJ, Montes C (1997) El cangrejo rojo de la marisma, una especie clave en los ecosistemas del Parque Nacional de Doñana. Sede para el estudio de los Humedales Mediterráneos SEHUMED, 11–12 (in Spanish)
Huner JV, Barr JE (1991) Red swamp crawfish. Biology and exploitation, 3rd edn. Louisiana sea grant collage programm, LSU, USA, Baton Rouge
Jiménez R (2000) Análisis del ambiente físico y la estructura y variación estacional de la comunidad macrobentónica de diversos humedales del área de Doñana. DEA thesis, Madrid: Departamento Interuniversitario de Ecología de Madrid. Universidad Autónoma de Madrid (in Spanish)
Lévêque C, Oberdorff T, Paugy D, Stiassny M, Tedesco P (2008) Global diversity of fish (Pisces) in freshwater. Hydrobiologia 595:545–567
Menetrey N, Oertli B, Sartori M, Wagner A, Lachavanne J (2008) Eutrophication: Are mayflies (Ephemeroptera) good bioindicators for ponds? Hydrobiologia 597(1):125–135
Millán A, Hernando C, Castro A, Ribera I, Aguilera P (2005) Los coleópteros acuáticos y semiacuáticos de Doñana: reconocimieto de su biodiversidad y prioridades de conservación. Boletin del la Sociedad Entomológica Aragonesa 36:157–164 (in Spanish)
Montes C, Bernués M (1991) Incidencia del flamenco rosa (*Phoenicopterus ruber roseus*) en el funcionamiento de los ecosistemas acuáticos de la marisma del Parque Nacional de Doñana (SW España). In: Pintos MR (ed) Reunión técnica sobre la situación y problemática del flamenco rosa en el mediterráneo occidental y África noroccidental. Junta de Andalucía, Agencia de Medio Ambiente, Sevilla, pp 103–110 (in Spanish)
Montes C, Borja F, Bravo MA, Moreira JM (coordinators) (1998) Reconocimiento Biofísico de Espacios Naturales Protegidos. Doñana: Una Aproximación Ecosistémica. Consejería de Medio Ambiente. Junta de Andalucía, Sevilla, 311 pp (in Spanish)

Rivas-Martínez S, Costa M, Castroviejo S, Valdés E (1980) La vegetación de Doñana (Huelva, España). Lazaroa 2:5–190 (in Spanish)

Rosenberg DM, Resh VH (eds) (1992) Freshwater biomonitoring and benthic macroinvertebrates. Chapman & Hall, London, 488 pp

Saura J, Bayán B, Casas J, Ruiz de Larramendi A, Urdiales C (2001) Documento Marco para el Desarrollo del Proyecto Doñana 2005. Regeneración Hídrica de las cuencas y cauces vertientes a las marismas del Parque Nacional de Doñana. Ministerio de Medio Ambiente, Madrid, 201 pp (in Spanish)

Serrano L, Reina M, Martín G, Reyes A, Arechederra A, León D, Toja J (2006) The aquatic systems of Doñana (SW Spain): Watersheds and frontiers. Limnetica 25(1–2):11–32

Usseglio-Polatera P, Bournaud M, Richoux P, Tachet H (2000) Biomonitoring through biological traits of benthic macroinvertebrates: How to use species trait databases? Hydrobiologia 422(423):153–162

Vicioso C (1946) Notas sobre la Flora Española. Anales del Jardín Botánico de Madrid 6(2):5–92 (in Spanish)

Chapter 30
Endangered Waterbirds at Doñana Natural Space

Manuel Máñez, Luis García, Fernando Ibáñez, Héctor Garrido, Juan Manuel Espinar, José Luis Arroyo, José Luis Del Valle, Alfredo Chico, Antonio Martínez, and Rubén Rodríguez

Background Information

The declarations of both Doñana Biological Reserve in 1964 and Doñana National Park in 1969 were based mainly on the presence of the Iberian lynx *Lynx pardinus* and the Spanish imperial eagle *Aquila adalberti*: two species endemic to the Iberian Peninsula that have been very scarce for some time (see Valverde 1960). Of equal importance were the many species of waterbirds that use the area (see García *et al.* 2000; Garrido *et al.* 2004), many of which are nationally or internationally threatened species (Madroño *et al.* 2004; BirdLife International 2008).

In 1989, the area was declared Doñana Natural Park, and, in 1999, Doñana Natural Space (DNS from here on) came into existence – a legal entity that includes the Natural Park, National Park and its protection zones (Fig. 30.1). The area is also designated as a Biosphere Reserve (UNESCO), a Wetland of International Importance of the Ramsar Convention, an Important Bird Area (IBA), a World Heritage Site and a Site of Community Interest.

DNS occupies 108,087 ha, with natural, temporary marshes that flood each autumn–winter (fed by rainwater and streams) forming 47.5% of the area: these start reducing in depth and area through spring, and dry up in summer (see Valverde 1960). However 1,160 ha have been converted into industrial saltpans, and 2,971 ha into ponds for extensive aquaculture, located mainly at Veta la Palma. Neither of these is dependent on rainfall, both acquire water from the Guadalquivir River. In addition, in the northern zone, 1,398 ha of wetlands have been converted into rice fields.

M. Máñez (✉), L. García, F. Ibáñez, H. Garrido, J.L. Arroyo, J.L. Del Valle, A. Chico, A. Martínez, and R. Rodríguez
Equipo de Seguimiento de Procesos Naturales, Estación Biológica de Doñana-CSIC, Avda. Américo Vespucio s/n, Isla de la Cartuja, Sevilla, 41092, Spain
e-mail: mmanez@ebd.csic.es (M. Máñez)

J.M. Espinar
Espacio Natural de Doñana, Centro Administrativo El Acebuche, Matalascañas, 21760, Spain

C. Hurford et al. (eds.), *Conservation Monitoring in Freshwater Habitats: A Practical Guide and Case Studies*, DOI 10.1007/978-1-4020-9278-7_30,

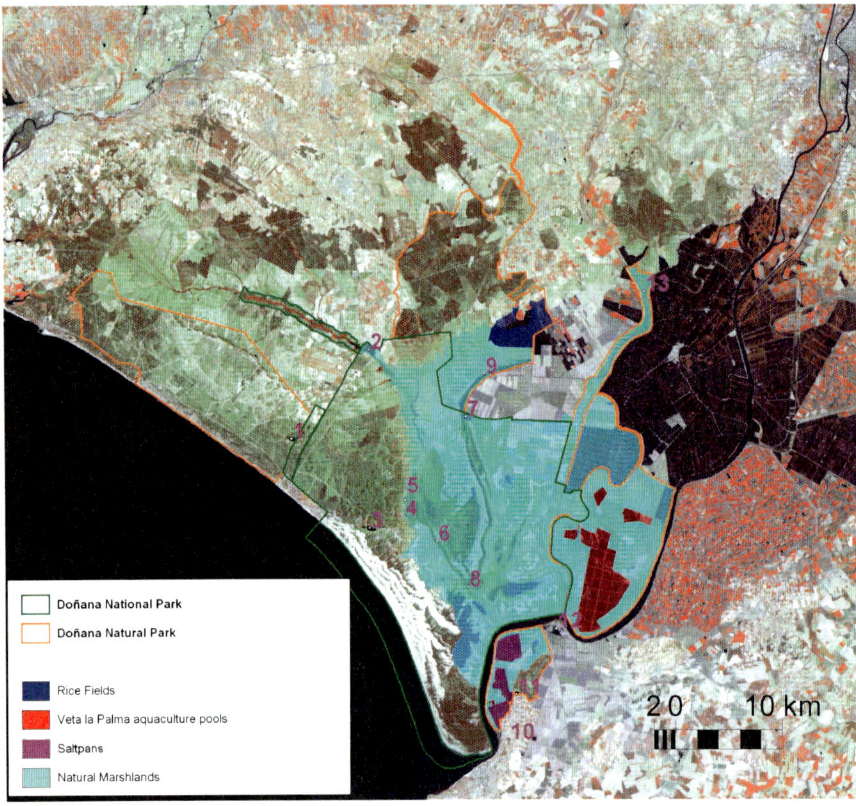

Fig. 30.1 Map of the Doñana Natural Space. The numbers indicate the approximate location of places cited in the text: 1 – El Acebuche lagoon complex; 2 – El Rocío Marshes; 3 – interdunal lagoons; 4 – Doñana Palace; 5 – Pajarera; 6 – Caño (marsh channel) of La Madre; 7 – Lucios (large depressions in the lower marshes) of FAO; 8 – Juncabalejo; 9 – Caño (marsh channel) of Guadiamar; 10 – Bonanza lagoons; 11 – Tarelo lagoon; 12-Codo de la Esparraguera; 13 – Ford of Don Simón

The remaining area of DNS represents aeolian deposits (Montes *et al.* 1998), the so-called "cotos", where Mediterranean scrub predominates, in a mosaic with extensive areas of cork oak *Quercus suber* and pine forest *Pinus pinea* and a large number of lagoons (Bravo and Montes 1993). While the majority of these lagoons are temporary, the interdunal lagoons are exceptional for being large and either permanent or semi-permanent.

Other lagoons of interest in DNS are the Tarelo lagoon and the lagoon complex at El Acebuche. The former is a permanent artificial lagoon, where water from the aquifer reaches the surface in an area where aggregates were mined during the 1980s. By contrast, the El Acebuche complex is a lagoon-marsh system that, after suffering a process of drainage in the 1950s, was restored during the 1970s and 1980s.

In this chapter, in addition to DNS, we cover two small wetlands located very close to the Cadiz sector of the Natural Park (Fig. 30.1): the Codo de la Esparraguera

(Trebujena), formed by marshland that has been converted into aquaculture ponds, and three lagoons at Bonanza (Sanlúcar de Barrameda), which have the same origin as the Tarelo lagoon.

The hydrological status of the natural wetlands depends basically on rainfall. The Palacio de Doñana meteorological station maintains a complete record of the total rainfall (in mm) during the crop or hydrometeorological year (from 1st September to 31st August of the following year). The meteorological station also records the rainfall registered during winter months (from 1st September to 31st March): most of the annual rainfall needed to fill the marshes occurs in this period. As illustrated in Fig. 30.2, this varies greatly, and determines the degree of wetland flooding and, by association, the availability of habitat for waterbirds. In order to evaluate the recorded rainfall, we have compared the values with the historical annual average (551.57 mm, n = 28) and historical average during the autumn–winter months (420.15 mm, n = 28).

This paper focuses on the seven species of waterbird that are classified as "Endangered": this category carries the maximum threat of extinction from a legal perspective in Spain (according to the National Catalogue of Threatened Species). These seven species are marbled teal *Marmaronetta angustirostris*, ferruginous duck *Aythya nyroca*, white-headed duck *Oxyura leucocephala*, bittern *Botaurus stellaris*, squacco heron *Ardeola ralloides*, black stork *Ciconia nigra* and red-knobbed coot *Fulica cristata*.

The three ducks species are also threatened at a global level, although in different categories: ferruginous duck is classified as "Near Threatened", marbled teal as "Vulnerable" and white-headed duck as "Endangered" (BirdLife International 2008).

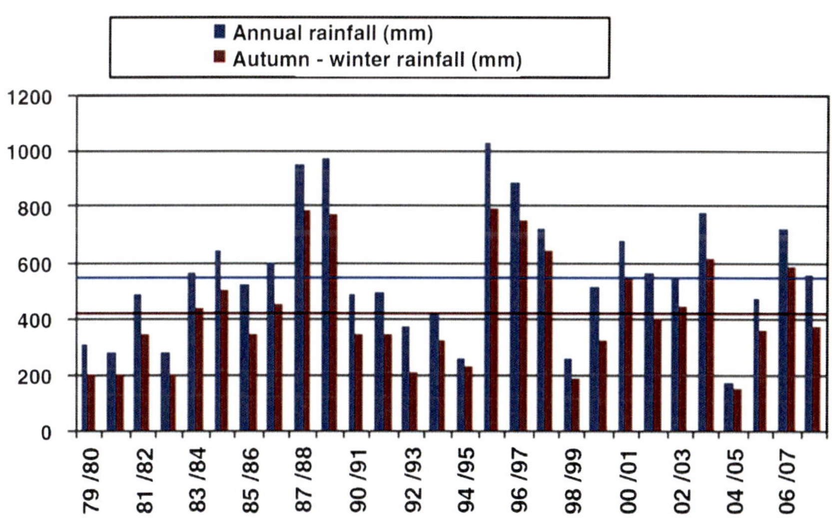

Fig. 30.2 Annual rainfall (*in blue*) and autumn–winter rainfall (*in red*) recorded at the Palacio de Doñana meteorological station from 1979/80 to 2007/08 (does not show the rainfall of 1988/89, as this was not measured in its entirety). This also shows the historical annual average (*blue line*) and historical average during the months of autumn–winter (*red line*)

The Methods

The main reason for monitoring these species is to determine the size of the population in DNS, especially during the breeding season, and to detect any trends that evolve over time. The method used is species-dependent, and is always based on a systematic survey of the entire natural wetland. This information is then supplemented with detailed surveillance of the nests, family groups (territorial males in the case of the bittern) and wintering birds. The marsh survey is carried out mainly on horseback and, in the case of the red-knobbed coot, using a "cajón" (a flat-bottomed boat propelled by poles). The census method involves a direct count of the bird flock settled in each of the wetlands: a method used for the majority of the waterbirds (Tellería 1986; Sutherland 1996). The count is performed from fixed locations (or by fixed route), depending on the size of the wetland and the existence of observation points.

The three duck species and the red-knobbed coot are also censused nationally in the middle of January, April, June, September and November. When pairs or nests of these species are detected, they are monitored throughout the breeding season.

The breeding population of the bittern is surveyed by locating 'booming' territorial males, systematically covering the entire wetland on horseback during the relevant period.

The breeding population of the squacco heron is censused by counting the nests or pairs in the breeding colonies from suitable locations.

Finally, we census the wintering population of black storks, taking account of data from the international census in January. This census covers all the wetlands on the right bank of the Guadalquivir river, both DNS and neighbouring areas that have been converted into rice fields or are used for the farming of other crops, and the Dehesa de Abajo Concerted Nature Reserve, as well as the main wetlands on the left bank.

The data are derived mainly from the archives of the Bird Section of the Natural Processes Monitoring Team (EBD-CSIC), which collaborates with the administrations managing DNS.

We have used TRIM (Trends & Indices for Monitoring Data; Pannekoek and Van Strien 1998) to analyse population trends: this is a statistical program designed for analysing data for wild populations. This program determines if the observed trend could be due to chance or not, and if not, determines the signal of the trend and the index of annual increase or decrease, taking the first year of sampling as the base year (100%).

Results by Species

Marbled Teal

Censuses show that the population trend from 2003/2004 to 2007/2008 has a similar pattern (low numbers in September, a peak in November, almost total absence in January, slight increase in April and larger increase in June), except 2006/2007 when very few ducks were seen. The Strait of Gibraltar is not a great barrier for the species, and flocks move between Morocco and Western Andalusia at any time of year, depending on changes in habitat availability (Green *et al.* 2004).

Marbled teal was the most abundant of the nesting ducks in these wetlands at the end of the nineteenth century. By the 1950s, however, only 100–200 pairs remained in years of average rainfall (Valverde 1960). During the 1970s, there was another marked decline, and by the end of that decade the breeding population was estimated at only 10–20 pairs.

Thereafter, following a drought in the early 1980s (Fig. 30.2), the breeding population recovered to over 100 pairs between 1984 and 1988, before falling again during a drought at the beginning of the 1990s, when the population reached its lowest level: in 1995 not a single pair was found in the Gualdalquivir Marshes (see Navarro and Robledano 1995; Green *et al.* 2004).

Between 1996 and 2008, estimates suggest that the maximum number of pairs in the area fluctuated between 10 and 83 (Fig. 30.3). The low numbers in 1996 can be explained by a severe drought that occurred over the previous 5 years, and the low numbers in 1999 and 2005 can be explained by extreme droughts in the previous winters (Fig. 30.2): the rainfall was not sufficient to flood the natural wetland and the Veta la Palma aquaculture pools and, during dry years, the salinity level is too high for breeding ducks (Green 2000).

Nevertheless, the low numbers since 2006 are worrying, since the predicted recovery after years of average and high rainfall (Fig. 30.2) has not materialised.

As far as broods are concerned, the estimates ranged from 2 to 31 (Fig. 30.3), with considerable differences between the numbers of pairs and the number of

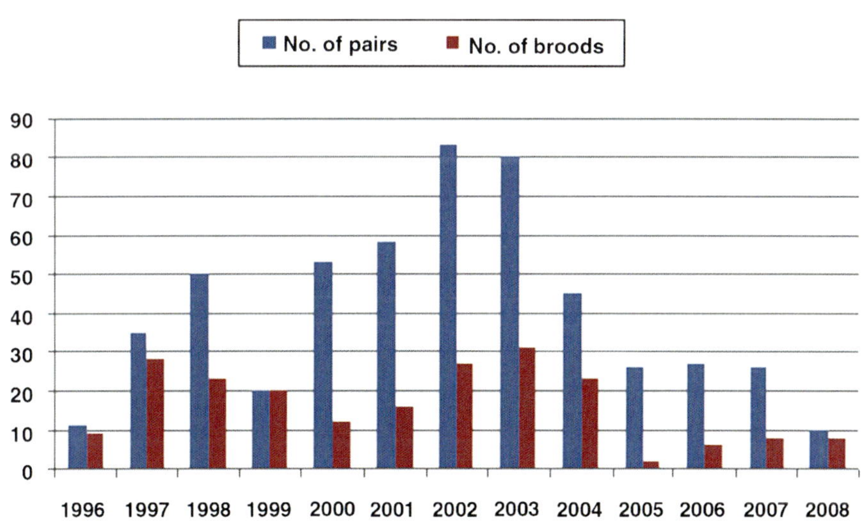

Fig. 30.3 Number of estimated pairs and number of estimated broods of marbled teal at Doñana Natural Space and other nearby wetlands over the period from 1996 to 2008

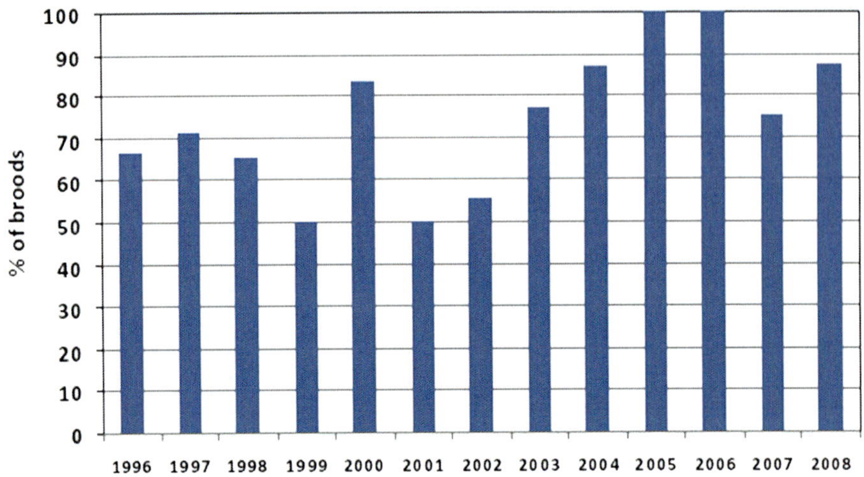

Fig. 30.4 Percentage of broods of marbled teal at Veta la Palma aquaculture poools to the total study area over the period from 1996 to 2008

broods seen. One of the main causes for this is the high predation rate, mainly by rats *Rattus* spp. in the natural marshes (Navarro and Robledano 1995; Green 1998), and by red fox *Vulpes vulpes* in Veta la Palma.

In keeping with the trend for pairs, the trend in the number of broods shows a slight decrease. Critically, however, between 50% and 100% of the broods have been found in Veta la Palma aquaculture pools every year (Fig. 30.4), with this percentage increasing significantly during the study period ($p = 0.0033$). This means that this area of man-made habitat, where broods are able to complete their development (marbled teal nests later than other duck species in the area, see Green 1998) is becoming crucial to the survival of the species in DNS.

The Spanish population of marbled teal is an important part of the Western Mediterranean population (Green *et al.* 2004), and these data show that DNS is, after the wetlands of the Southern Alicante region (Gómez *et al.* 2006), the second most important area for the species in the country.

Ferruginous Duck

Valverde (1960) reckoned that about 500 ferruginous duck pairs bred in the Guadalquivir Marshes at the beginning of the last century, compared to just a dozen at the end of the 1950s.

Between 1970 and 1992, breeding was confirmed in only three seasons, in 1984 (two broods), 1987 (one brood) and in 1989 (some fledglings) (Green 2004). During the 5-year period 1992–1996, more than 125 captive birds were released: first in the El Acebuche lagoon complex (DNS) and later in Portil lagoon, in the province of Huelva (Green 2004). Since then, breeding was confirmed at El Acebuche in 1993 (three pairs), 1994 (two broods) and 1996 (one brood).

Despite annual sightings of males and females at El Acebuche, the next confirmed breeding was in 1999, when a female with four ducklings was seen in one of the interdunal lagoons. Breeding was also confirmed in 2000 at El Acebuche, with the sighting of a female with five small ducklings in the middle of August, and an adult female with two juveniles (male and female) in October, presumably the survivors.

Although there have been breeding season sightings in 2001, 2002 and 2003, and winter sightings since then, there have been no subsequent breeding records at DNS. These data suggest that the releases undertaken in the 1990s have not aided the recovery of the population in Doñana and the Guadalquivir Marshes, the most important in all of Spain (Green 2004). This could be because the main cause of the decline (habitat degradation) still persists. Ferruginous ducks seem to require wetlands that are rich in emergent, floating vegetation and submerged macrophytes (Green 2004).

White-Headed Duck

The coordinated censuses from 2003/2004 to 2007/2008 have shown that this species is currently much more abundant in DNS and its surrounding areas in autumn and winter than during the breeding season. The main wintering locations are the fishponds of Veta la Palma and the Tarelo lagoon.

The first Spanish records for white-headed ducks were at Doñana and the Guadalquivir Marshes (see review in Torres Esquivias *et al.* 1986). According to Amat and Sánchez (1982), some 200 pairs bred in this area in the 1950s. However, the species stopped breeding in Doana and the Guadalquivir Marshes in the late 1960s (Ree 1973), and there were no further breeding records until 1985, when between three and 10 broods seem to survive (García *et al.* 1986; Clarita 1986). The following year at least one brood hatched in one of the interdunal lagoons (García *et al.* 1989). After this, there was no evidence of breeding in the natural marshes until 2007, though 11 ducklings were seen at Tarelo lagoon in June 1993, with successful breeding also confirmed there in 1994 and 1995 (Torres Esquivias and Moreno-Arroyo 2000).

The number of estimated broods in DNS and surrounding areas from 1996 to 2008 are shown in Fig. 30.5. Due to the low numbers, it is not possible to establish a statistically significant trend. Note, however, that this population represents only a small percentage of the total Spanish population of around 2,300 birds (Torres Esquivias 2004).

These data suggest that Tarelo lagoon has diminished in importance as a breeding site for the species over the last 3 years. There was no nesting there in 2002 and 2003, which coincides with a reduction in benthic chironomid biomass related to the hypereutrophy and episodes of anoxic conditions (Serrano *et al.* 2004),

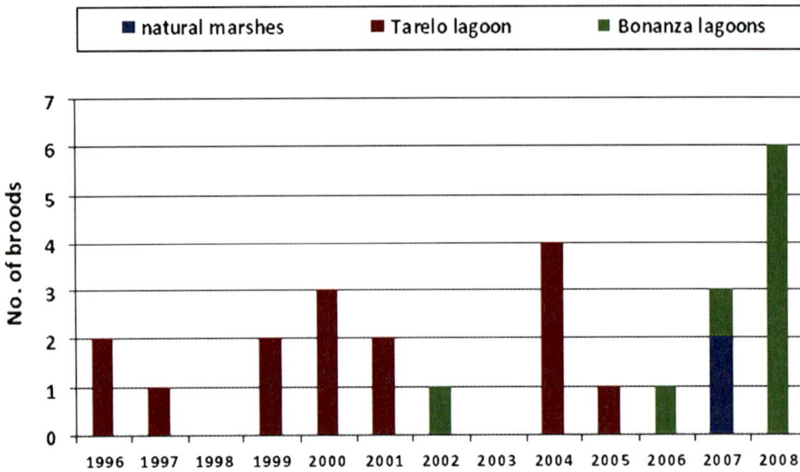

Fig. 30.5 Number of estimated broods of the white-headed duck in Doñana Natural Space (natural marsh and Tarelo lagoon) and Bonanza lagoons over the period from 1996 to 2008

which may also have happened subsequently. It seems unlikely that this species will return to nest in the natural marshes on a regular basis, given that its disappearance is due to the lack of deep and semi-permanent waters suitable for breeding (Green and Figuerola 2003).

For this reason, urgent measures are needed to protect the Bonanza lagoons, where the species has been breeding in recent years, as these lagoons are located in an area of intensive farming, where the risks of pollution and degradation due to hyper-eutrophication are high. Similarly, illegal hunting and disturbance by domestic animals in the lagoons pose a serious threat to the breeding population.

Bittern

According to Valverde (1960), the bittern was a scarce but well-distributed nesting species in the Marshes to the late 1950s. According to Urdiales (unpublished report), it certainly bred in what today represents the National Park until the beginning of the 1980s, when large areas occupied by the species and dominated by bulrush *Typha* sp. disappeared; meanwhile, in the marshes to the north of the present DNS, it seems that it bred until the middle of the 1980s, when all the northern "lucios" were drained.

In 1991, this last author detected eight territories in the Guadalquivir Marshes (five of them located in the "Eastern Branch" of the Guadalquivir River, where bitterns were heard every year). Thereafter, there were no records during a cycle of drought that lasted from 1992 until 1995 (Fig. 30.2).

Subsequently, despite abundant rainfall in several years, with the exception of one bird seen near El Rocío on 17th April 1999 (Arce and Sal 2001), the only observations were outside the breeding season. There was no evidence of breeding until 2002, when four territorial males were present in the natural marshes throughout the breeding season. As nests are very difficult to locate, there is no evidence that the species bred, or evidence to the contrary.

In the 2003 season, 12 territorial males were present, and breeding was confirmed in the natural marshes of the National Park and in Veta la Palma (Ibáñez *et al.* 2004). In 2004, however, 11 territorial males were recorded, but there was no evidence of fledged young and at least three of the four nests were lost, two reoccupied by purple heron *Ardea purpurea* and one raided by wild boar *Sus scrofa* (Espinar and Ibáñez 2004).

The drought of 2005 prevented breeding in Doñana, while in 2006 five territorial males were seen but there was no evidence of breeding.

In the 2007 season, nine territorial males were reported and five nests located, two of which were reoccupied by purple heron and one raided by wild boar. The others were successful in producing young, though only one was seen to fledge.

In 2008, eight territorial males were reported at the beginning of the breeding season, but these had stopped calling by the end of April, suggesting that they all left the area earlier than normal because of the low water level in the natural marshes. This year the total annual rainfall was close to the historical average (Fig. 30.2), but for the first time the rainiest month was April, in the middle of the spring.

Although not statistically significant, the trend in the number of territorial males since the species began to nest again in DNS, shows a slight increase. As a consequence, DNS is now the second most important breeding area for this species in Spain after the middle Ebro river valley, which supports between 12 and 17 territorial males (Bertolero and Soto-Largo 2004). However, it must be born in mind that a long drought could threaten the survival of this species in DNS.

One problem we have mentioned is the reoccupation of bittern nests by purple herons, which happens in areas of wetland fenced to exclude livestock and encourage regeneration of common reed *Phragmites australis*. This fencing has resulted in dense areas of reedbed where colonies of purple heron have proliferated, and where bitterns have also settled. The problem does not occur when bitterns nest in wetland dominated by alkali bulrush *Bolboschoenus maritimus*, though they can be easy prey for wild boar in these areas if the water level is not sufficiently high.

Squacco Heron

According to Valverde (1960), the 60–80 pairs of squacco heron that nested in the Guadalquivir Marshes during the 1950s in the famous "Pajarera" (a big colony of herons and spoonbills *Platalea leucorodia*) of Doñana, was the most important population in Western Europe. During the 1960s and 1970s, their numbers in this colony declined, until in 1980 only two pairs were seen (Aguilera and Sañudo 1986).

Meanwhile at La Rocina, near El Rocío, an estimated 100 to 200 pairs was present during the early 1970s (Ree 1973), but this colony disappeared in 1974 (Castroviejo 1993). Fortunately, since 1974, the species has taken refuge on the edge of the Guadalquivir River, first in La Isleta and later in Los Olivillos (La Puebla del Río, Sevilla). In 1974, some 50 pairs were present (Castroviejo 1993) and, since then, between 5 and 70 pairs have bred in the area every year.

Although one pair was reported in the "Pajarera" in 1990, the species really came back to breed in DNS in 1996 (Table 30.1), after the long period of drought of the early 1990s (Fig. 30.2). This time, the species settled in the Casas "lucio" of FAO, where it formed a mixed colony with purple herons and glossy ibis *Plegadis falcinellus*. This is an area protected from large herbivores and terrestrial predators, which can be supplied with water from a well, and where the species nests in both bulrush *Typha dominguensis* and African tamarisk *Tamarix africana*.

Since then, this colony has consolidated (Table 30.1) and, except in years of severe drought, such as 1999 and 2005 (Fig. 30.2), these three species have bred. Currently, this is one of the most important colonies for Ciconiiforms in the whole of Europe. Also in 1996, a pair bred successfully in a mixed colony in a pine forest by the El Acebuche lagoon (Máñez and Garrido 1997); but this was an isolated incident.

In 1998, the squacco heron returned to the "Pajarera" as a breeding species, but in 1999 it did not nest at all within DNS due to the drought. The following year, the number of pairs in DNS was low, no doubt affected by the drought of the previous year. However in 2001, the number of pairs in the FAO and the "Pajarera" increased substantially, and the species formed a third colony in DNS, in Juncabalejo, an area of common reed fenced to protect it from herbivores (Fig. 30.1). However, due to the low water table in this area, a wild boar raided eight nests and an Egyptian mongoose *Herpestes ichneumon* ate some of the chicks from the remaining nests.

In 2002, the number of pairs breeding in the three colonies increased within DNS, while in 2003 and 2004 this increase was only seen in the FAO. The severe drought of 2005 practically prevented this species from breeding in DNS, and only two pairs were reported in a new colony located on some tamarisks in Entremuros, close to Ford of Don Simón (Natural Park), although successful breeding could not be confirmed. In 2006, the breeding population in DNS recovered and reached 84 pairs, of

Table 30.1 Minimum number of estimated breeding pairs of squacco heron at the colonies of Doñana Natural Space over the period from 1996 to 2008. No squacco herons bred in 1999

Colony/year	96	97	98	00	01	02	03	04	05	06	07	08
Pajarera	0	0	5	5	22	29	26	17	0	1	31	0
Acebuche	1	0	0	0	0	0	0	0	0	0	0	0
FAO	130	100	45	15	70	160	331	389	0	80	100	100
Juncabalejo					25	67	3	24	0	0	40	0
Ford of Don Simón									2	2	0	2
Tarelo lagoon										1	0	6
Total	131	100	50	20	117	256	360	430	2	84	171	108

which 95% where in the FAO area. There was also confirmation, for the first time, of breeding (one fledgling) in a colony of small herons in Tarelo lagoon.

In 2007, the breeding population in DNS doubled by comparison with 2006, but this population increase was not sustained in 2008, when low water levels contributed to a lack of colonies in the inner National Park (Pajarera and Juncabalejo).

Despite the trend in number of pairs from 1996 to 2008 showing a slight decrease, the DNS population is the second most important in Spain, second only to that of the Ebro delta, in Tarragona (Pérez-Aranda *et al.* 2003). It seems that excluding large herbivores from areas of marshland vegetation may have contributed to this.

Black Stork

Valverde (1960) refers to concrete evidence of black storks (Fig. 30.6) nesting within the limits of the present DNS, and the species is included in the list of birds that nested in Santa Olalla, the main interdunal lagoon in 1774. There has been no other confirmed breeding in the area that today corresponds to DNS, although in 2007, there was a late attempt to nest in a pine forest in Doñana National Park, but the birds left before egg-laying.

The monitoring of this species, mainly summer resident in Spain, has detected the wintering in the study area since the early eighties, which has grown from a few individuals to over 300 in January 2007, although this number dropped by 60% over the subsequent two winters, returning to numbers similar to those in 2004 and 2005.

Fig. 30.6 A black stork at Doñana Natural Space. *Photo by J.A. Sencianes/EBD-CSIC*

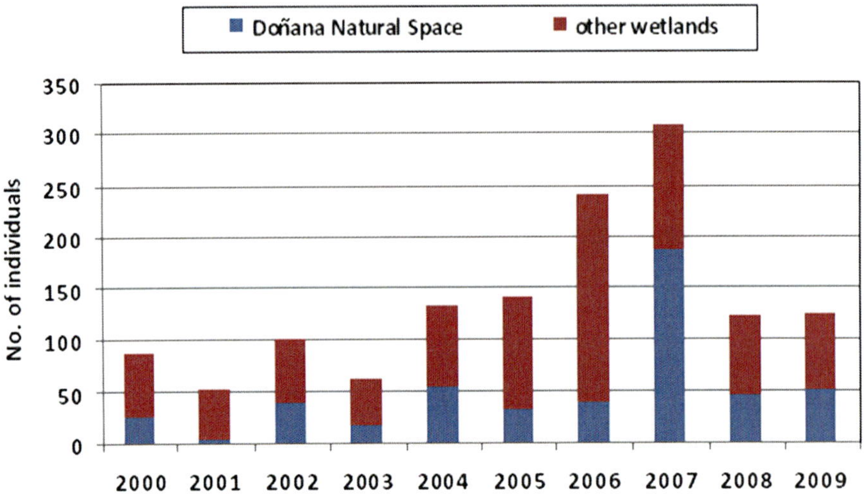

Fig. 30.7 Number of individuals of black stork in the censuses of January 2000 to 2009 at Doñana Natural Space and other wetlands of the Guadalquivir Marshes

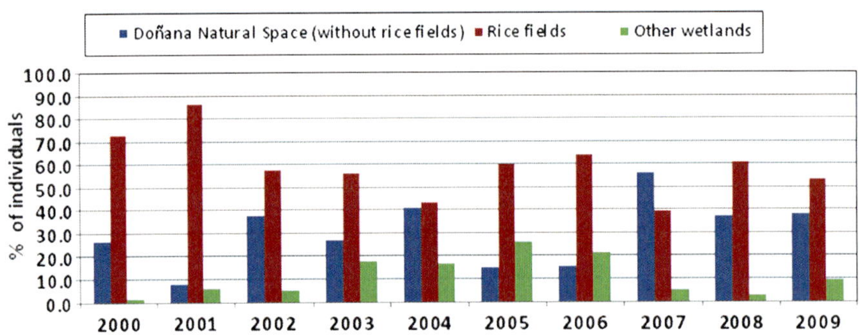

Fig. 30.8 Percentage of individuals of black stork observed from census in January 2000 to 2009 in Doñana Natural Space (excluding rice fields), rice fields and the other wetlands in the Guadalquivir Marshes

However, these numbers mean that the Gualdalquivir Marshes are the main wintering ground of the species in Spain.

Overall the trend in numbers of wintering black stork between the years 2000 and 2009 (Fig. 30.7) shows a slight increase; this corresponds to significant increment in DNS ($p < 0.05$), 18.6% annually, while not being statistically significant across the rest of the wetlands.

If we break down the census taking into account sightings within DNS (except the small rice field zone within it, which occupies 1.3% of the area), in all the rice fields and in the rest of wetlands censused (Fig. 30.8), the highest count every year (except one) is in the rice fields. This is of great importance from a conservation perspective, since at least 93% of the rice field area that is cultivated every year (ranging from a minimum of 22,000 ha to a maximum of 37,000 ha from 2000 and 2007) is not protected.

Red-Knobbed Coot

Valverde (1960) stated that, according to professional egg collectors, the red-knobbed coot (Fig. 30.9) was formerly common in the wetlands and nested in a 1:10 ratio with common coot *Fulica atra*. By the end of the 1950s, however, the species nested only in very small numbers, raising fears of rapid extinction in the area. Although still rare, the species continued nesting until the end of the 1980s, with a breeding population of between 10 and 20 pairs and a ratio of 1 red-knobbed coot to every 630–710 common coots (Máñez 1991).

In 1990, a year of abundant rainfall, it was estimated that there were only three pairs in the area, and during the 3-year period between 1996 and 1998, also with abundant rainfall, very few birds were recorded during the breeding season and only one nest (in 1996) was found.

However, at the end of the winter of 2000/2001, sightings increased, and at least 37 pairs were seen in DNS in spring 2001 (32 in the National Park and five in the marsh channel of Guadiamar), six of these broods were monitored. The reasons for this spectacular increase are not certain, but it seems logical to assume that it was influenced by habitat degradation at some of the best breeding locations for the species in Morocco (Green *et al.* 2002), and augmented by the release of birds from captive breeding programs in Andalusia and Valencia (Amat and Raya 2004).

Fig. 30.9 Adult red-knobbed coot in breeding plumage. *Photo by Héctor Garrido/EBD-CSIC*

In 2002, estimates suggested there were 50–54 pairs in DNS, of which 80% were in the National Park, and the rest in the marsh channel of Guadiamar. In 2003, 73–82 pairs were detected in DNS, of which 90–93% were in the National Park, 5.5–7% in the marsh channel of Guadiamar and the rest in Veta la Palma. Finally, in 2004, there were between 91–111 pairs in DNS, of which >95% were in the National Park and the rest in the marsh channel of Guadiamar. These surveys indicated that DNS was again the most important breeding area for the species in Spain, after the barren 1990s, when the species found refuge mainly in the Cádiz lagoons (Amat and Raya 2004). Moreover, within the Guadalquivir Marshes, the National Park is the most important area, especially the marsh area close to Juncabalejo, where the two main marsh channels meet.

Subsequently, in 2005, a very dry year, there was no sign of breeding, and the following year only one nest was detected in the whole of DNS, in the natural marshes of the National Park. As no chicks were seen in the area, which dried up soon after, we can assume that breeding was not successful.

However, the population partially recovered in the spring of 2007, where an estimated 64 pairs gathered mainly in the marsh area close to Juncabalejo, where they bred fairly successfully. Only four pairs were studied outside this area, three in the marsh channel of Guadiamar, where no chicks hatched, and one at El Acebuche lagoon, a new location for the species, where two chicks fledged.

The poor hydrological conditions during 2008 prevented the species from settling in the natural marshes of the National Park, nor were there sightings of pairs in the marsh channel of Guadiamar. Only two pairs were seen: one that raised a chick at El Acebuche lagoon (F. Robles, pers. comm.) and the other at El Rocío Marshes.

The trend in the number of pairs of red-knobbed coot since its return as a breeding species shows a slight decrease: this, however, is undoubtedly influenced by the years when numbers were very low and when the species was absent.

Conclusions

The long-term surveillance of the endangered breeding species suggests that the three duck species have declined in DNS by comparison with the populations in the 1950s. Ferruginous duck rarely breeds any more, and the others have found refuge in artificial habitats, marbled teal in the aquaculture pools of Veta la Palma and white-headed duck in man-made lagoons.

These population declines are related to changes in the natural marshes, with the disappearance of deep water and semi-permanent areas, suitable for the two diving species, and a shorter annual period of flooding, affecting all three species (Green and Figuerola 2003).

In the case of the marbled teal, predation by rats and disturbance (by humans and livestock) in the natural wetland led to the creation of a "Reserve Zone" located in the southern marsh in 2005. These problems are yet to be fully eliminated.

The remaining species are faring considerably better. The bittern, which was extinct as a breeding species in the area during the 1980s and 1990s, has recovered part of its original population in the early years of the twenty-first century, and the squacco heron population is larger than it was in the 1950s, and with a substantial increase in the number of colonies in DNS. A few consecutive years of adequate rainfall, combined with the various measures taken by the Management Body of DNS, such as (a) the reduction of livestock in the National Park, (b) the fencing of areas in the natural marshes (to exclude livestock and allow regeneration of marsh vegetation), and (c) improved exchange of water with the estuary (allowing fish to enter from the Guadalquivir River), appear to have aided this recovery.

The populations of red-knobbed coot are higher than those of the 1950s, although much lower than previously. As this species seems to require fresh water wetlands and clear water with a high diversity and coverage of submerged plants (Green *et al.* 2002), it can be construed that some areas of the natural marshes have been in optimum condition during recent years. Nevertheless, during the open season for hunting in Andalusia, hunting common coots should be banned at sites that provide shelter for red-knobbed coots, a measure that is already in place in some areas.

In the case of the black stork, wintering depends to a large extent on the existence of rice fields, a crop of great importance for many species of wintering waterbird in the Guadalquivir Marshes (Toral and Figuerola, unpublished report). Therefore its replacement by any other crop or use, with the exception of the restoration of the natural wetland, would be detrimental for this species and many others.

In summary, survey and monitoring of the bird populations is a basic and essential tool for both researchers and managers. The researchers can then study the relationships between the changes in the bird community and the temporary changes in the wetland conditions (Green and Figuerola 2003). They can also assess the adequacy of the management measures implemented by managers, and may recommend that these measures are reviewed within the proposed conservation objectives.

Acknowledegments This work would not have been possible without the help and collaboration of different administrations that have managed the current DNS during these years of research. Likewise, colleagues in Doñana Biological Station and in DNS, as well as a good number of volunteers have contributed information and collaborated in the work. Ricardo Diaz-Deldado produced Fig. 30.1. Thanks also to Olga Ceballos and Clive Hurford for correcting the drafts. Since 2003, many of the monitoring tasks are performed through an agreement with the Administration of DNS (Autonomous Body for National Parks of the National Ministry of Environment until 2006, and the Regional Environment Authority of Andalusia since 2007), and censuses of waterbirds have been funded through a contract with Environmental Management Company (EGMASA), of the Regional Environment Authority of Andalusia.

References

Aguilera E, Sañudo J (1986) Pasado y presente de las colonias de Ciconiformes en Doñana y Odiel. Bios 2:11–13 (in Spanish)

Amat JA, Raya C (2004) Focha moruna, *Fulica cristata*. In: Madroño A, González C, Atienza JC (eds) Libro Rojo de las Aves de España. Dirección General para la Biodiversidad-SEO/ BirdLife, Madrid, pp 199–202 (in Spanish)

Amat JA, Sánchez A (1982) Biología y ecología de la malvasía (*Oxyura leucocephala*) en Andalucía. Doñana Acta Vertebrata IX:251–320 (in Spanish)

Arce LM, Sal V (2001) Noticiario Ornitológico 2001 (1). Avetoro común *Botaurus stellaris*. Ardeola 48(1):137 (in Spanish)

Bertolero A, Soto-Largo E (2004) Avetoro común, *Botaurus stellaris*. In: Madroño A, González C, Atienza JC (eds) Libro Rojo de las Aves de España. Dirección General para la Biodiversidad-SEO/BirdLife, Madrid, pp 65–69 (in Spanish)

BirdLife International (2008) Species factsheet. Available at http://www.birdlife.org (accessed on 3 March 2009)

Bravo MA, Montes C (1993) Inventario de las formaciones palustres del manto eólico del Parque Nacional de Doñana (SW España). Actas del VI Congreso Español de Limnología Granada, pp 31–43 (in Spanish)

Castroviejo J (1993) Memoria del mapa del Parque Nacional de Doñana. Consejo Superior de Investigaciones Científicas y Agencia de Medio Ambiente de la Junta de Andalucía (in Spanish), Sevilla

Clarita P (1986) Noticiario Ornitológico. Malvasía (*Oxyura leucocephala*): Nidificación comprobada en las marismas del Guadalquivir. Ardeola 33(1–2):206 (in Spanish)

Espinar JM, Ibáñez F (2004) Noticiario Ornitológico. Avetoro común (*Botaurus stellaris*). Ardeola 51(2):543–544 (in Spanish)

García L, Calderón J, Castroviejo J (1986) Informe sobre la reproducción de las aves del Parque Nacional de Doñana en 1985. Estación Biológica de Doñana, Sevilla (in Spanish)

García L, Calderón J, Castroviejo J (1989) Las aves de Doñana y su entorno. Estación Biológica de Doñana (CSIC) – Cooperativa Marismas del Rocío (in Spanish)

García L, Ibáñez F, Garrido H, Arroyo JL, Máñez M, Calderón J (2000) Prontuario de las Aves de Doñana. Anuario Ornitológico de Doñana, nº 0, Diciembre 2000. Estación Biológica de Doñana y Ayuntamiento de Almonte, Almonte (Huelva) (in Spanish)

Garrido H, Arroyo JL, García L, Ibáñez F, Máñez M, Vázquez M (2004) Anuario Ornitológico de Doñana, nº 1 (Septiembre 1999 – agosto 2001). Estación Biológica de Doñana y Ayuntamiento de Almonte, Almonte (Huelva) (in Spanish)

Gómez JA, Dies JI, Vilalta M (2006) Las aves acuáticas de la Comunitat Valenciana: censos y evolución de las poblaciones (1984–2004). Conselleria de Territori i Habitatge, Generalitat Valenciana, València (in Spanish)

Green AJ (1998) Clutch size, brood size and brood emergence in the Marbled Teal *Marmaronetta angustirostris* in the Marismas del Guadalquivir, southwest Spain. Ibis 140:670–675

Green AJ (2000) The habitat requirements of the Marbled Teal (*Marmaronetta angustirostris*), Ménétr., a review. In: Comín FA, Herrera JA, Ramírez J (eds) Limnology and aquatic birds: Monitoring, modelling and management. Proceedings of the 2nd SIL international congress. Universidad Autónoma del Yucatán, Mérida, pp 131–140

Green AJ (2004) Porrón pardo, *Aythya nyroca*. In: Madroño A, González C, Atienza JC (eds) Libro Rojo de las Aves de España. Dirección General para la Biodiversidad-SEO/BirdLife, Madrid, pp 108–110 (in Spanish)

Green AJ, Figuerola J (2003) Aves acuáticas como bioindicadores en los humedales. In Paracuellos M (ed) Ecología, manejo y conservación de los humedales. Instituto de Estudios Almerienses, Almería, pp 47–60 (in Spanish)

Green AJ, El Hamzaoui M, El Agbani MA, Franchimont J (2002) The conservation status of Moroccan wetland with particular reference to waterbirds and to changes since 1978. Biol Conserv 104:71–82

Green AJ, Echevarría JL, Ferrández M (2004) Cerceta pardilla, *Marmaronetta angustirostris*. In Madroño A, González C, Atienza JC (eds) Libro Rojo de las Aves de España. Dirección General para la Biodiversidad-SEO/BirdLife, Madrid, pp 100–104 (in Spanish)

Ibáñez F, Espinar JM, Máñez M (2004) El retorno del avetoro a Doñana. Quercus 225:14–19 (in Spanish)

Madroño A, González C, Atienza JC (eds) (2004) Libro Rojo de las Aves de España. Dirección General para la Biodiversidad-SEO/BirdLife (in Spanish), Madrid

Máñez M (1991) Estado actual en el Parque Nacional de Doñana de las especies de aves incluidas en la "Lista Roja de los vertebrados de España" en las categorías de "En peligro" y

"Vulnerable". Jornadas de Zonas Húmedas Andaluzas. Fuente de Piedra, 20-22 Abril 1990, pp 41–49 (in Spanish)

Máñez M, Garrido H (1997) Noticiario Ornitológico. Garcilla cangrejera *Ardeola ralloides*. Ardeola 43(2):240–241 (in Spanish)

Montes C, Borja F, Bravo MA, Moreira JM (1998) Reconocimiento Biofísico de Espacios Naturales Protegidos. Doñana: Una Aproximación Ecosistémica. Consejería de Medio Ambiente. Junta de Andalucía, Sevilla (in Spanish)

Navarro JD, Robledano F (1995) *La cerceta pardilla* (Marmaronetta angustirostris) en España. ICONA Colección Técnica (in Spanish)

Pannekoek J, Van Strien A (1998) TRIM 3 manual (trends & indices for monitoring data). JM Voorburg, Statistcs Netherlands

Pérez-Aranda D, Ibáñez F, García L, Jiménez M, Garrido H, Máñez M (2003) Garcilla Cangrejera, *Ardeola ralloides*. In: Martí R, Del Moral JC (eds) Atlas de las Aves Reproductoras de España. Dirección General de Conservación de la Naturaleza – Sociedad Española de Ornitología, Madrid, pp 110–111 (in Spanish)

Ree V (1973) Dagens avifaunistiske situasjon i Las Marismas i Sør-Spania. Sterna 12(4):225–268 (in Norwegian)

Serrano L, Reina M, Arechederra A, Casco MA, Toja J (2004) Limnological description of the Tarelo lagoon (SW Spain). Limnetica 23(1–2):1–10

Sutherland WJ (1996) Ecological census techniques: A handbook. Cambridge University Press, Cambridge

Tellería JL (1986) Manual para el censo de los Vertebrados Terrestres. Editorial Raíces, Madrid (in Spanish)

Torres Esquivias JA (2004) Malvasía cabeciblanca, *Oxyura leucocephala*. In Madroño A, González C, Atienza JC (eds) Libro Rojo de las Aves de España. Dirección General para la Biodiversidad-SEO/BirdLife, Madrid, pp 111–113 (in Spanish)

Torres Esquivias JA, Moreno-Arroyo B (2000) La recuperación de la malvasía cabeciblanca (*Oxyura leucocephala*) en España durante el último decenio del siglo XX. Oxyura 10(1):5–51 (in Spanish)

Torres Esquivias JA, Arenas R, Ayala JM (1986) Evolución histórica de la población española de malvasía (*Oxyura leucocephala*). Oxyura 3(1):5–17 (in Spanish)

Valverde JA (1960) Vertebrados de las Marismas del Guadalquivir (introducción a su estudio ecológico). Archivos del Instituto de Aclimatación IX:1–168 (in Spanish)

Chapter 31
Monitoring Marsh Dynamics Through Remote Sensing

Ricardo Díaz-Delgado, David Aragonés, Iban Ameztoy, and Javier Bustamante

The Value of Remote Sensing Images

The first synoptic images from space were provided by TIROS spacecraft with a video camera onboard in 1960 (Susskind *et al.* 1984), these images revealed a huge spheroid fully covered by oceans and clouds. There was still a long way to travel from simply picturing Earth to the current technological achievements of remote sensing science. However, the main scientific purposes of using remote sensing images have been maintained since those days. The very earliest uses of remote sensing are now much more focused and widespread, and often driven by military objectives. Indeed, remote sensing owes its rapid development almost exclusively to military initiatives (Barret and Curtis 1999). From a military perspective, targets should be differentiated as accurately as possible, and this requirement has resulted in the increased spatial resolution of remote sensors. Spectral resolution – the number of sensor bands imaging at different wavelengths – has been cultivated to help discriminate between false positives and true targets.

Space technology has also played an essential role in improving the potential of remote sensing – mostly by enhancing satellite platform designs, accurate orbiting and the launch of more satellites.

However, it was at the beginning of the 1970s when a new concept for remote sensing was introduced and that the new technology drifted towards a more applied discipline: Earth Observation Systems (EOS). While hundreds of satellites were launched for telecommunication, military uses or spatial prospecting, a few were also designed simply to monitor the Earth surface under the EOS concept. This new

R. Díaz-Delgado (✉)
Head of Landscape & Remote Sensing Monitoring, Doñana Biological Station-CSIC, c/Américo Vespucio s/n. Isla de la Cartuja, Sevilla, 41092, Spain
e-mail: rdiaz@ebd.csic.es

D. Aragonés, I. Ameztoy and J. Bustamante
Remote Sensing & GIS Laboratory (LAST), Doñana Biological Station-CSIC, c/Américo Vespucio s/n. Isla de la Cartuja, Sevilla, 41092, Spain

C. Hurford et al. (eds.), *Conservation Monitoring in Freshwater Habitats: A Practical Guide and Case Studies*, DOI 10.1007/978-1-4020-9278-7_31,

generation of Earth sentinels aimed to systematically capture images from all over the world. The first two missions with this aim were the NOAA-AVHRR (Advanced Very High Resolution Radiometer) mission and the Landsat mission (originally named ERTS from Earth Resources Technology Satellite). The former was basically designed for oceanography and meteorology and launched in 1979. The latter, with the MSS (Multi-Scanner Sensor) onboard, was launched in 1972 and designed primarily to image Earth land cover. Both missions are ongoing and carry completely new sensors (Table 31.1). The main differences between the two are the spatial and temporal resolutions. NOAA-AVHRR was able to acquire 2,580 km swath of 1.1 × 1.1 km^2 pixel size images while Landsat captured 185 × 185 km^2 scenes of 59 × 79 m^2 pixel size images. NOAA-AVHRR missions have increased our knowledge of weather patterns, sea surface temperature, global vegetation dynamics, wildfire monitoring, snow cover, etc., whereas Landsat missions have contributed more to understanding land cover dynamics and change, such as deforestation, flood mapping, fire scar and fire risk mapping.

Since 2000, hundreds of Earth Observation Systems have been launched with significantly improved spatial resolutions of up to 60 cm (Quickbird panchromatic), improved spectral resolution (prototype hyperspectral missions are already providing 200 bands images such as Hyperion), improved radiometric resolution (up to 1,024 bits such as MODIS Terra/Aqua) and improved temporal resolution (geostationary satellites such as MSG-2 provide images every 15 min). These technological achievements have successfully contributed to the development of remote sensing, and forestry, geology, agriculture, urban planning, biology, ecology, climatology, geography, archaeology and hydrology, are just some of the disciplines that have benefited from the use of remote sensing images.

However, there is also an added value attached to the use of remote sensing images. Some sensors have been capturing images of the same locations for 36 years, giving end-users a powerful multi-temporal time series of images that can be acquired retrospectively to study any natural process and its dynamics, as well as historical

Table 31.1 Comparison of the spectral, temporal and spatial resolutions of Landsat MSS, TM and NOAA-AVHRR imagery. Modified from Ozesmi and Bauer (2002)

Resolution	Landsat MSS	Landsat TM	AVHRR
Band 1		0.45–0.52	0.58–0.68
Band 2		0.52–0.60	0.72–1.10
Band 3		0.63–0.69	3.55–3.93
Band 4	0.5–0.6	0.76–0.90	0.77–0.86
Band 5	0.6–0.7	1.55–1.75	11.5–12.5
Band 6	0.7–0.8	10.4–12.5	
Band 7	0.8–1.1	2.08–2.35	
Radiometric (bits)	64	256	1,024
Temporal	18 days	16 days	12 h
Spatial	59 × 69 m^2	30 m, 120 m TM	1.1 km at nadir
First launched	1972	1982	1979

changes, by means of digital image analysis. Hopefully, many of the historical sensors will remain active so that we can expand these time series into the future.

Remote Sensing Limitations

Although many studies provide evidence to support the use of remote sensing images for mapping land use and land cover, as well as the different structural and biophysical parameters characterising them, we are also aware of many limitations. These restrictions increase when our aim is to analyse a time series of images, for example:

1. High resolution satellite images are expensive
2. There are intellectual property restrictions
3. There is limited availability and usability of optical images as a result of:
   - Cloud cover
   - Sensor or acquisition failure
   - Acquisition planning and scheduling
   - Mission continuity
4. The need for geometric and radiometric consistency among scenes to allow consistent removal of atmospheric effects
5. The restricted spectral range of nominal bands
6. Insufficient revisiting time
7. The need for ground-truth data (data collected in the field)
8. Extensive data processing
9. Computer limitations
10. Subscription rates for future image acquisitions

The following example illustrates how these limitations can impact on a project. In 2002, the Remote Sensing & GIS Laboratory (LAST-EBD) from Doñana Biological Station had the opportunity to acquire a long time-series of Landsat images (MSS, TM and ETM+) for the Doñana region (Landsat scene 202/34). We ordered all the available scenes for this path and row through the European distributor Eurimage. At that time, 239 scenes were available, covering the period from 1975 to 2001. The MSS scenes cost 200 € each, while TM and ETM+ were on special offer at 700 € each, giving a total cost of 133,000 €. From the available dates we requested, about 30 scenes were under cloud cover and another 56 were rejected due to different acquisition failures. We then entered a contract for a periodical research subscription for future scene acquisitions, priced at 500 € per image. To date, LAST-EBD has been under subscription for 5 years, but in 2004 the subscription rules changed. Initially, there was a chance to replace up to two cloud-covered images with new requests. Nowadays, there is no option to reject fully cloud-covered scenes despite the images being totally useless.

Before processing the time-series, we set up a protocol that included metadata retrieval, geometric correction, co-registration, radiometric correction and normalisation and backup for every scene. This process can be time consuming (up to 4 h per image) but recently we have reduced this by up to 1 h per image.

One of the first applications of the Landsat time series was to map flooding levels in the Doñana marshland and reconstruct the historic flooding regime, looking at both temporal and spatial patterns. After several trials to identify true wetland, cross-validated with ground-truth data, TM band 5 (1.55–1.75 m) was selected as the best variable to mask water bodies by simple thresholding. We then computed flood masks, i.e. binary images of flooded pixels (value 1) and non-flooded pixels (value 0) for every single date and calculated the hydroperiod (duration of flooding per hydrological cycle). However, the unavailability of certain scenes prevented us from estimating this valuable variable for every cycle. The secondary goals of the research project were to map aquatic vegetation tightly linked to the hydroperiod and then to analyse plant dynamics and changing trends. Different approaches have been trialled to this end, but with no satisfactory results to date, mostly due to the lack of thin spectral bands in the short wave infrared region that might help to discriminate the critical plant species.

In 1999, Landsat 7 was launched with the introduction of the ETM+ sensor, an Enhanced Thematic Mapper that added a new panchromatic band of 15 × 15 m^2 pixel size and increased the thermal band spatial resolution up to 60 m. Due to its longevity, the Landsat mission has suffered many major setbacks in its time, including a failure to reach orbit (Landsat 6 in 1993) and unexpected sensor malfunction, such as the Scan Line Corrector failure of ETM+ in 2003 that yields a zigzag pattern across the satellite ground track. However, astonishingly, Landsat 5, which was launched in 1984, is still orbiting and has been reprogrammed to capture new images up to the present time. Most recently, in April 2008, USGS (US Geological Survey) who operate the Landsat mission, announced plans to provide all archived Landsat scenes at no charge to users. Today, the historical dataset for any place in the world can be requested and downloaded from Glovis (http://glovis.usgs.gov) or Earth Explorer (http://earthexplorer.usgs.gov).

As such, many of the cited constraints associated with using remote sensing as the main tool for long-term monitoring have been resolved (within reason), increasing its suitability for monitoring large or inaccessible wetlands.

Remote Sensing for Wetlands

The first remote sensing applications targeting wetlands dealt with basic delineation of wetland boundaries. In addition to eventual flood and floodplain mapping, wetlands have been mapped and inventoried with remote sensing images since the 1970s. Remote sensing appears to be the best tool for mapping large and relatively inaccessible areas (Engman and Gurney 1991). The rationale behind the remote sensing approach to delimiting water bodies relies on the fact that water has a relatively low reflectance, especially in the near-infrared region, from 0.7 to 3.0 m (Gardiner and Díaz-Delgado 2007). Therefore, the accuracy of detection is based on the reliable identification of water and the spatial resolution of the sensor. Interestingly, cloud shadows, dark soils, closed canopy forests and urban areas may show similar reflectance values. Image interpretation was used as mapping tool long before the development of digital image analysis techniques. The human eye may easily integrate information from colour, texture, intensity and contextual characteristics allowing quick delineation of water bodies despite being shallow or deep, turbid or pristine, plant covered or bare water. However, this approach is always subject to individual interpretation and can be very time-consuming. Digital image analysis offers, through the application of validated algorithms and classification rules, an objective way to proceed with a multi-temporal series of images.

Long-Term Monitoring of Doñana Marshland Through Remote Sensing

Doñana marshes occupy almost 500 km^2 at the mouth of the Guadalquivir River (SW Spain). Protected as a National Park since 1968, it is the largest wetland in Europe and vast numbers of waterfowl breed or use the marshes as a stop-over site during migration. Natural inundation takes place between October and March, mostly fed by rain in the drainage watershed. Since an artificial levée was built in 1984, tidal influence is no longer a significant factor. The levée retains water inside the marsh for as long a time as possible, operating as a dump and preventing saline water from entering. In 1998, toxic runoff from the Aználcollar mine into the Guadiamar River made it necessary to dam the affected river to avoid a pollution of the Doñana marshlands, which caused an artificial increase of the marshland hydroperiod. As such, the historical inundation regime has changed dramatically as a consequence of management decisions. Today, Doñana encompasses an ambitious restoration project called Doñana 2005, which aims to restore the natural inundation process by applying widely accepted decisions based on data gathered over the past 10 years on a point station basis.

In this context, more than ever, it is necessary to know the historical inundation patterns, either spatial or temporal, and their relationships with natural variability and human modifications. For many years, details of the inundation process, such as inter-annual and seasonal variation, both spatial and temporal, as well as the influence of human transformations have been required by decision-makers in order to apply a scientific-based or adaptive management regime to Doñana marshes. To date, the hydrological management has been conducted on an "event-reaction" basis, which has led to temporary solutions eventually becoming a part of new problems. Hence, to date, the long-term monitoring protocols for Doñana wetlands include hydroperiod mapping, i.e. the duration of flooding within the hydrological cycle; spatially explicit flood mapping, i.e. to reconstruct the flooding regime; water turbidity; water depth and water temperature. Remote sensing is the main tool used for monitoring these protocols.

Flood Mapping, Hydroperiod and Flooding Regime

As stated previously, water bodies have low reflectivity, especially in Near Infrared and Mid Infrared bands (bands 4, 5 and 7 of TM and ETM+), and several procedures have been developed to identify flooded areas based on the low reflectivity of water in these spectral regions. Some indices have been proposed to automatically determine the inundation level in Landsat scenes. Ángel-Martínez (1994) suggests the CEDEX index to discriminate continental waters:

CEDEX = (TM4/TM3) – (TM4/TM5)

Where TM4 accounts for either TM or ETM band 4. According to Castaño *et al.* (1999) CEDEX values below 0.4 are inundated areas. Domínguez Gómez (2002) suggests the Normalized Difference Water Index (McFeeters 1996) as being useful to discriminate oligotrophic waters:

NDWI = (TM2 – TM4) / (TM1 + TM4)

Domínguez Gómez (2002) argues that the NDWI fails when applied to images collected by the airborne ATM-Daedalus sensor and suggests that it is better to apply simple band thresholding (density slicing) on the B4 histogram. Kyu-Shun *et al.* (2001), working in wetlands with different turbidity levels, show that TM5 is less sensitive to sediment charged waters and therefore the best to delineate the borders between water and soil in turbid waters.

Finally, several authors have argued that the third component of Tasseled Cap transformation for TM (Crist and Cicone 1984; Baker *et al.* 2006), known as 'Wetness', is tightly related to soil wetness. We also compute this index with the coefficients for Landsat 5 and 7 for reflectivity values in order to evaluate the capability for discriminating inundation levels.

Doñana marshlands are highly variable. They may appear as dry soil in summer, as bare water pools during the flooding season, or as pools densely covered by emergent and floating plants; in both cases with patchy turbidity patterns due to differences in suspended sediment concentrations (Fig. 31.1). In order to capture this variability, we carried out seven ground-truth sampling campaigns simultaneously with the acquisition of six Landsat scenes (4 TM and 2 ETM+).

Several transects were located across heterogeneous inundated areas and every 60 m we recorded information on water turbidity (measured in NTU with a nephelometric turbidimeter), depth, bare ground cover, plant cover, open water cover, dominant and most abundant plant species and inundation level.

Among several bands and radiometric indices, band TM5 was selected as the best flooding level discriminator for all of the cases. Single image slicing using a regression tree technique enabled us to map four flooding levels (dry soil, wet, damp and inundated). Regression trees identify threshold values to assign spectral classes to thematic categories. Subsequent grouping of the dry soil and wet pixels, and damp soil and inundated pixels, allowed us to automatically produce flooding inundation masks for every available Landsat scene covering the period 1984–2007 (Fig. 31.2). A similar process was applied retrospectively to band 7 of the Landsat MSS time series of images, which covered the period 1975–1984.

Fig. 31.1 Pictures showing the variability of Doñana marshlands

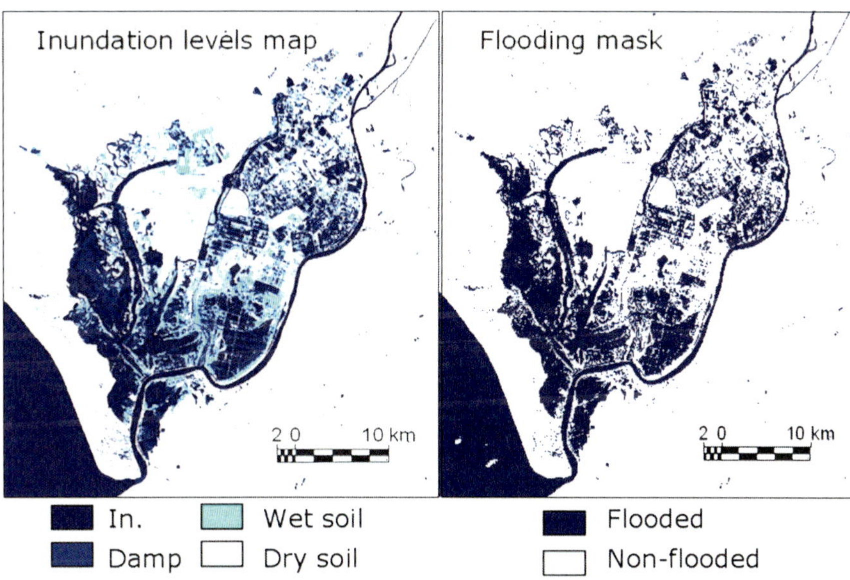

Fig. 31.2 An example of the inundation levels map and flooding mask for the same date image, 7th January 2002. Note that the dry and wet soils are classified as non-flooded areas in the flooding mask

We estimated the hydroperiod for the whole time series by applying two different approaches to the masks. The first method provides a synthetic image in which every pixel shows the number of days it remained inundated over a complete hydrological cycle (from 1st September to 31st August). This method assumes that flooding has been maintained between two consecutive scenes if flooding has been detected in both. This first approach consisted of assigning the cycle day number (number corresponding to image date being 1st September valued as 1 and 31st August valued 365) to flooded pixels of every inundation mask.

Thus, final images give information on the number of days a pixel remained inundated per cycle, named annual hydroperiod (Díaz-Delgado *et al.* 2006a). We also generated decadal hydroperiod composites by adding every available date per month and subsequently averaging the results per decade. This latter method was designed to allow inter-decadal comparison after important transformations in the marshland, while the former aims to detect quantitative local changes in hydroperiod trends when subjected to natural variability and human transformations (Fig. 31.3). Both automatic mapping procedures are also systematically applied to the new acquired scenes by Landsat TM and ETM+.

Water Turbidity and Depth

Turbidity is a measure of how much of the light travelling through water is scattered by suspended particles. The scattering of light increases with increasing suspended

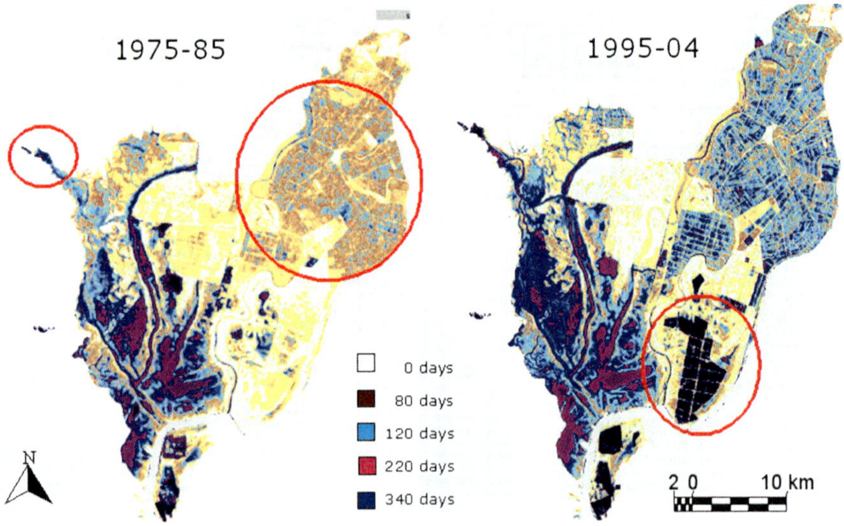

Fig. 31.3 A comparison of the decadal hydroperiod in the 1975–1985 and 1995–2004 decades. The legend shows the average number of days per decade inundated (from 0 to 365). *Red circles* highlight large differences in the hydroperiod between decades, such as the Veta la Palma fisheries, rice fields and main water income areas

solid and plankton content, i.e. turbidity is a tight surrogate of SSC (Suspended Sediments Concentration) and chlorophyll content.

In order to map the turbidity and depth of Doñana marshlands we fit, in a step-by-step mode, a generalized additive model (GAM) to ground-truth data by using the best Landsat sensor bands and indices as predictors (Bustamante *et al.* 2009). The best GAM model selected with the step-by-step procedure included bands TM3, TM5 and the ratio B1/B4. The model indicated a positive curvilinear relationship with TM3, a negative curvilinear relationship with TM5 and a negative linear relationship with the ratio B1/B4. The model explained 40% of the variance in turbidity (Fig. 31.4a).

Water turbidity is best estimated in situations where bottom soil reflectance and aquatic vegetation do not interfere, but it is still possible to build a predictive model for situations like those of the Doñana marshes – with shallow waters and abundant aquatic vegetation

Unlike for turbidity, the best predictors bands for the depth model were TM1, TM5, the ratio TM2/TM4, and bottom soil reflectance in band TM4 (from an image in September, when the marsh is completely dry) and the ratio between the B4 reflectance and the B4 reflectance for the same pixel in September. As expected, we found that band TM1 was the most informative wavelength as it is inversely related to depth. The model explained 75.42% of the variance. Once applied to the images we produced more than 240 maps of the water depth in Doñana marshlands showing the historical changes in its bathymetric characteristics (Fig. 31.4b).

On the basis of these results and newly available image acquisitions, we are currently monitoring water turbidity and depth by producing quantitative maps and identifying the main trends for both parameters (Díaz-Delgado *et al.* 2006b).

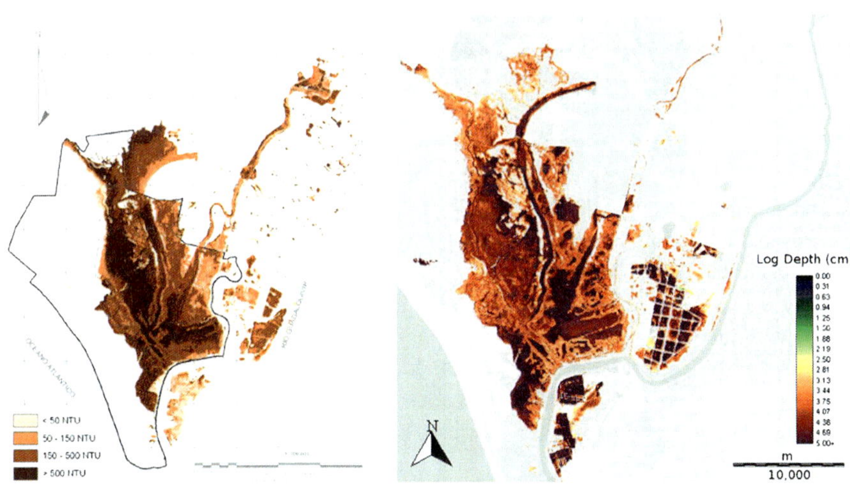

Fig. 31.4 (**a**) A reclassified water turbidity map of the Doñana marshes for the Landsat TM image of 14th January 1990. (**b**) The water depth map of the flooded area in the Doñana marshes in logarithmic scale for the Landsat TM image of 25th March 2004

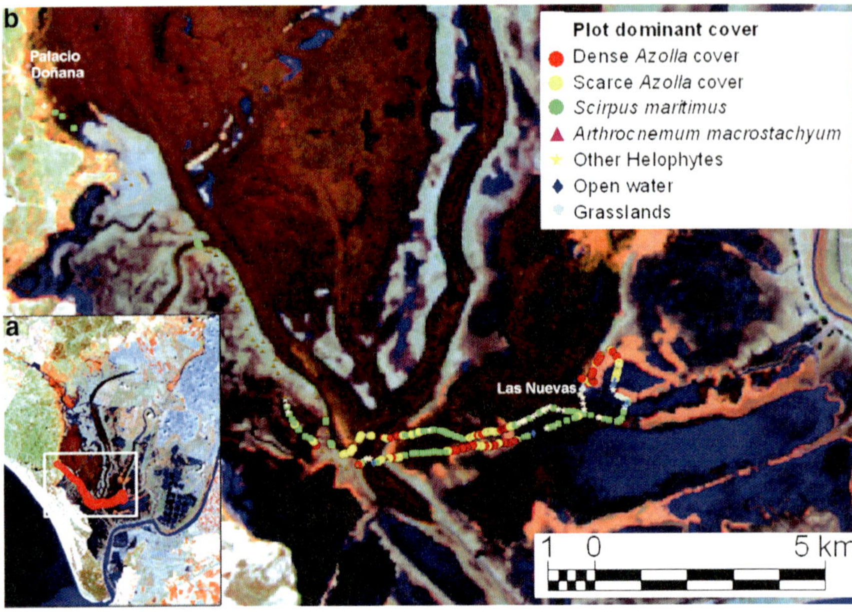

Fig. 31.5 (**a**) Landsat TM image of Doñana acquired on 5th May 2007 (false colour composite RGB bands 4-5-3) showing simultaneous to image acquisition ground-truth transect points (**b**) enlargement of sampled area showing the dominant species at each point. Densely covered *Azolla* points are clearly identified by *bright red circles*

Monitoring the Spread of an Aquatic Alien Species

The aquatic fern *Azolla filiculoides* is a small floating plant originally from America and largely naturalized in Northern Europe. Its physiology and growth pattern limits access to light for the submerged vegetation.

Azolla tends to deplete phosphorous, generates anoxic bottoms under the dense carpets it forms, and prevents the development of other submerged or floating vegetation species. This pteridophyte was first detected in Doñana marshlands in 2000, and has been found recurrently during the flooding season, invading larger and larger areas. In Doñana it is invading open marshes (open water bodies), marshland covered by helophytes and macrophytes, "*lucios*" (shallow lagoons) and "*caños*" (old tidal water courses). As Doñana marshlands occupy up to 25 000 ha, that become virtually inaccessible when flooded, it is a difficult task to map *Azolla* distribution through conventional methods. This fern grows and spreads at the beginning of the flooding season (January) and finally dies off during the drying out period (from May to June).

Usually, the invaded wetlands appear fully covered and this modifies their reflective characteristics. The alien fern shows a characteristic spectral signature that is very different from the rest of aquatic plants, and this makes it possible to map it even amid other dominant species such as *Scirpus litoralis*, *Scirpus maritimus* or *Arthrocnemum macrostachyum*. Since 2003, the LAST-EBD has been systematically collecting ground-truth data in Doñana marshlands, recording the presence and cover (%) of *Azolla* across transects simultaneously to Landsat TM and ETM+ acquisitions. Areas densely covered by *Azolla* appear in the images as highly reflective in the Near Infrared region (700–1,100 nanometers): this suggests high photosynthetic activity, much higher than that shown by native vegetation in flooded areas (Fig. 31.5).

Discussion

Traditional monitoring at the plot scale is strongly complemented by monitoring using remote sensing, as the latter provides information integrated at the landscape scale that can easily be combined with ground-truth data. While probes and sensors may be reporting local changes for a parameter, remote sensing images give the opportunity to extrapolate the information and quantitatively map the targeted variable and its gradients in a spatially explicit way. Underlying processes mapped at the landscape scale may reveal natural trends, induced changes, sudden threats or blooms, thus providing a valuable synoptic tool for adaptive wetlands management (Gardiner and Díaz-Delgado 2007). Furthermore, long time-series of images enhance long-term monitoring by helping to reconstruct historical processes that would otherwise remain uncertain. Retrospective mapping of synoptic parameters allows us to assess land cover changes in comparison to historical trends (anomalies from the average). Restoration projects may also benefit by using historical remote sensing images and the resulting maps as a baseline reference.

Landscape monitoring of Doñana marshlands is helping to integrate plot data on species richness, abundance, presence, water quality or breeding success with inundation level, annual hydroperiod or water turbidity. Other remote sensing monitoring protocols to be soon implemented will focus on: water temperature, evapotranspiration, net primary production, plant cover, and helophyte and macrophyte distribution mapping.

However, the techniques that we have implemented to monitor the processes mentioned above might not be easily extrapolated to other wetlands.

Specific characteristics of the Doñana marshlands have led us to apply specific and empirical remote sensing methods not necessarily valid for other wetlands. However, while there is likely to be a site-specific element to the development of protocols for other sites, many of the general approaches we have developed could be readily adapted elsewhere.

Acknowledgements This study was funded by the Doñana National Park administration (Spanish Ministry of Environment) through the research project "Reconstruction of bird population dynamics during the last three decades", by the Spanish Ministry of Science and Education through the research project HYDRA (# CGL2006-02247/BOS) and by help from the National Environmental Remote Sensing Network (# CGL2007-28828-E/BOS). Doñana National Park and Natural Park provided permits for field work in protected areas with restricted access. A. Travaini, H. Le Franc, D. Paz, A. Polvorinos, C. Rodríguez, and I. Román helped with field work. J.C. Gilabert, J.L. Pecharromán, L. M. Campoy and P.L. Porta helped with image processing. The authors want also to thank Clive Hurford for reviewing the manuscript and for his suggestions to improve the readability of the chapter.

References

Ángel-Martínez MC (1994) Aplicación de la teledetección en la localización de superficies de agua. Unpublished CEDEX Technical Report, 120 pp

Baker C, Lawrence R, Montagne C, Patten D (2006) Mapping wetlands and riparian areas using landsat ETM+ imagery and decision-tree-based models. Wetlands 26:465–474

Barret EC, Curtis LF (1999) Introduction to environmental remote sensing, 3rd edn. London, Routledge 457 pp

Bustamante J, Pacios F, Díaz-Delgado R, Aragonés D (2009) Predictive models of turbidity and water depth in the Doñana marshes using Landsat TM and ETM+ images. J Environ Manage 90:2219–2225

Castaño S, Mejuto MF, Vela A, Quintanilla A, Ruiz JR (1999) Monitoring of wetlands evolution. In: Montesinos S, Castaño S (eds) Application of space techniques to the integrated management of river basin water resources (ASTIMwR). European Commission, Brussels, pp 27–49

Crist EP, Cicone RC (1984) A physically-based transformation of thematic mapper data: The TM Tasseled Cap. IEEE Trans Geosci Remote Sens GE-22:256–263

Díaz-Delgado R, Bustamante J, Pacios F, Aragonés D (2006a) Hydroperiod of Doñana marshes: Natural or anthropic origin of inundation regime? In: ESA & Ramsar Convention (eds) Proceedings of the 1st GlobWetland Symposium. 19–20 October, Frascatti, Italy

Díaz-Delgado R, Bustamante J, Aragonés D, Pacios F (2006b) Determining water body characteristics of Doñana shallow marshes through remote sensing. In: IGARSS 2006 (ed) IEEE international conference on geoscience and remote sensing symposium, July 31–August 4, Denver, pp 3662–3663

Domínguez Gómez JA (2002) Estudio de la calidad del agua de las lagunas de gravera mediante teledetección. PhD thesis, Alcalá de Henares, Universidad de Alcalá, 320 pp

Engman ET, Gurney RJ (1991) Remote sensing in hydrology. Chapman & Hall, London, 225 pp

Gardiner N, Díaz-Delgado R (2007) Trends in selected biomes, habitats and ecosystems: Inland waters. In: Strand H, Höft R, Strittholt J, Miles L, Horning N, Fosnight E (eds) Sourcebook on remote sensing and biodiversity indicators. Secretariat of the Convention on Biological Diversity, Montreal, pp 83–102

Kyu-Shun L, Tae-Hoon K, Yeo-Sang Y, Sang-Ming S (2001) Spectral characteristics of shallow turbid water near the shoreline on inter-tidal flat. Korean J Remote Sens 17:131–139

McFeeters SK (1996) The use of the Normalized Difference Water Index (NDWI) in the delineation of open water features. Int J Remote Sens 17:1425–1432

Ozesmi SL, Bauer ME (2002) Satellite remote sensing of wetlands. Wetlands Ecol Manage 10:381–402

Susskind J, Rosenfield J, Reuter D, Chahine M (1984) Remote sensing of weather and climate parameters from HIRS 2/MSU on TIROS-N. J Geophys Res Atmos 89:4677–4697

Chapter 32
New Technologies for Long-Term Biodiversity Monitoring

Data Quality and Real-Time Availability

Hugues Lefranc, Regino Núñez, and Justin Steventon

Introduction

Now, more than ever before, environmental data are needed to inform efficient environmental management. Information on biodiversity and ecosystems is essential, for governments as well as the scientific and educational communities. Much of this information, however, is not readily accessible, yet scientists need access to high quality data for their research; planners need readily available and reliable information to make appropriate decisions for the management of environmental resources; and we should not forget the need to satisfy the growing interest in biological information from the general public. Enabling access to this information is of vital importance in the face of the rapid decline in biological diversity. The purpose of this chapter is to describe the solution developed to automate the process of data collection and facilitate the use of long-term monitoring information by the Natural Processes Monitoring Team (Equipo de Seguimiento de Procesos Naturales, ESPN) of the Doñana Natural Park.

Obtaining Quality Data

Data are of high quality if they are suitable for use in operations, decision-making and planning (Juran 1964 in Chapman 2005). It is important for the organisations that plan to make their data available to others to have a policy with respect to the

H. Lefranc (✉)
Natural Processes Monitoring Team, Doñana Biological Station (CSIC), Seville, Spain
e-mail: hugo@ebd.csic.es

R. Núñez
Advanced Technology Systems (SATEC), Seville, Spain

J. Steventon
Steventon Consulting, Redmond, WA 98052, USA

C. Hurford et al. (eds.), *Conservation Monitoring in Freshwater Habitats: A Practical Guide and Case Studies*, DOI 10.1007/978-1-4020-9278-7_32,

quality of data. Projects for the evaluation of biodiversity require large amounts of information on species in order to obtain significant results. These projects are very expensive to run, and data collection, particularly gathering information on species, is invariably the most time-consuming activity. However, data quality principles must be applied at all the stages of the process of data management (capture, digitalisation, storage, analysis, presentation and use). Any loss in quality at any of these stages reduces the applicability, usage and validity of the data.

Traditionally, during fieldwork, observers either follow a standard protocol and fill in the appropriate forms or make field notes recording the place, date, time, location, etc. The forms are then handed in to the office for their storage and subsequent analysis. This method, although widely used, has some problems both in the field and in the office.

Data collection with pen and paper is not an ideal method for guaranteeing the quality of the information. Handwritten material could, at the time of digitalisation, be badly interpreted due to the handwriting or even discarded for being illegible. The drifting attention of the observer due to the repetitive nature of the work, or due to bad weather, can lead to errors in form completion, or even the loss of forms. The absence of a structure or uniform format in the field notebooks makes the transfer of the information to a database even more complex and time-consuming, not to mention the tedious process of transcribing handwritten data into a digital format that also can lead to errors.

The key factors for improving the quality of the data are prevention and validation. The prevention of errors is closely related to the information gathering and database entry phases. The development of a simple and easy-to-use interface for data input can be a way of minimising errors, because there are fewer opportunities for them to occur. Although considerable effort should be directed towards the prevention of errors, it is also important to consider the validation of data (error detection plus correction). And one of the best ways to reveal errors is to allow data to be seen (Edwards 2004).

New Technologies

Biodiversity informatics (Bisby 2002) is an emerging discipline born from the need to find more efficient methods for capturing and processing environmental information. It draws on collaborations between biologists and software and network engineers. The objective is to acquire with precision, represent, communicate, integrate, analyse and apply the information extracted from natural systems. It is supported by the application of new technologies to guarantee the fulfilment of protocols, and the introduction of data integrity and quality controls. This system will facilitate the rapid presentation of reports and data in real or quasi-real time, with all of this based on a powerful device for database management.

New technologies have the potential to revolutionise the process of data gathering. Personal Digital Assistants (PDA) and handheld electronic diaries have the same functionality and capabilities as microcomputers and, owing to their portability, allow observations to be digitised in the field. For example, the Global Positioning System (GPS) allows users to accurately register the location of specific sites; over the Internet, via WiFi (wireless Internet) or using a mobile telephone network, users can send observations from a remote database in the field to database servers; and, finally, the data can easily be made available to analysts and managers through the Web.

One of the most revolutionary of these new technologies is the free software CyberTracker (www.cybertracker.co.za). This is a program for data capture widely used in environment and wildlife conservation. It is a very efficient interface for collecting large quantities of geo-referenced data. CyberTracker can incorporate positional data from a GPS - if available (some e-diaries have built-in GPS, and others allow connection to small GPS receptors via Bluetooth). The original concept behind its development was to create an interface that was very user-friendly and could be used by fieldworkers. The tool is based on the principle of using the touch screen of the PDA or electronic diary to store information. Computerising data *in situ* is considerably more efficient than digitising the data manually.

The application is very simple and allows for personalisation in its use. A user can design and edit a database, developing screen sequences with different interfaces (Fig. 32.1). For each methodological protocol for data acquisition, a sequence is created in which, step-by-step, all the possible options and elements needed for describing field observations should appear on the screen. The sequence, which is the same for all the observers, thereby enables a standardisation of the methodology, leaving little leeway for personal interpretation of protocols by fieldworkers.

With CyberTracker, the design of a database with sequential screens does not require knowledge of programming. The creation of elements for each screen allows the automatic generation of a structured database.

Once the data are downloaded onto the PC, they are transferred into a database for permanent storage, analysis, visualisation and export. Data are viewed in tables but can also be projected over maps and aerial photographs, allowing the verification of the entries before their final validation. The information can also be rapidly exported into other programs, such as spreadsheets or GIS (Geographical Information System) software, for advanced analysis. In this way, CyberTracker integrates the traditional gathering of environmental data and the disciplines of spatial information analysis.

In the context of a set of manual procedures with little or no automation, the ESPN of Doñana Natural Park began to consider the idea of incorporating this new technology for the process of capture, classification and exploitation of data deriving from its own work on long-term monitoring programmes. Due to the diversity of protocols, the vast number of observers and the annual deadlines for submitting reports, the use of this technology has meant a radical change in the way of acquiring data in the field, leaving aside the traditional field book, in exchange for the modern PDA.

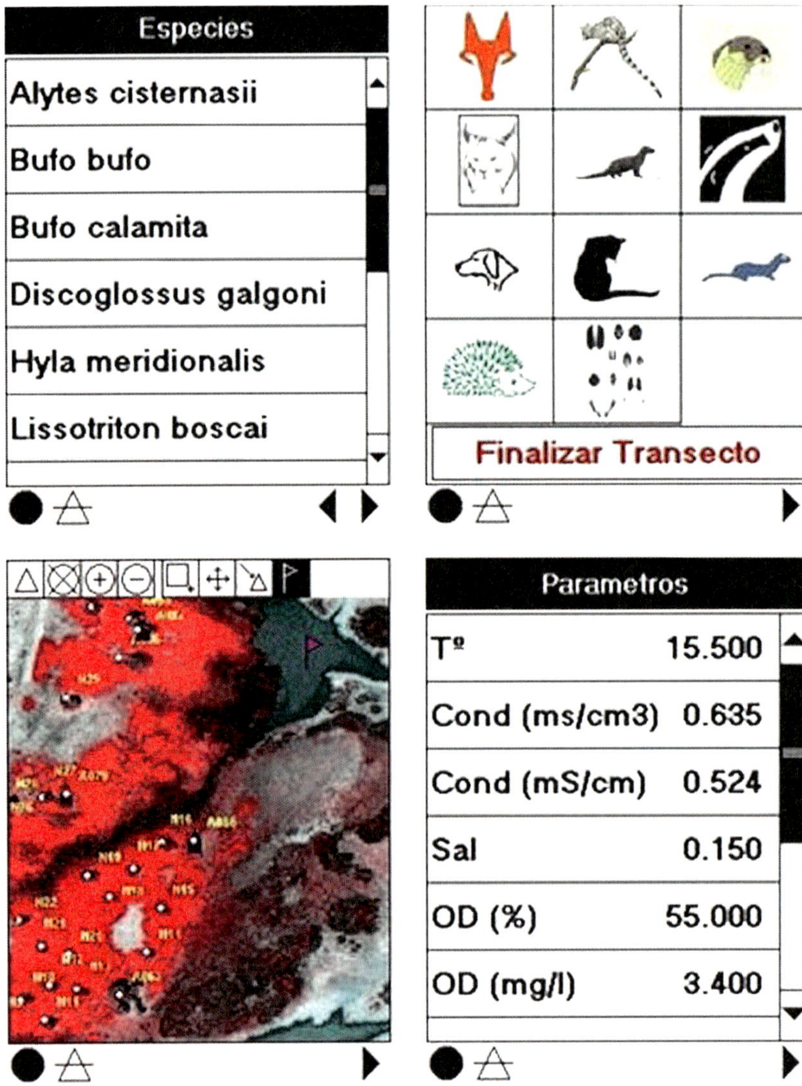

Fig. 32.1 Examples of PDA screens used for collecting field data

The Starting Point

In October of 2007 a new work area was developed in the ESPN, the area of data quality. The objective was to introduce IT tools to improve data collection, preserve the integrity of the information and offer end-users the information more rapidly.

From March 2007, we started to use the CyberTracker software for the acquisition of data in more than 50 protocols for long-term ecological monitoring in Doñana Natural Park.

The ESPN was born with the objective of generating basic, reliable and verifiable information about the state of the physical and biological environment of Doñana. This information is used by the managers of Doñana Natural Park, as well as the scientific community, to detect changes in the fauna, flora and environmental quality of the protected area.

The main aim of the ESPN consists of the implementation of an open-ended monitoring programme, to evaluate the spatial and temporal tendencies of the biological diversity and the ecosystems of Doñana. Its work is supervised by a scientific committee composed of researchers both within the Biological Station and from elsewhere. Every year since 2002, the ESPN and its 18 technical staff have applied more than 50 methodological protocols for capturing information in fields as diverse as water quality, hydrological dynamics and geomorphological processes; land use; dynamics of the aquatic and terrestrial vegetation communities; herbaceous plants and bush productivity; changes in the surface coverage of plant matter; distribution and population dynamics of invertebrates, fish, amphibians, reptiles, mammals, and of threatened, key and invading species. Some of the data collection process was already computerised, with data from PDA's downloaded onto non-shared databases and/or spreadsheets that had to be subsequently homogenised and filtered. Finally, after technical validation, the information was exported to permit its publication on the web, with this process being carried out by hand.

The introduction of these new technologies in the ESPN has involved a major investment in training to extend their use across the workforce. Interestingly, some of the members of the ESPN, who were initially reluctant to use the new IT, were later among the most convinced and adept users of the PDAs and CyberTracker. This updating process has also meant an improvement in protocols, as it entailed a profound revision of the data collection procedures. After the initial implementation, it can be stated that the most important benefit has been the improvement in data quality, reflected in the complete compliance with protocol requirements, the disappearance of omitted data and semantic homogeneity. In addition, the experience has allowed the maximum automation of the chain of data capture, reducing the minimum number of users necessary, for the processes of acquisition and digitalisation, to just one. Thus, CyberTracker has shown itself to be an optimum program to comply with most of the criteria for data quality.

This process has returned naturalists to a leading role on the staff in place of the technical experts. It is a very democratic program in the sense that with knowledge of nature, and a little practice, anybody can collect data with this system with no previous expertise. Thanks to CyberTracker, we have made the most of the time spent on the fieldwork. For example, in the case of the monitoring of the aquatic ecosystems of Doñana, the limnological sequence allows the user to perform 18 different protocols with the same sequence (Table 32.1). Once the data have been downloaded, all the information can be interrogated through the same database.

Table 32.1 Wetland monitoring protocols used in Doñana

Habitat	Objectives
Aquatic fauna	Quantitative estimation of fish, amphibians, reptiles and macroinvertebrates in temporary and semi-temporary ecosystems Specific monitoring of *Procambarus clarkii* (alien species) Thirty-eight locations: twice a year
	Quantitative estimation in shallow tidal waters of the marshes Three locations: every other month
	Trachemys scripta elegans (exotic species) Four locations: monthly
	Population dynamics of native turtles Five locations: every 4 years
	Survey of amphibians in temporary ponds Thirty locations: monthly checks when there is water
Aquatic vegetation	General description of coverage Specific description of the coverage, phenology and location Estimation of biomass Thirty-eight locations: twice a year
	Non-systematic monitoring of protected and alien species
Water	Estimation of water quality (T°, Cond., DO, pH, NO_3, NH_4, Chlorophyll and turbidity) in temporary and semi-temporary ecosystems Thirty-eight locations: twice a year
	Estimation of water quality in ephemeral ponds Estimation of the area submerged and duration Thirty locations: monthly checks when there is water
	System for early alert of algal blooms At least four locations: at least twice a week in high risk periods

Another very important benefit of the software is the considerable saving of time dedicated to the digitising process. This gain allows the observer more continuity in their fieldwork, while at the same time, affording more time for error detection. The possibility of seeing the data directly, the same day, in a table and projected over a map, makes it easier to detect errors in numbers and in geographic positioning or failed data.

However, CyberTracker, despite all the improvements that it has given to the ESPN, has not provided, to date, a reduction in the time between data acquisition and delivery to the final users, i.e. the scientific personnel, managers and the general public. As the program does not amass all of the consolidated data in a single platform or unified scheme that would allow independent access to the process, it has not solved the problem of distributing the information. If the data cannot be downloaded centrally onto a single PC, the information manager will have the laborious task of compiling the data, and the more PDA users there are collecting the same sequence of data, the more dispersed the information will be. Updating CyberTracker was also difficult: if we wanted to change a data sequence, we would have to gather in the PDAs and do this manually. Furthermore, the software did not, originally, allow us to make the information gathered by the technical staff of the

ESPN available in real time. Therefore, we proposed a new software development that would benefit both CyberTracker and the ESPN. This was necessary to centralise data and to deal with incorporating the data from all the monitoring protocols (more than 50) and from the technical workforce (18 people).

With the financial help of the Spanish government programme for the recognition of Unique Scientific and Technical Infrastructures (ICTS), and in collaboration with the consulting company SATEC (specialists in communication systems), the ESPN asked the CyberTracker programmer, Justin Steventon, to carry out some modifications and improvements to the software. Basically, the technical request consisted of making it possible to use CyberTracker in a network, with all the sequences and databases migrating to a single server. The synchronisation, the transfer of sequences, the downloading of data and the visualisation of data were to be performed remotely from the server using an internet connection. As a consequence of these improvements, it would be possible to download data collected with the PDA to the server, from any location with wireless coverage or by using the mobile telephone network.

Planning for Data Transfer to a Central Server

The first step for optimising the process of data capture and storage is to develop a structured diagram showing how the processes of capture, consolidation and validation, and finally exploitation of data can be separated on different levels, based on a classical three-level diagram with a client–server structure.

Data Capture

At this level we needed to generate the sequences necessary for allowing data capture by PDA users. The data can then be transferred either in real time (if there is wireless or mobile phone coverage) or delayed, and then transferred en masse when the mobile devices are connected to PC workstations. The basic tools for this process are:

- PDA with Wireless connection
- An updated version of CyberTracker, for data transmission through the web
- Observation sequence (methodological protocol)

Data Consolidation

This is facilitated using a network relational database, MySQL. This database (DB) will store data in double format: on the one hand, it is a repository of the CyberTracker database itself, and, on the other hand, due to the proprietary way in which CyberTracker stores its data (encrypted), it also stores certain types of data in a non-encrypted format, so that they are accessible directly by third party applications. The basic components of this level are:

- FTP Server
- CyberTracker
- MySQL DB Server

Data Exploitation

This has two methods of consulting and/or exploitation, due to the double format in which data are stored. The first one is for the CyberTracker application itself, to enable researchers to perform complex tasks such as making advanced filters for consultations; and the second method is to enable the general public to access these data en masse, through the ICTS of the Doñana website. The components of this third level are:

- CyberTracker
- ICTS Web (http:/icts.ebd.csic.es)

Figure 32.2 shows the physical diagram outlining the implementation of the new processes for monitoring with CyberTracker in the Doñana Biological Reserve (DBR).

Observations made can be sent to a FTP/HTTP server on the DBR network from the PDAs, making use of the WiFi Access Points to the Reserve's data net-

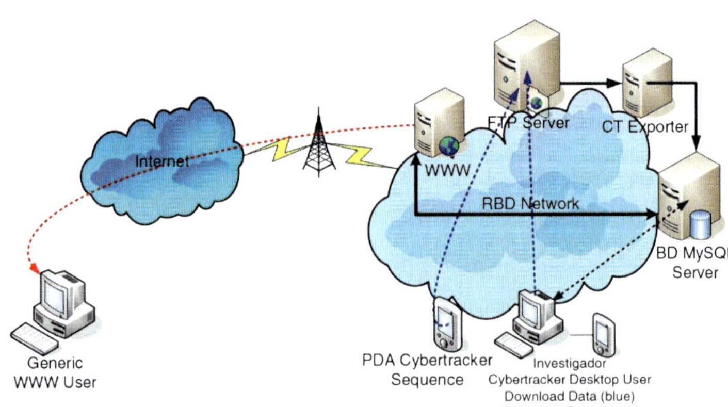

Fig. 32.2 A diagram illustrating the network infrastructure in Doñana Natural Park

work or via an Internet connection directly from the PDA, and from a PC, connected to the Internet and synchronised with the PDA. The data on the FTP/HTTP server are transferred onto the MySQL DB using scheduled processes of CyberTracker and stored in both non-encrypted and CyberTracker proprietary formats. Both generic users and researchers can access the data, the former via the Internet, by consulting the website of the ICTS, and the latter, by access through the CyberTracker application that is installed on their personal equipment and making use of the centralised database. The stages in the new monitoring scheme are as follows:

The observer loads the CyberTracker sequence corresponding to the appropriate monitoring protocol onto the PDA. Data collection is initiated following the chosen sequence step-by-step. The CyberTracker configuration allows the instantaneous transmission of the data when desired if there is wireless or GSM coverage, or subsequently, when the aforementioned connections become available or on instigation by the user. The data are sent to an FTP server in the CyberTracker proprietary exchange format. A process launched by a Windows scheduler downloads the data from the FTP server to the MySQL DB in CyberTracker format, while, in parallel, another independent process exports the CyberTracker data to tables previously designed for each type of sequence/protocol, to ensure that these data can be used by applications other than CyberTracker. This flow of data between users and the servers is illustrated in Fig. 32.3. Researchers can use the CyberTracker software from their own equipment through the central DB. Generic users can view the data in a standard format through a web interface using the application for data exploitation on the ICTS website.

The Application of Our Design in International Monitoring Networks

One of the major difficulties in forming global strategies for evaluating biodiversity is data comparability. Reliable measurement of trends and changes in biodiversity requires standardised data collection and analysis in order to be comparable. Therefore, if we want to evaluate data both at a local scale and global scale, we will need a coherent system for data collection and communication.

The technology incorporated in the long-term ecological monitoring programme of Doñana Natural Park has recently been put forward as a technological reference for the regional network LTER-Europe (European Long-Term Ecosystem Research Network, www.lter-europe.net). The aim is to extend the use of the CyberTracker software as a tool for the normalisation and harmonisation of data acquisition protocols. This would improve the quality, and quasi-real time availability, of data for end-users.

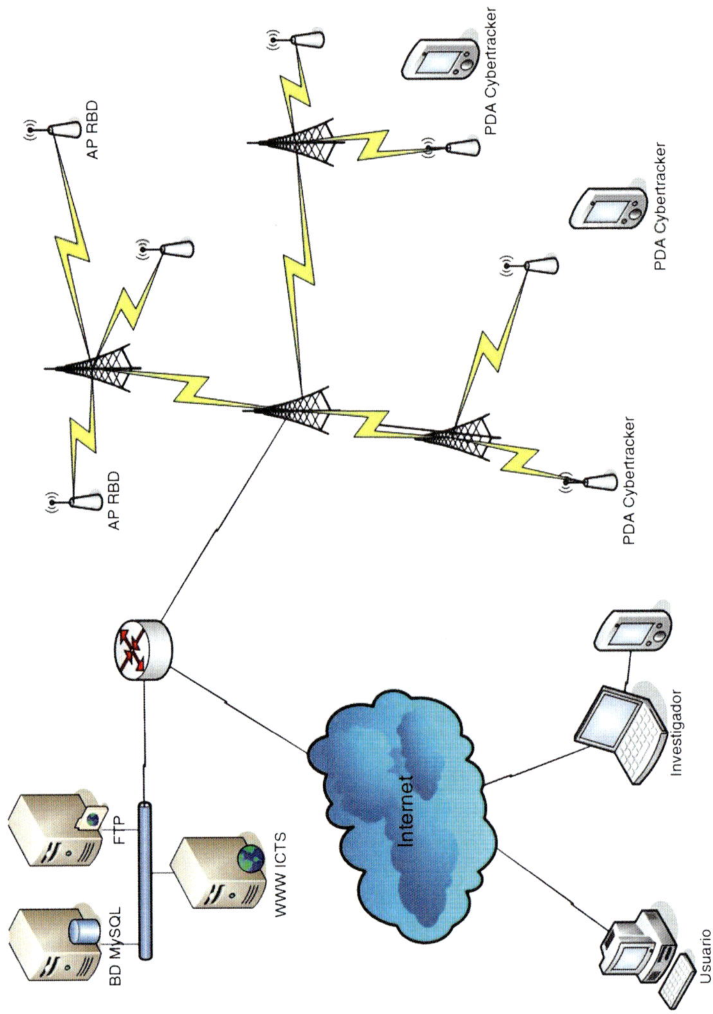

Fig. 32.3 A diagram illustrating the flow of monitoring data in Doñana Natural Park

References

Bisby FA (2002) The quiet revolution: biodiversity informatics and the internet. Science 289:2309–2312

Chapman AD (2005) Principles of data quality. Report for the global biodiversity information facility 2004. GBIF, Copenhagen

Edwards JL (2004) Research and societal benefits of the global biodiversity information facility. BioScience 54(6):485–486

Section 7
Appendix

Glossary

Accuracy: how close a single measurement, or group of measurements, is to the true value of a parameter.

Alien species: a species occurring outside its natural past or present distributional range, where its presence and dispersal is due to intentional or unintentional human action.

Area of search: the area searched at the scale of an individual sample point: in a river, the area of search for a single sample might be 5000 m^2. By contrast, in terrestrial grassland it might be 2 m^2.

Area of study: the total area that we look at in the field, including every point which might be included as part of a sample.

ASPT: Average score per taxon.

Attribute: measurable characteristic of a habitat or species.

Baltic: the area surrounding the Baltic Sea.

Bathymetric: of, or relating to, measurements of the depths of oceans or lakes.

Bathyscope: a device for looking below the surface of water.

Benthic: of or relating to, or happening on the bottom under a body of water, e.g. bottom-dwelling aquatic plants and animals.

Biomass: the total mass of living matter within a given unit of area.

Blanket-weed algae: dense surface scum of filamentous algae – commonly *Cladophora* species.

BMWP method: the Biological Monitoring Working Party (BMWP) procedure for measuring water quality using macro-invertebrates to create quality indices.

Brackish: water with a higher salinity than fresh water but a lower salinity than seawater.

By-catch: animals caught unintentionally while fishing for other species.

Carrying capacity: the population size of a species that the environment can sustain in the long term.

Catchment: the area of land that collects rainfall, which then flows into a waterway, or into the groundwater system.

Charophytes: a division of the plant kingdom that includes the stoneworts.

Circumpolar: found in areas immediately surrounding the North Pole.

Cryptogams: a group of plants that reproduce using spores, not seeds.

CSM: Common Standards Monitoring.

DAFOR scale: Dominant, Abundant, Frequent, Occasional, Rare – a subjective cover/abundance scale for estimating the amount of plant material present.

Derogations: the addition of exemptions, special conditions or partial repeal of a law, which in some way diminishes the original intent or scope.

Detection rate: the frequency at which a species is detected by different observers.

Diffuse pollution: Pollution resulting from scattered or dispersed sources that are collectively significant but to which effects are difficult to attribute individually.

DNS: Doñana Natural Space.

DVT: Dominant Vegetation Type – a stand of homogenous vegetation classified by the most dominant component.

EOS: Earth Observation Systems.

ESPN: Natural Processes Monitoring Team (Equipo de Seguimiento de Procesos Naturales) of Doñana Natural Park, Spain.

EUNIS: A comprehensive pan-European habitat classification system.

Eutrophication: an increase in compounds containing nitrogen and/or phosphorous, usually associated with pollution. It is the process by which a lake or other water body becomes rich in dissolved nutrients and often deficient in oxygen. Eutrophication can occur either as a natural stage in lake maturation or can be artificially induced by human activities (e.g. organic waste, fertilizers, sewage run-off from the land). Excessive nutrient-enrichment is often referred to as hyper-eutrophication.

Evapotranspiration: the term used to describe the sum of evaporation and plant transpiration from the earth's land surface to the atmosphere.

Favourable condition: this is used to describe a habitat or species that is in a desired or acceptable state. This state is defined in the condition indicator table.

Feature (interest feature): a general term sometimes used in this book to mean a habitat or species designated as of conservation importance on a given site.

Filamentous algae: single algae cells that form long visible chains, threads, or filaments. These filaments intertwine forming a mat that can resemble a blanket of wet wool. Filamentous algae start growing along the bottom in shallow water or attached to structures in the water (including other aquatic plants). Often filamentous algae float to the surface forming large mats, commonly referred to as pond scum.

Fossorial: living underground.

Generalized Additive Model (GAM): a statistical model developed for blending properties of generalised linear models with additive models.

GIS (Geographical Information Systems): sophisticated computer systems able to manipulate, analyse and display spatial information.

GPS (Global Positioning System): a highly accurate satellite navigation system. The receiver calculates a position on the ground by timing signals that are broadcast from a series of special GPS satellites.

HELCOM: the Helsinki commission, the governing body of the Helsinki Convention.

Helophyte: the phytosociologic definition of a biennial or herbaceous plant of which only the buds survive a harsh period, such as winter.

Heredability: a measurement of the relative importance of the genetic variance (mainly genetic additive and dominance variance) in the phenotypic variability found for a certain quantitative trait within a population. It scores from 0 to 1.

Hibernacula: sites where bats congregate to hibernate during the winter months.

Hydroacoustic: the study or use of sound in water to remotely obtain information about the physical characteristics of the water body, such as the bed of a lake or vegetation structure.

Hydromorphology: describes the hydrological and geomorphological processes and attributes of surface water bodies. For example, for rivers, hydromorphology describes the form and function of the channel as well as its connectivity (up and downstream and with groundwater) and flow regime.

Hydroperiod: the length of time that a wetland holds water.

Indicator assemblage: a suite of co-existing species that will respond either positively or negatively to factors expected to impact on the condition of a habitat.

Indicator species: a species known to respond, either positively or negatively, to factors expected to impact on the condition of a habitat – or that of associated species.

Insolated: exposed to sunshine for most of the day.

Introduction: direct or indirect movement by human agency of an organism outside its past or present natural range.

Invasive species/invader: an established alien species that is rapidly extending its range in a new region. This term is usually associated, although not necessarily, with an alien species causing significant harm to biological diversity, ecosystem functioning, socio-economic values and human health in the region of introduction.

JNCC: Joint Nature Conservation Committee.

Lairs: caves in the snow layer on the sea ice that ringed seals use.

Land uplift: the rebound of an area after an ice shield has disappeared.

Landsat: the Landsat Program is a series of Earth-observing satellite missions jointly managed by NASA and the U.S. Geological Survey. Since 1972, Landsat satellites have collected information about Earth from space.

LEAFPACS: a recording method for river macrophytes, recommended for Water Framework Directive reporting in the UK.

Limnology: a discipline that concerns the study of fresh waters, specifically lakes, ponds and rivers (both natural and manmade), including their biological, physical, chemical, and hydrological aspects.

Lotic: of, relating to, or living in moving water.

LTER: Long-Term Ecological Research. It usually refers to research programmes carried out on natural sites belonging to a network.

LTER-Europe: European Long-Term Ecosystem Research Network.

LTSER: Long-Term Socio-Ecological Research. It usually refers to socio-economic and ecological research programmes carried out on platforms (several sites) belonging to a network.

Macrophytes: plants (including vascular plants, bryophytes, algae and lichens) that are visible to the naked eye and nearly always identifiable in the field, typically used in reference to aquatic vegetation.

Magnopotamion or **Hydrocharition-type vegetation**: these are aquatic plant communities protected under Natura 2000 legislation. Hydrocharition are free-floating surface communities comprising species such as duckweeds, bladderwort and water solider. Magnopotamion are large pondweeds such as floating leaved and fine-leaved pondweeds (*Potamogeton* spp.).

Marl-lake: a lake rich in natural marl deposits - Marl or Marlstone being a calcium carbonate or *lime*-rich mud or mudstone, formed under freshwater conditions in deposits of lake sediments.

Mesotrophic: the prefix "meso" means mid-range. The mesotrophic state is defined as having a moderate supply of nutrients and therefore moderate biological productivity.

Mid Infrared: the mid-wave infrared region of the electromagnetic spectrum (from 3,000 to 5,000 nm).

MIS: Macrophyte Index Scheme.

MNA: Minimum Number Alive – Minimum number of water voles alive in any one month based on trapping data that month (or animals tagged previously and captured thereafter).

Monitoring protocol: a predefined written procedural method in the design and implementation of long-term monitoring.

Monitoring: for the purposes of this book, monitoring is defined as assessing the condition of a habitat or species against a predetermined standard.

MS: macrophyte survey.

MTR: Mean Trophic Rank.

Near Infrared: the near infrared region of the electromagnetic spectrum (from about 800 to 2,500 nm).

Necropsy: the surgical examination of a dead animal.

Nephelometric turbidimeter: an instrument for determining the concentration or particle size of suspensions by means of transmitted or reflected light.

NNR: National Nature Reserve.

NVC: National Vegetation Classification (NVC). This is a comprehensive floristic classification of terrestrial and freshwater vegetation types in the UK. It recognises roughly 400 separate plant communities, many further divided into sub-communities.

NWCW: National Wetland Centre Wales – a National Key Site for water voles in Wales.

Oligo-mesotrophic lake: lakes and pools with waters moderately rich in dissolved bases. The water is very clear (pH often above 7.5) and poor to moderate in nutrients; commonly the water is oxygen rich and low in turbidity.

Oligotrophic: describes ecosystems that are nutrient-poor and have a low productivity. Often referring to rivers and lakes that have clear water and low biological productivity (oligo = little; trophic = nutrition).

Optimal condition: used to describe when a habitat or species of conservation priority on a site is in the best condition that we could hope to achieve, as defined in the condition indicator table. The term 'favourable condition' suggests something less than this, and is often interpreted as an 'acceptable' state.

ORL: Observed Range Length – estimates of the minimal range of an animal calculated as the distance between the two furthest capture positions following the contours of the watercourse on the satellite map.

Otter holt: a place where otters rest and breed.

Parr: a juvenile fish preparing to leave the fresh water of its natal habitat.

PCB: polychlorinated biphenyls, artificial organic compounds that are classified as persistent organic pollutants that are toxic and accumulate in animals.

PDA: Personal Digital Assistant, a handheld computer.

Phenology: the study of periodic plant and animal life cycle events and how these are influenced by seasonal and interannual variations in climate.

Phreatic: refers to matters relating to ground water below the static water table.

Physiology: the study of the mechanical, physical, and biochemical functions of living organisms.

Phytoplankton: the term phytoplankton, or algae, encompasses all microorganisms that make energy from sunlight in water. Phytoplankton serves as the base of the aquatic food web.

Piezometer: small diameter water well used to measure the hydraulic head of groundwater in aquifers.

Point Source: a stationary location or fixed facility from which pollutants are discharged; or any single identifiable source of pollution, such as a pipe, ditch or channel.

Post-glacial: after the ice age.

Precision: how close repeated measurements of the same attribute are to one another, though not necessarily to the true value of the attribute: we may not know the true value.

Public Service Agreement: the British Government's Public Service Agreement (PSA) target to have 95% of the SSSI area in favourable or recovering condition by 2010.

PVI: Percent Volume Infested – the percentage of the underwater space occupied.

RCS: River Corridor Survey.

Red Data Book: the IUCN (International Union for the Conservation of Nature and Natural Resources) maintains an international list, published as the Red Data Book. Red Data Book species are classified into different categories of perceived risk. Each Red Data Book usually deals with a specific group of animals or plants (for instance, reptiles, insects or mosses).

Remote sensing: broadly used to describe any measurement taken of an object at some distance, rather than by direct contact.

Rhizomes: underground stems of a plant.

RSPB: Royal Society for the Protection of Birds (UK).

SAC: Special Area of Conservation – a site protected under Natura 2000 legislation.

Saxicolous: living on rocks or stones.

SCUBA survey: underwater dive survey.

SCV: Species Cover Value – a component of Mean Trophic Rank analysis.

Secchi disk: a disk, divided into black and white quarters, used to gauge water clarity by measuring the depth at which it is no longer visible from the surface.

Sediment: matter that has been deposited by some natural process; any particulate matter that can be transported by fluid flow, and which eventually is deposited.

Semi-aquatic: spending at least part of its existence living in water.

Silt: silt is soil or rock derived granular material of a grain size between sand and clay. Silt may occur as a soil, or as suspended sediment in a surface water body.

Site: used in this book to describe the area of conservation interest under discussion. In general, it has been used to refer to any area protected for nature conservation.

SMHI: Swedish Meteorological and Hydrological Institute.

Smolt: a stage in the life cycle of salmonid fish, where they become physiologically adapted to salt water and when they move down to the sea.

Special Areas of Conservation: Natura 2000 sites that are designated under the Habitats Directive.

SPOT: Système pour l'observation de la Terre. A sensor used for Earth Observation.

SSSI: Site of Special Scientific Interest. A site of national conservation importance protected by legislation in the UK.

Stolons: horizontal stems which grow at the soil surface or below ground.

Stonewort (Charophyte): Stoneworts are classified along with green algae and are simple multi-celled organisms descended from some of the earliest life forms that appeared on Earth. The plant consists of a series of "giant cells" up to several centimetres in length.

STR: Species Trophic Rank – a component of Mean Trophic Rank analysis.

Surveillance: typically, a series of repeat surveys used to detect or track trends of habitats or species. Differs from monitoring by not measuring against a predetermined standard.

Survey: a set of standard observations, usually obtained with a standard method and within a restricted time period, typically a one-off exercise, e.g. mapping a habitat or compiling a species list.

Thematic map: a map that displays the spatial pattern of a theme or series of attributes, showing statistical rather than topographical information about a place. Thematic maps emphasise spatial variation of one or a small number of geographic distributions.

TIROS: Television Infrared Observation Satellite.

Total phosphorus (TP): the total concentration of phosphorus found in the water. Phosphorus is a nutrient essential to the growth of organisms, and is widely believed to be the limiting factor in the primary productivity of surface water bodies.

Turbid water: turbidity of water can be caused by suspended particles of sediment or algae that prevent sunlight from entering the water.

Water body: (Water Framework Directive) a manageable unit of surface water, being the whole (or part) of a stream, river or canal, lake or reservoir, transitional water (estuary) or stretch of coastal water. A ‘body of groundwater’ is a distinct volume of underground water within an aquifer.

Water Framework Directive: European Union legislation – Water Framework Directive (2000/60/EC) – establishing a framework for European Community action in the field of water policy.

WiFi: wireless Internet.

Xeric: relating or adapted to an extremely dry habitat. Succulents such as cacti, aloes, and agaves are xeric plants.

Index